T0213763

Lecture Notes in Computer Science 10304

Commenced Publication in 1973
Founding and Former Series Editors:
Gerhard Goos, Juris Hartmanis, and Jan van Leeuwen

More information about this series at http://www.springer.com/series/7407

Pascal Weil (Ed.)

Computer Science –
Theory and Applications

12th International Computer Science Symposium in Russia, CSR 2017
Kazan, Russia, June 8–12, 2017
Proceedings

 Springer

Editor
Pascal Weil
CNRS and University of Bordeaux
Talence
France

ISSN 0302-9743   ISSN 1611-3349   (electronic)
Lecture Notes in Computer Science
ISBN 978-3-319-58746-2   ISBN 978-3-319-58747-9   (eBook)
DOI 10.1007/978-3-319-58747-9

Library of Congress Control Number: 2017939718

LNCS Sublibrary: SL1 – Theoretical Computer Science and General Issues

Printed on acid-free paper

This Springer imprint is published by Springer Nature
The registered company is Springer International Publishing AG
The registered company address is: Gewerbestrasse 11, 6330 Cham, Switzerland

Preface

The 12th International Computer Science Symposium in Russia (CSR 2017) was held during June 8–12, 2017 in Kazan, hosted by Kazan Federal University. It was the 12th event in the series of regular international meetings, following CSR 2012 in Nizhny Novgorod, CSR 2013 in Ekaterinburg, CSR 2014 in Moscow, CSR 2015 in Listvyanka, and CSR 2016 in St. Petersburg.

The opening lecture was given by Thierry Coquand (Chalmers) and five other invited plenary lectures were given by Javier Esparza (Munich), Elham Kashefi (Paris and Edinburgh), Andrew McGregor (Amherst), Ronitt Rubinfeld (MIT), and Marc Zeitoun (Bordeaux). We would like to extend our thanks to all of them.

This volume contains the accepted papers and those sent by the invited speakers. The scope of the proposed topics for the symposium is quite broad and covers a wide range of areas in theoretical computer science and its applications. We received 44 papers in total, from which the Program Committee selected 22 papers for presentation at the symposium and for publication in the proceedings. The committee met electronically over a full week and the chair wants to thank all the committee members for their availability during that intensive time and for interesting discussions. Our work would not have been possible without the generous help of over 80 additional external reviewers. To avoid conflicts of interest, the Program Committee members were asked to refrain from submitting papers to the conference.

The reviewing process relied in an essential manner on the *EasyChair* conference system created by Andrei Voronkov: It greatly improved the efficiency of the committee's work and of the logistics of producing this proceedings volume, and we are grateful for its availability.

We appreciate the support of *Kazan Federal University* and of *Yandex*. As in previous editions of the conference, Yandex generously provided the Best Paper Awards. The recipients of these awards were selected by the Program Committee.

- The Best Paper Award was split between two papers: by Alexei Miasnikov, Svetla Vassileva, and Armin Weiß, *"The Conjugacy Problem in Free Solvable Groups and Wreath Products of Abelian Groups is in* TC^0*"*; and by Lukas Fleischer and Manfred Kufleitner, *"Green's Relations in Finite Transformation Groups."*
- The Best Student Award goes to the paper by Alexey Milovanov, *"On Algorithmic Statistics for Space-Bounded Algorithms."*

The workshop *Computation and Cryptography with qu-bits* (CCQ 2017) was co-located with CSR 2017.

Finally, we thank the local organizers for their impeccable work: Farid Ablayev (chair), Aida Gainutdinova, Anton Marchenko, Daniil Musatov, Alina Petukhova, Alexander Vasiliev, Valeria Volkova, Mansur Ziatdinov, and Marsel Sitdikov.

April 2017   Pascal Weil

Organization

Program Committee

Farid Ablayev	Kazan State University, Russia
Ittai Abraham	VMware Research, Israel
Isolde Adler	University of Leeds, UK
Frédérique Bassino	LIPN, Université Paris 13, France
Véronique Bruyère	University of Mons, Belgium
Maike Buchin	Ruhr Universität Bochum, Germany
Hubie Chen	Universidad del País Vasco and Ikerbasque, Spain
Anuj Dawar	University of Cambridge, UK
Stéphanie Delaune	CNRS, IRISA, France
Anna Frid	Aix-Marseille Université, France
Elena Grigorescu	Purdue University, USA
S. Krishna	IIT Bombay, India
K. Narayan Kumar	Chennai Mathematical Institute, India
Frédéric Magniez	CNRS, Université Paris Diderot, France
Meena Mahajan	The Institute of Mathematical Sciences, Chennai, India
Grigory Marshalko	Technical Committee for Standardisation TC26, Russia
Catuscia Palamidessi	Inria, France
Victor Selivanov	Institute on Informatics Systems, Russia
Kavitha Telikepalli	Tata Institute of Fundamental Research, Mumbai, India
Thomas Thierauf	HTW Aalen, Germany
Szymon Toruńczyk	University of Warsaw, Poland
Hélène Touzet	CNRS, University of Lille and Inria, France
Mikhail Volkov	Ural State University, Russia
Dorothea Wagner	Karlsruhe Institute of Technology, Germany
Pascal Weil	LaBRI, CNRS and Université de Bordeaux, France

Additional Reviewers

Alagic, Gorjan	Bulín, Jakub	Gacs, Peter
Aschieri, Federico	Capelli, Florent	Gainutdinova, Aida
Atig, Mohamed Faouzi	Costa, Alfredo	Gandikota, Venkata
Barth, Lukas	Czerwiński, Wojciech	Glasser, Christian
Batra, Jatin	Dupuis, Frédéric	Gogacz, Tomasz
Berger, Ulrich	Durand, Bruno	Griffin, Christopher
Bhandari, Siddharth	Fages, François	Haase, Christoph
Bollig, Beate	Fernique, Thomas	Heunen, Chris
Buchhold, Valentin	Gabbasov, Bulat	Hoshino, Naohiko

Huang, Chien-Chung
Inenaga, Shunsuke
Jerrum, Mark
Kao, Mong-Jen
Kaplan, Marc
Kiefer, Sandra
Klasing, Ralf
Kocman, Radim
Konnov, Igor
Kozma, Laszlo
Kołodziejczyk, Leszek
Kumar, Akash
Kuske, Dietrich
Lavado, Giovanna
Lhote, Loick
Lhoussaine, Cedric
Limaye, Nutan
Lin, Young-San

Liu, Yanhong A.
Lohrey, Markus
Manuel, Amaldev
Markey, Nicolas
Mengel, Stefan
Merkle, Wolfgang
Misra, Neeldhara
N.P., Swaroop
Natarajan, Abhiram
Nimbhorkar, Prajakta
Oum, Sang-Il
Panolan, Fahad
Place, Thomas
Pournin, Lionel
Qian, Chen
Radermacher, Marcel
Romanovski, Nikolay
Roos, Yves

Rudskoy, Vladimir
Salfelder, Felix
Salvy, Bruno
Saurabh, Nitin
Schmitz, Heinz
Schnoebelen, Philippe
Shilov, Nikolay
Shukla, Anil
Shur, Arseny
Sijben, Stef
Subrahmanyam, Venkata
Talbot, Jean-Marc
Vandenberghe, Lieven
Vandin, Andrea
Zhou, Samson
Ziatdinov, Mansur

Contents

Type Theory and Formalisation of Mathematics

Thierry Coquand[1,2(✉)]

[1] Department of Computer Science and Engineering,
Göteborg University, Gothenburg, Sweden
thierry.coquand@cse.gu.se
[2] Department of Computer Science and Engineering,
Chalmers University of Technology, Gothenburg, Sweden

It is difficult to overestimate the importance of *modularity* for specifying and reasoning about software [1], or for checking large and complex mathematical arguments [8–10]. The goal of this presentation is to explain in what way a recent development in type theory, the formulation of the *axiom of univalence*, addresses these modularity issues.

1 Equality and Collections in Mathematics

It will be convenient to start with a quick description of how structures are represented in mathematics. There is a natural stratification: at the first level, we have collections representing algebraic or ordered structures, like groups, rings, lattices, and at the next level, we have collections of such structures (categories), and so on.

At the first level, a structure is a *set* equipped with some operations and/or relations. This is the level considered by Bourbaki in his *théorie des structures* [3]. One important modularity principle is then that two *isomorphic* structures should be considered to be the *same*. In particular, two isomorphic structures should satisfy the same properties. For instance, if G and H are two isomorphic groups, and G is abelian (resp. solvable) then so is H. This is not quite true in set theory however since we can formulate properties that are not invariant by isomorphisms (for instance the real number π may be in G and not in H). For this reason, Bourbaki formulated the notion of *transportable* or «structural» properties, that are exactly the properties invariant by isomorphisms[1].

At the second level, where we consider collection of structures, we find a new notion of «being the same collection». A typical example is given by the collection of all families of sets over a given (base) set B. One representation is SET^B which is the collection of all family of sets X_b, $b \in B$. Another representation is SET/B the collection of all pairs Y, f where Y is a set and f a function $Y \to B$. For a mathematician, these two representations are «structurally the same». We have two canonical maps $F : \mathsf{SET}^B \to \mathsf{SET}/B$ and $G : \mathsf{SET}/B \to \mathsf{SET}^B$ for changing representations: if $X = (X_b)$ is a family of sets over B then $F(X)$ will

---

[1] One important fact will be that, in contrast with set theory, all statements expressible in type theory are transportable.

© Springer International Publishing AG 2017
P. Weil (Ed.): CSR 2017, LNCS 10304, pp. 1–6, 2017.
DOI: 10.1007/978-3-319-58747-9_1

be the collection $\Sigma(b \in B)X_b$ of pairs (b, u) with u in X_b together with the first projection. If $f : Y \to B$ then $G(Y, f)$ will be the family $X_b = \{y \in Y \mid f(y) = b\}$. However this does not define an isomorphism: $G(F(X))$ and X are only isomorphic (and not equal as sets in general). One way to understand the situation is that we should consider the two collections SET^B and SET/B as *groupoids* and the pair F, G define an *equivalence* (and not an isomorphism) between these groupoids. This gives us a *new* way to identify collections, and hence a new (powerful) modularity principle[2]. This new notion of identification comes from mathematical practice, but one can argue that it is not (contrary to the first level) so well represented in set theory (for instance, a notion of «transportable» property at this level would be quite complex, and does not seem to have been formulated).

At the next level, where we consider collections of groupoids, or collections of categories, we have the notion of 2-groupoids, then n-groupoids, with more and more complex notions of *equivalences*, and where it is less and less clear when a property is transportable. As we are going to see, type theory, together with the axiom of univalence, provides a system with a (formally) *simple* and general notion of equivalence, and where every expressible statement is invariant w.r.t. equivalences.

2   Dependent Type Theory

The notion of *type*, which comes from logic [6] has been found convenient in the theory of programming languages. Each type represents a collection of values. In particular, we have *base* types, such as the type *nat* of natural numbers, or the type *bool* of Boolean values, and if A and B are two types, we can form the type $A \to B$ which is the type of programs representing functions from the collection A to the collection B, and the type $A \times B$ which is the type of pairs of elements of type A and of type B.

Dependent type theory extends this «simple» [6] type theory by adding a *universe* type U, or a type of «small» types, and the following type: given a_0, a_1 element of type A we can form the *identification* type $\mathsf{Id}\ A\ a_0\ a_1$ of all possible identifications of a_0 and a_1. For instance, if A is a type representing a collection of some algebraic structures, e.g. groups, and a_0 and a_1 represent two groups, then $\mathsf{Id}\ A\ a_0\ a_1$ should represent the type of all possible identifications of a_0 and a_1. The introduction of these two new type formations introduces *dependent* types: if F is of type $A \to U$, then $F\ x$ represents a family of (small) types indexed by x of type A and if x and y are of type A then $\mathsf{Id}\ A\ x\ y$ represents a family of types indexed by x and y. It is then natural, if $B(x)$ is a family of types indexed by x of type A, to introduce the types $\Pi(x : A)B$ and $\Sigma(x : A)B$ which generalize respectively the function type and the binary product type.

---

[2] To give another (maybe more surprising) example: the collection of all linear orders with a fixed finite number of elements is a large collection, but it has no non trivial automorphisms, and should be considered to be the «same» as the groupoid with one object and only the identity morphism.

Interestingly, this notion of dependent types was already introduced by the mathematician N.G. de Bruijn as a good system of notations for representing mathematical arguments [4]. A further idea was to represent a proposition as a type of its proofs. Then $A \to B$ will represent implication: a proof that A implies B can indeed be seen as a function transforming a proof of A to a proof of B, while $A \times B$ similarly represents conjunction. One can then think of $\Pi(x : A)B$ as representation universal quantification, and $\Sigma(x : A)B$ as representation of (a strong form of) existential quantification. For instance, one could consider the two types

1. $\Pi(a : A)\Sigma(x : A)\mathsf{Id}\ A\ a\ x$
2. $\Sigma(a : A)\Pi(x : A)\mathsf{Id}\ A\ a\ x$

The first type has only one inhabitant, since we can choose x to be the same as a and we should have a trivial identification in $\mathsf{Id}\ A\ a\ a$, while the second type, which is usually abrviated as $\mathsf{isContr}\ A$, expresses that the type A contains exactly one element. One obtains in this way an elegant formalism for expressing proofs of mathematical theorems, which works entirely by type declarations.

While N.G. de Bruijn was representing a proposition as the type of its proofs, he noticed that these types should have a special property of *proof-irrelevance*. He gave the following example [5]. If we have a type R of real numbers with a family of type $x > 0$ expressing the property of being strictly positive, then the logarithm function ln will be represented as an element of the type $\Pi(r : R)(r > 0) \to R$, since it expects both a real number r *and* a proof that this real number is strictly positive. We expect also that $ln\ r\ p_0$ is the same as $ln\ r\ p_1$ if we have two proofs p_0 and p_1 of $r > 0$. This is the case if we have a proof of $\mathsf{Id}\ (x > 0)\ p_0\ p_1$, which expresses proof-irrelevance. In univalent foundations, this is used as a definition of when a *type A* represents a *proposition*, this is the case if we can prove (i.e. we can find an inhabitant of) the type[3]

$$\mathsf{isProp}\ A\ =\ \Pi(x_0\ x_1 : A)\mathsf{Id}\ A\ x_0\ x_1$$

We can then define when a type A is a *set* by being such that each identification type $\mathsf{Id}\ A\ x_0\ x_1$ is a proposition (that is, there is at most one identification of two elements of A, like it is for a set in set theory). This is represented by the type

$$\mathsf{isSet}\ A\ =\ \Pi(x_0\ x_1 : A)\mathsf{isProp}\ (\mathsf{Id}\ A\ x_0\ x_1)$$

We can use the universe and dependent sums to form the type of *structures*. For instance the type

$$\Sigma(X : U)X \times (X \to X)$$

represents the type of element $A, (a, f)$ with $a : A$ and $f : A \to A$, so the type os structures with one constan and one unary function. The type of *semi-groups* will be represented by the type

$$SG\ =\ \Sigma(X : U)sG\ X$$

---

[3] We write $\Pi(x\ y : A)B$ for $\Pi(x : A)\Pi(y : A)B$.

where $sG\ X$ is the type of semigroup structures on X

$$sG\ X\ =\ \text{isSet}\ X \times \Sigma(f : X \to X \to X)\Pi(x_0\ x_1\ x_2 : X)\text{Id}\ X\ (f\ (f\ x_0\ x_1)\ x_2)\ (f\ x_0\ (f\ x_1\ x_2))$$

We expect identifications for this type SG to correspond to isomorphisms of semigroups, and so, we don't expect SG to be a set: in general there may be several identifications between two given semigroups.

3   Logical Laws of Identifications

What is remarkable is that, when expressed in this formalism of dependent type theory, the general laws of identifications can all be listed in a short way. The two first logical laws are that there should be a trivial identification 1_a of type Id $A\ a\ a$ for any a in A, and that we have a general transport principle

$$\text{Id}\ A\ a_0\ a_1 \to (B(a_0) \to B(a_1))$$

whenever $B(x)$ is a family of types indexed over x of type A. Notice that this gives not only the principle of substitution of equal by equal (if $B(x)$ is a family of propositions) but also, if A is a type of structure, and $B(x)$ a general family of types, the fact that we can use an identification of a_0 and a_1 to *transport* any element of $B(a_0)$ to an element of $B(a_1)$. This provides a vast generalization of Bourbaki's notion of *transport of structures*.

P. Martin-Löf noticed, in [11], that these laws should be complemented by the following principle, which can be expressed by the fact that, in order to prove $C(x, p)$, for $x : A$ and $p : $ Id $A\ a\ x$ it is enough to prove $C(a, 1_a)$. This principle can be seen as a *new* logical law of identifications and has several remarkable consequences. For instance, using the two first logical law we can define a composition operation

$$\text{Id}\ A\ a_0\ a_1 \times \text{Id}\ A\ a_1\ a_2\ \to\ \text{Id}\ A\ a_0\ a_2$$

which reflects the transitivity of equality. A consequence of the third identification law is that this composition is associative, and that 1_a acts like a neutral element of this composition.

4   Mathematical Law of Identifications: The Axiom of Univalence

What is missing at this point is a law describing the equality in the type U. This law, the axiom of univalence, is a generalization of the (extensionality) principle that two logically equivalent propositions are equal.

We first need a description of the notion of equivalence. We define, for f of type $A \to B$

$$\text{isEquiv}\ f\ =\ \Pi(y : B)\text{isContr}\ (\Sigma(x : A)\text{Id}\ B\ (f\ x)\ y)$$

which expresses roughly that each fiber of f is a singleton. While formally simple, this generalizes in a uniform way all possible notions of equivalence (logical equivalence of propositions, isomorphisms of structure, categorical equivalence between groupoids or categories) that occur in mathematics. We define then the type of equivalences Equiv A B as the type Σ $(f : A \to B)$isEquiv f.

As a special case of the general transport principle, we have a transport function

$$\text{Id } U \ A \ B \to (A \to B)$$

which transforms an idenfication p in a transport function $t(p) : A \to B$. Using Martin-Löf's law, one can show that $t(p)$ is an equivalence, and we get a map Id U A $B \to$ Equiv A B. The *axiom of univalence* can be expressed as the fact that this map has a *section*, so that any equivalence can be seen as a transport map of an identification.

An unexpected consequence of this axiom is the principle of *function extensionality*: two functions can be identified if they can be pointwise identified. This is an important *modularity* principle: if two functions have the same input-output behavior we can in any context replace one by the other.

The axiom of univalence can be seen as a generalization of this modularity principle. It implies for instance that if two structures are *equivalent* (e.g. two groups are isomorphic, or two categories are equivalent) we can replace in any context one by the other. In particular, not only they share the same properties, but we can transport any notion for one structure to the other.

Let us now give a further example of modularity that we get by expressing all notions in an «invariant» way. If $S(X)$ is a notion of structure, any equivalence $f : A \to B$ provides automatically a transport of structures $f^* : S(A) \to S(B)$. In this way, any given structure s on A can be transported to a structure f^*s on B. If $t : S(B)$ is a structure on B, we can then define an equivalence $f : A \to B$ to be a *morphism* from A, s to B, t if we can show Id $S(B)$ (f^*s) t.

Let us now assume that we have defined a general notion (it can be a property or a structure) $Z(X, s) : X \to U$ for $X : U$, $s : S(X)$. It follows then from the general laws of identifications that for any equivalence $f : A \to B$

$$\text{Id } (A \to U) \ (Z(B, f^* \ s) \circ f) \ Z(A, s)$$

where $g \circ f$ is the composition $(g \circ f) \ x = g \ (f \ x)$. In particular, if f is a automorphism of the structure A, s that is f^*s can be identified to s, we get

$$\text{Id } (A \to U) \ (Z(A, \ s) \circ f) \ Z(A, s)$$

which expresses that anything expressible about a structure is automatically invariant by any automorphism of this structure[4].

---

[4] For a trivial, but significant [7] example, the center of group is automatically invariant by any automorphism of a group, and in particular, it is a normal subgroup.

5 Actual Formalization of Mathematics

As we have tried to explain, dependent type theory with the axiom of univalence provides a formalism in which we can only express notions invariant by equivalence, with a notion of equivalence which captures in a uniform way several notions of equivalence in mathematics, and we have argued that this can be essential for having good modularity properties.

One can however wonder if using such a formalism will not put too many constraints on the user. While it is too early to be sure, preliminary experiments [12,13] seem to indicate that this is not the case. In particular, basic notions of homological algebra (abelian, triangulated categories) can be expressed quite elegantly, and several results that require a strong form of the axiom of choice when formulated in set theory can be expressed effectively in this framework.

References

1. Appel, A.: Modular verification for computer security. In: CSF 2016: 29th IEEE Computer Security Foundations Symposium, June 2016
2. Bishop, E.: Mathematics as a numerical language 1970 Intuitionism and Proof Theory (Proc. Conf., Buffalo, N.Y.), pp. 53–71 (1968)
3. Bourbaki, N.: Éléments de mathématique. Chapitre 4: Structures. Actualités Sci. Ind. no. 1258 Hermann, Paris (1957)
4. de Bruijn, N.G.: The mathematical language AUTOMATH, its usage, and some of its extensions. In: 1970 Symposium on Automatic Demonstration. Lecture Notes in Mathematics, vol. 125, pp. 29–61
5. de Bruijn, N.G.: A survey of the project AUTOMATH. In: To H. B. Curry: essays on combinatory logic, lambda calculus and formalism, pp. 579–606 (1980)
6. Church, A.: A formulation of the simple theory of types. J. Symbolic Logic **5**, 56–68 (1940)
7. Interview of P. Deligne. https://www.simonsfoundation.org/science_lives_video/pierre-deligne/
8. Gonthier, G.: A computer-checked proof of the Four Colour Theorem. Microsoft report (2005)
9. Gonthier, G., et al.: A machine-checked proof of the odd order theorem. In: Blazy, S., Paulin-Mohring, C., Pichardie, D. (eds.) ITP 2013. LNCS, vol. 7998, pp. 163–179. Springer, Heidelberg (2013). doi:10.1007/978-3-642-39634-2_14
10. Th. Hales. Developments in formal proofs. Astérisque No. 367–368, Exp. No. 1086, pp. 387–410 (2015)
11. Martin-Löf, P.: Constructive mathematics and computer programming. In: Logic, methodology and philosophy of science, VI (Hannover, 1979). Stud. Logic Found. Math., vol. 104, pp. 153–175 (1982)
12. Voevodsky, V.: An experimental library of formalized mathematics based on the univalent foundations. Math. Structures Comput. Sci. **25**(5), 1278–1294 (2015)
13. Library UniMath. https://github.com/UniMath/UniMath/tree/master/UniMath

Advances in Parameterized Verification
of Population Protocols

Javier Esparza[✉]

Technische Universität München, Munich, Germany
esparza@in.tum.de

Abstract. Population protocols (Angluin et al. *PODC*, 2004) are a formal model of sensor networks consisting of identical mobile devices. Two devices can interact and thereby change their states. Computations are infinite sequences of interactions satisfying a strong fairness constraint.

A population protocol is well specified if for every initial configuration C of devices, and every computation starting at C, all devices eventually agree on a consensus value depending only on C. If a protocol is well specified, then it is said to compute the predicate that assigns to each initial configuration its consensus value.

While the computational power of well-specified protocols has been extensively studied, much less is known about how to verify their correctness: Given a population protocol, is it well specified? Given a population protocol and a predicate, does the protocol compute the predicate? Given a well-specified protocol, can we automatically obtain a symbolic representation of the predicate it computes? We survey our recent work on this problem.

Population protocols [3,4] are a model of distributed computation by anonymous, identical finite-state agents. While they were initially introduced to model networks of passively mobile sensors [3,4], they capture the essence of distributed computation in diverse areas such as trust propagation [16] and chemical reaction networks [27], a popular model for theoretical chemistry.

The Model. A protocol has a finite set of states Q, and a set of transitions of the form $(q, q') \mapsto (r, r')$, where $q, q', r, r' \in Q$. At each computation step of a population protocol a scheduler selects two agents, observes their current states, say q_1 and q_2, selects a rule with (q_1, q_2) on the left-hand side, say $(q_1, q_2) \mapsto (q_3, q_4)$, and updates the states of the agents to q_3 and q_4[1]. Since agents are anonymous and identical, it is irrelevant which agent moves to which state. Further, the global state of a protocol is completely determined by the number of agents at each state, i.e., by a mapping $Q \to \mathbb{N}$. Such a mapping is called a *configuration*.

Without loss of generality, one can assume that for every two states q_1, q_2 the protocol has at least one transition of the form $(q_1, q_2) \mapsto (q_3, q_4)$ for some

---

[1] Since I am often asked this question, let me mention that extensions to k agents for some $k > 2$ can be easily simulated by protocols with $k = 2$.

© Springer International Publishing AG 2017
P. Weil (Ed.): CSR 2017, LNCS 10304, pp. 7–14, 2017.
DOI: 10.1007/978-3-319-58747-9_2

states q_3, q_4 (perhaps the transition $(q_1, q_2) \mapsto (q_1, q_2)$ that does not change the current configuration). Then the protocol cannot block, and all computations have an infinite number of steps. Schedulers are requested to satisfy a fairness condition: if a configuration C appears infinitely often in the execution, then every transition enabled at C is taken infinitely often in the execution. This condition approximates the behavior of a probabilistic scheduler that selects agents at random, either uniformly or according to some fixed probability distribution.

Computing with Population Protocols. Population protocols are machines for the distributed computation of predicates $\mathbb{N}^n \to \{0, 1\}$. In the simplest version, a predicate $P(x_1, \ldots, x_n)$ of arity n is computed by a protocol with *initial states* $q_1, \ldots, q_n$. Given $k_1, \ldots, k_n \in \mathbb{N}^n$, the value $P(k_1, \ldots, k_n)$ is computed by preparing the initial configuration C_0 given by $C_0(q_i) = k_i$ for every $1 \leq i \leq n$ and $C_0(q) = 0$ for every non-initial state q, and letting the protocol run. Intuitively, $P(k_1, \ldots, k_n) = b$ holds if in *every* fair execution starting at C_0, all agents eventually agree to b and do not ever change their mind—so, loosely speaking, population protocols compute by reaching consensus. But what does it mean that the agents agree to b? For this, we define a mapping that assigns to every state $q \in Q$ an *opinion* $O(q) \in \{0, 1\}$, with the intended meaning that agents at state q have opinion $O(q)$. Therefore, we have $P(k_1, \ldots, k_n) = b$ if every fair computation $C_0 C_1 C_2 \ldots$ reaches a configuration C_i such that for every $j \geq i$ and for every state q satisfying $C_j(q) > 0$, the state q satisfies $O(q) = b$.

Observe that not every protocol computes a predicate. It is easy to construct protocols such that for some initial configuration C_0 some fair computation never agrees to a value, or for which two fair computations agree to two different values. We call such protocols *ill specified*, and say that a protocol is *well specified* if it is not ill specified.

Computing Power. Most of the work on population protocols carried out by the distributed computing community has concentrated on characterizing the predicates computable by well-specified protocols. In particular, Angluin et al. [3,4] gave explicit well-specified protocols to compute every predicate definable in Presburger arithmetic, the first-order theory of addition, and showed in a later paper (with a different set of authors) that they cannot compute anything else, i.e., well-specified population protocols compute exactly the Presburger-definable predicates [5,7]. There is also extensive work on the computational power and runtime of protocols. For example, protocols for any Presburger-definable predicate satisfying a certain runtime guarantee are shown in [6], while other papers have designed particularly efficient protocols for specific predicates (see e.g. [2]).

Verification Problems for Population Protocols. The verification community is interested in other questions. Given a population protocol, is it well specified? Given a population protocol and a Presburger predicate (represented by a Presburger formula), does the protocol compute the predicate? Given a well-specified protocol, can we automatically obtain a symbolic representation of the predicate it computes? It is also interested in properties concerning the executions of a protocol, seen not as a device for the distributed computation of a predicate, but as a concurrent system. In particular, we are interested in checking if a protocol satisfies a given temporal logic specification expressed in, for example, Linear Temporal Logic, one of the standard specification languages. All these problems are challenging because of their *parameterized* nature. The semantics of a population protocol is an *infinite* family of finite-state transition systems, one for each possible input. Therefore, any of the problems above is actually a question of the form: Do all members of the family satisfy a certain property?

Deciding the property for one single member requires to inspect only one finite transition system, and can be done automatically using a model checker. This is the approach that verification researchers investigated first [11,12,28,31], but it only proves the correctness of a protocol for a finite number of inputs. Alternatively, one can also formalize a proof of well specification in a theorem prover [15], but this approach is not automatic: a human prover must first come up with a proof for each particular protocol. Can we go beyond this? The verification community has studied parameterized verification problems for a long time (see [9,17]), and so at first sight one could feel optimistic. Unfortunately, the techniques developed in this area cannot be applied to the analysis of population protocols. It is worth to spend some lines explaining why.

Classical Parameterized Verification Techniques Do Not Work. Many simple parameterized models of computation have Turing power, and so all interesting analysis problems are undecidable for them [8]. For example, given a Turing machine M and an input x, we can easily construct a little finite-state program that simulates a tape cell. The program has a boolean variable indicating whether the head is on the cell or not, a variable storing the current tape symbol, and a third variable storing the current control state when the head is on the cell. A agent running the program communicates with its left and right neighbors by message passing. If M accepts x, then it does so using a finite number N of tape cells. Therefore, the instance of the system containing N agents eventually reaches a configuration in which the value of the control-state variable of a agent is a final state of M. On the contrary, if M does not accept x, then no instance, however large, ever reaches such a configuration. So the reachability problem for parameterized programs is undecidable. However, this proof sketch contains the sentence "the program communicates with its left and right neighbors". This is achieved by giving agents an *identity*, typically a number in the range $[1..N]$. This number appears as a parameter i in the code, and so it is not the case that all agents execute *exactly* the same code, but the code where the parameter is instantiated with the agent identity. Since in population protocols agents have no identity, the argument above does not apply. So, are parameterized verification questions for systems consisting of completely identical agents decidable, and, if this is the case, can these results be applied to population protocols?

The answer to the first question is positive but, unfortunately, the second is not. Many questions about systems of identical agents can be proved decidable using the theory of well-quasi orders [1,21], which have become the standard mathematical tool for parameterized verification. Loosely speaking, the theory can be applied when the specification to be checked satisfies a monotonicity property: if the instance consisting of N agents satisfies the property, then for every K the instance with $N + K$ agents will also satisfy it. However, the central verification questions for population protocols, like well-specification, are not monotonic. This is due to the notion of fairness inherently present in population protocols. While an execution of the instance of the protocol running with N agents is also an execution of the instance with $N + K$ agents (the K additional agents just do not participate in the execution), the same is not true for *fair* executions, because the new K processes cannot be indefinitely left out. This problem can be overcome for some simple notions of fairness, but not for the particularly demanding one required by population protocols.

Parameterized Verification of Population Protocols. In 2014 I started an initiative to attack the parameterized verification of population protocols, which has been (and continues to be) quite successful. Together with my colleagues and students Michael Blondin, Pierre Ganty, Stefan Jaax, Jérôme Leroux, Rupak Majumdar, and Philipp J. Meyer, we have

- obtained decidability and undecidability results that establish the limits of what is algorithmically achievable, and several complexity bounds showing that many questions have very high complexity; and
- initiated the study of population protocols that are *good for verification*: classes of protocols with the same expressive power as the general class, but with more tractable verification problems.

In the rest of this note I summarize our results.

Decidability of Central Verification Questions. In [18], a paper published at CONCUR 2015, we showed that the three central verification problems for population protocols are all decidable:

(a) Given a population protocol, is it well specified?
(b) Given a population protocol and a Presburger predicate (represented by a Presburger formula), does the protocol compute the predicate?
(c) Given a well-specified protocol, can we automatically obtain a symbolic representation of the predicate it computes?

We call (b) the *fitting problem* (whether a given protocol *fits* the specification), and (c) the *tailor problem* (*tailoring* a protocol that fits the specification). The proofs relied on breakthrough results by Jérôme Leroux on the theory of Petri nets, a model very close to population protocols [22–25]. Shortly after submitting this paper we were able to substantially improve our results: We showed that questions (a) and (b) are recursively equivalent to the reachability problem for Petri nets, eliminating the need for the most advanced and complicated of Leroux's results (the ones of [22]). These improvements are contained in the journal version of [18], recently published in Acta Informatica [20] (I do not recommend to read [18] any more!).

The reductions from the reachability problem show that there is little hope of finding reasonably efficient algorithms for arbitrary protocols, even small ones: The reachability problem is known to be EXPSPACE-hard, and all known algorithms for it have non-primitive recursive complexity (actually, the first explicit upper bound for the problem was only obtained very recently, see [29] for a recent survey). In particular, there are no stable implementations of any of these algorithms, and they are considered impractical for nearly all applications.

To solve problem (c), we introduce a notion of certificate of well-specification for a protocol. We provide algorithms that, given a protocol and an advice string decide if the string is a certificate, and extract from it a Presburger formula of the predicate computed by the protocol. The overall algorithm for problem (c) just enumerates all advice strings, checks if they are a certificate, and if so computes a formula. However, this algorithm may not terminate if a protocol happens to have no certificates. So we also show that this is not the case: every well-specified protocol has at least one certificate. Since this certificate approach only provides two semi-decision algorithms for problem (c), we know even less concerning its complexity, no upper bound has been given yet.

Decidability of Model Checking Problems. Population protocols have been used to model systems beyond their initial motivation in distributed computing. In particular, they can also model trust propagation [16], evolutionary dynamics [26], or chemical reaction systems [27,30]. These systems do not always aim at the computation of predicates, or, if they do, they do not compute them in the way defined by Angluin et al. [3]. With more diverse applications of population protocols comes also new properties one would like to reason about. For instance, Delporte-Gallet et al. [14] studied privacy in population protocols. They proved (by hand) different properties of specific protocols, like "the system can reach a good configuration without any interaction involving a distinguished agent p_0". For these reasons, in [19], published at FSTTCS 2016, we studied the general model checking problem for population protocols against probabilistic linear-time (LTL) specifications. We use the probabilistic semantics in which the scheduler selecting the agents at each step proceeds stochastically. We assume that each transition carries a label that can be observed whenever it occurs. This assigns to each computation an infinite word over the set of labels. We can then speak of the probability of the set of computations satisfying a given formula ϕ of LTL. We show that the qualitative model checking problem (i.e., deciding if ϕ holds with probability 1) is decidable, while the quantitative problem (deciding if the probability that ϕ holds exceeds a given probability) is undecidable.

Efficiently Verifiable Protocols. The theoretical results of [18,20] on the well-specification problem can be reformulated as: the membership problem for the class WS of well-specified protocols is decidable but at least as hard as the reachability problem for Petri nets. This motivates the search for a class of protocols properly contained in WS and satisfying the following three properties:

(a) *No loss of expressive power*: the class should compute all Presburger-definable predicates.
(b) *Naturality*: the class should contain most protocols discussed in the literature.
(c) *Feasible membership problem*: membership for the class should have reasonable complexity.

The class WS obviously satisfies (a) and (b), but not (c). In our most recent work, currently under submission and available online in [10], we introduce a new class WS^3, standing for *Well-Specified Strongly Silent* protocols. We show that WS^3 still satisfies (a) and (b), and then prove that the membership problem for WS^3 is in the complexity class DP; the class of languages L such that $L = L_1 \cap L_2$ for some languages $L_1 \in$ NP and $L_2 \in$ coNP. This is a dramatic improvement with respect to the EXPSPACE-hardness of the membership problem for WS.

Our proof that the problem is in DP reduces membership for WS^3 to checking (un)satisfiability of two systems of boolean combinations of linear constraints over the natural numbers. This allows us to implement our decision procedure on top of the constraint solver Z3 [13], yielding the first software able to automatically prove well-specification for *all* initial configurations. We have tested our implementation on the families of protocols studied in [11,12,28,31]. These papers prove correctness for some inputs of protocols with up to 9 states and 28 transitions. Our approach proves correctness for all inputs of protocols with up to 20 states in less than one second, and protocols with 70 states and 2500 transitions in less than one hour. In particular, we can automatically prove well-specification for *all* inputs in less time than previous tools needed to check one single large input.

Acknowledgments. I want to thank my colleagues Pierre Ganty, Jérôme Leroux, and Rupak Majumdar, my postdoc Michael Blondin, and my PhD students Stefan Jaax and Philipp J. Meyer for agreeing to put their brains to work on these questions. Thank you also to Dejvuth Suwimonteerabuth for implementing the little simulator I use in my talks on population protocols.

References

1. Abdulla, P.A., Cerans, K., Jonsson, B., Tsay, Y.-K.: General decidability theorems for infinite-state systems. In: Proceedings of the 11th Annual IEEE Symposium on Logic in Computer Science, LICS 1996, pp. 313–321. IEEE Computer Society (1996)
2. Alistarh, D., Gelashvili, R., Vojnovic, M.: Fast and exact majority in population protocols. In: Proceedings of the ACM Symposium on Principles of Distributed Computing, PODC 2015, pp. 47–56. ACM (2015)

3. Angluin, D., Aspnes, J., Diamadi, Z., Fischer, M.J., Peralta, R.: Computation in networks of passively mobile finite-state sensors. In: PODC 2004, pp. 290–299. ACM (2004)
4. Angluin, D., Aspnes, J., Diamadi, Z., Fischer, M.J., Peralta, R.: Computation in networks of passively mobile finite-state sensors. Distrib. Comput. 18(4), 235–253 (2006)
5. Angluin, D., Aspnes, J., Eisenstat, D.: Stably computable predicates are semilinear. In: PODC 2006, pp. 292–299. ACM (2006)
6. Angluin, D., Aspnes, J., Eisenstat, D.: Fast computation by population protocols with a leader. Distrib. Comput. 21(3), 183–199 (2008)
7. Angluin, D., Aspnes, J., Eisenstat, D., Ruppert, E.: The computational power of population protocols. Distrib. Comput. 20(4), 279–304 (2007)
8. Apt, K.R., Kozen, D.C.: Limits for automatic verification of finite-state concurrent systems. Inf. Process. Lett. 22(6), 307–309 (1986)
9. Bloem, R., Jacobs, S., Khalimov, A., Konnov, I., Rubin, S., Veith, H., Widder, J.: Decidability of parameterized verification. In: Synthesis Lectures on Distributed Computing Theory. Morgan & Claypool Publishers (2015)
10. Blondin, M., Esparza, J., Jaax, S., Meyer, P.: Towards efficient verification of population protocols. CoRR, abs/1703.04367 (2017)
11. Chatzigiannakis, I., Michail, O., Spirakis, P.G.: Algorithmic Verification of Population Protocols. In: Dolev, S., Cobb, J., Fischer, M., Yung, M. (eds.) SSS 2010. LNCS, vol. 6366, pp. 221–235. Springer, Heidelberg (2010). doi:10.1007/978-3-642-16023-3_19
12. Clement, J., Delporte-Gallet, C., Fauconnier, H., Sighireanu, M.: Guidelines for the verification of population protocols. In: ICDCS 2011, pp. 215–224 (2011)
13. Moura, L., Bjørner, N.: Z3: an efficient SMT solver. In: Ramakrishnan, C.R., Rehof, J. (eds.) TACAS 2008. LNCS, vol. 4963, pp. 337–340. Springer, Heidelberg (2008). doi:10.1007/978-3-540-78800-3_24
14. Delporte-Gallet, C., Fauconnier, H., Guerraoui, R., Ruppert, E.: Secretive birds: privacy in population protocols. In: Tovar, E., Tsigas, P., Fouchal, H. (eds.) OPODIS 2007. LNCS, vol. 4878, pp. 329–342. Springer, Heidelberg (2007). doi:10.1007/978-3-540-77096-1_24
15. Deng, Y., Monin, J.-F.: Verifying self-stabilizing population protocols with Coq. In: TASE 2009, pp. 201–208. IEEE Computer Society (2009)
16. Diamadi, Z., Fischer, M.J.: A simple game for the study of trust in distributed systems. Wuhan Univ. J. Nat. Sci. 6(1–2), 72–82 (2001)
17. Esparza, J.: Keeping a crowd safe: on the complexity of parameterized verification (invited talk). In: STACS. LIPIcs, vol. 25, pp. 1–10. Schloss Dagstuhl - Leibniz-Zentrum fuer Informatik (2014). Corrected version available in CoRR, abs/1405.1841
18. Esparza, J., Ganty, P., Leroux, J., Majumdar, R.: Verification of population protocols. In: Proceedings of the 26th International Conference on Concurrency Theory (CONCUR), pp. 470–482 (2015)
19. Esparza, J., Ganty, P., Leroux, J., Majumdar, R.: Model checking population protocols. In: FSTTCS. LIPIcs, vol. 65, pp. 27:1–27:14 (2016)
20. Esparza, J., Ganty, P., Leroux, J., Majumdar, R.: Verification of population protocols. Acta Inf. 54(2), 191–215 (2017)
21. Finkel, A., Schnoebelen, P.: Well-structured transition systems everywhere! Theor. Comput. Sci. 256(1–2), 63–92 (2001)
22. Leroux, J.: The general vector addition system reachability problem by Presburger inductive invariants. In: LICS 2009, pp. 4–13. IEEE Computer Society (2009)

23. Leroux, J.: Vector addition system reversible reachability problem. In: Katoen, J.-P., König, B. (eds.) CONCUR 2011. LNCS, vol. 6901, pp. 327–341. Springer, Heidelberg (2011). doi:10.1007/978-3-642-23217-6_22
24. Leroux, J.: Presburger vector addition systems. In: LICS 2013, pp. 23–32. IEEE Computer Society (2013)
25. Leroux, J.: Vector addition system reversible reachability problem. Log. Methods Comput. Sci. **9**(1) (2013)
26. Moran, P.A.P.: Random processes in genetics. Math. Proc. Camb. Philos. Soc. **54**(1), 60–71 (1958)
27. Navlakha, S., Bar-Joseph, Z.: Distributed information processing in biological and computational systems. Commun. ACM **58**(1), 94–102 (2014)
28. Pang, J., Luo, Z., Deng, Y.: On automatic verification of self-stabilizing population protocols. In: Proceedings of the 2nd IEEE/IFIP International Symposium on Theoretical Aspects of Software Engineering (TASE), pp. 185–192 (2008)
29. Schmitz, S.: The complexity of reachability in vector addition systems. SIGLOG News **3**(1), 4–21 (2016)
30. Soloveichik, D., Cook, M., Winfree, E., Bruck, J.: Computation with finite stochastic chemical reaction networks. Nat. Comput. **7**(4), 615–633 (2008)
31. Sun, J., Liu, Y., Dong, J.S., Pang, J.: PAT: towards flexible verification under fairness. In: Bouajjani, A., Maler, O. (eds.) CAV 2009. LNCS, vol. 5643, pp. 709–714. Springer, Heidelberg (2009). doi:10.1007/978-3-642-02658-4_59

Verification of Quantum Computation
and the Price of Trust

Alexandru Gheorghiu[1], Theodoros Kapourniotis[2], and Elham Kashefi[1,3(✉)]

[1] School of Informatics, University of Edinburgh, Edinburgh, UK
ekashefi@inf.ed.ac.uk
[2] Department of Physics, University of Warwick, Coventry, UK
[3] CNRS LIP6, Université Pierre et Marie Curie, Paris, France

This is an extended abstract of the presentation made at CSR 2017. Complete details can be found in our upcoming paper [1].

Quantum computers promise to efficiently solve not only problems believed to be intractable to classical computers [2], but also problems for which verifying the solution is considered intractable [3]. In particular, there are problems in the complexity class BQP, i.e. solvable in polynomial time by a quantum computer, that are believed to be outside of NP, the class of problems for which checking the solution can be performed in polynomial time by a classical computer. This raises the question of how one can verify whether quantum computers are indeed producing correct results. Answering this question leads to *quantum verification*, which has been highlighted as a significant challenge on the road to scalable quantum computing technology. Verification is pertinent to both medium-sized quantum computers, expected to be developed in under a decade, but also to future quantum cloud supercomputers used by remote users. It is also relevant for experiments of quantum mechanics, where the size of the system involved is beyond the regime of classical simulation. In this paper we attempt to categorize the different methods of quantum verification that have appeared in recent years. Since most of them are based on cryptographic primitives and treat quantum devices as untrusted entities, we highlight a general trade-off between trust assumptions and complexity.

The setting in which quantum verification has been studied extensively is that of *interactive proof systems*. This involves two distinct entities: a trusted party called the *verifier* (also known as *client*), tasked with verifying the correctness of a computation and an untrusted party called the *prover* (also known as *server*), who runs the computation and attempts to convince the verifier of the result. Formally, for some language $L \in$ BQP the verifier wants to know, for an input x, whether $x \in L$ or $x \notin L$. The prover is trying to convince the verifier that one of these statements is true usually by demonstrating that it has performed the correct quantum computation. To ensure this, in a typical run of a verification protocol, the verifier asks the prover to not only perform the quantum computation, but also a series of trials that will be used to test his behaviour. Cryptographic methods are applied so that the prover cannot distinguish the tests from the computation and try to cheat selectively. This class of protocols constitutes the majority of verification protocols developed so far. For this reason, in our paper, we will primarily review these types of approaches.

P. Weil (Ed.): CSR 2017, LNCS 10304, pp. 15–19, 2017.
DOI: 10.1007/978-3-319-58747-9_3

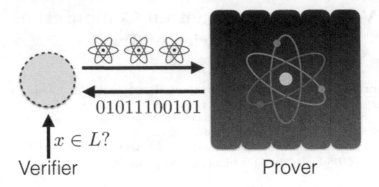

Fig. 1. Prepare and send verification protocol

It is worth mentioning that all techniques reviewed in this paper assume that the prover can deviate in any possible way that follows the laws of quantum mechanics.

Essential to the effectiveness of a verification protocol is the ascription of trust to some of the used devices. Ideally, one wants to restrict the trust to the classical computer which the verifier controls. However, all existing approaches require some extra trust assumptions on the quantum devices or the channels involved in the protocol. For instance, protocols in which the verifier interacts with a single quantum prover require the verifier to possess a trusted quantum device. If there is more than one prover, the verifier can indeed be fully classical, but then the provers are forbidden from interacting with each other. Our goal in this paper is to highlight the trade-off between the trust assumptions of each verification technique and the required resources to achieve the same level of confidence in the verification.

We proceed by first considering protocols which make use of cryptographic primitives and have *information-theoretic security*. These protocols are divided into two broad categories:

1. *Prepare and send/receive and measure protocols.* These are protocols in which the verifier and the prover exchange qubits through some quantum channel. As the name suggests, the verifier either prepares and sends qubits to the prover [4–9] or, alternatively, receives qubits from the prover and measures them [10–12]. In the first case, the verifier relies on the uncertainty principle and the no-cloning theorem to ensure that the prover cannot distinguish tests from computations. In the second case, the verifier uses a type of cut-and-choose technique to decide whether to test the prover or perform a computation using the received states. In both cases, the essential element is the fact that the prover is oblivious to some part of the delegated computation. This property is commonly referred to as *blindness* [13–20] and is a shared feature of most verification protocols. A schematic illustration of a prepare and send protocol is shown in Fig. 1.

2. *Entanglement-based protocols.* These are protocols in which entangled states are shared either between the verifier and the prover [21,22] or between multiple provers [23,24]. One of the main reasons for considering the entanglement-based setting is because it can lead to *device-independent* verification. In other words, because of the remarkable properties of non-local correlations, it is possible to verify a quantum computation in a situation in which all quantum devices are untrusted. It is, however, necessary to assume that the quantum devices sharing entanglement are not communicating throughout the protocol. In this case, the verifier needs to test not only the prover performing the computation, but also any other quantum device that is sharing entanglement. Depending on the trust assumptions about the shared entangled states as well as the measurement devices we notice different scalings for the communication complexity of the protocols as we show in the table below.

We then also consider protocols which are not based in cryptography, but are more akin to quantum state certification. These are known as *post-hoc* verification protocols [25–27] and can also be categorized as either receive and measure or entanglement-based. While the cryptographic protocols aim to test the operations performed by the prover(s) towards achieving universal quantum computation, post-hoc protocols simply check quantum witnesses for decision problems. In other words, deciding whether some input x belongs or not to a language $L \in \mathsf{BQP}$ reduces to performing a two-outcome measurement of a quantum witness state $|\psi\rangle$. The protocols either have the prover send this state to the verifier to be measured, or the verifier coordinates a set of entangled provers to prepare and measure $|\psi\rangle$.

Measurements	Entanglement		
	Trusted	Semi-trusted	Untrusted
Trusted	$O(N)$	$O(N^2)$	$O(N^{13}log(N))$
Untrusted	$O(N^2)$	$O(N^2)$	$O(N^{64})$

In both the previously mentioned cryptographic and post-hoc protocols, there are no limiting assumptions about the computational powers of the provers. In other words, even though we regard them as BQP machines, verification is possible even if the provers are computationally unbounded. Recently, however, verification protocols have been proposed for settings in which the provers are limited to a sub-universal model of quantum computations. The two that we review are for the *one-pure-qubit model* [28] and the *instantaneous quantum polynomial-time model* (or IQP) [29,30].

Lastly, we address the issue of *fault-tolerance* [31]. This entails the ability to perform verification in a setting where quantum devices and quantum states are subject to noise that scales with the size of the system. Achieving fault-tolerant verification is crucial for the practical applicability of these protocols and their use in near-future experiments of *quantum supremacy* (attempting to demonstrate the "supraclassical power" of quantum computing).

By categorizing and analysing the resources required in each protocol, while at the same time making the trust assumptions explicit, we illustrate the bigger picture of quantum verification in the delegated setting. This highlights the significant overlap between quantum computation, cryptography and complexity theory and can serve as a guide for the development and improvement of future protocols.

References

1. Gheorghiu, A., Kapourniotis, T., Kashefi, E.: Verification of quantum computation and the price of trust (2015)
2. Shor, P.W.: Polynomial-time algorithms for prime factorization and discrete logarithms on a quantum computer. SIAM Rev. **41**(2), 303–332 (1999)
3. Aharonov, D., Jones, V., Landau, Z.: A polynomial quantum algorithm for approximating the Jones polynomial. In: Proceedings of the Thirty-Eighth Annual ACM Symposium on Theory of Computing, pp. 427–436. ACM (2006)
4. Aharonov, D., Ben-Or, M., Eban, E.: Interactive proofs for quantum computations. In: Proceedings of Innovations in Computer Science 2010 (ICS 2010), p. 453 (2010)
5. Fitzsimons, J.F., Kashefi, E.: Unconditionally verifiable blind computation. arXiv preprint (2012). arXiv:1203.5217
6. Broadbent, A.: How to verify a quantum computation. arXiv preprint (2015). arXiv:1509.09180
7. Barz, S., Fitzsimons, J.F., Kashefi, E., Walther, P.: Experimental verification of quantum computation. Nat. Phys. **9**(11), 727–731 (2013)
8. Kapourniotis, T., Dunjko, V., Kashefi, E.: On optimising quantum communication in verifiable quantum computing. arXiv preprint (2015). arXiv:1506.06943
9. Kashefi, E., Wallden, P.: Optimised resource construction for verifiable quantum computation. arXiv preprint (2015). arXiv:1510.07408
10. Morimae, T.: Measurement-only verifiable blind quantum computing with quantum input verification (2016). arXiv preprint arXiv:1606.06467
11. Hayashi, M., Morimae, T.: Verifiable measurement-only blind quantum computing with stabilizer testing. arXiv preprint (2015). arXiv:1505.07535
12. Hayashi, M., Hajdusek, M.: Self-guaranteed measurement-based quantum computation. arXiv preprint (2016). arXiv:1603.02195
13. Arrighi, P., Salvail, L.: Blind quantum computation. Int. J. Quantum Inform. **4**(05), 883–898 (2006)
14. Childs, A.: Secure assisted quantum computation. Quant. Inf. Compt. **5**(6), 456 (2005)
15. Broadbent, A., Fitzsimons, J., Kashefi, E.: Universal blind quantum computation. In: 50th Annual IEEE Symposium on Foundations of Computer Science (FOCS 2009), pp. 517–526. IEEE (2009)
16. Mantri, A., Perez-Delgado, C.A., Fitzsimons, J.F.: Optimal blind quantum computation. Phys. Rev. Lett. **111**(23), 230502 (2013)
17. Perez-Delgado, C.A., Fitzsimons, J.F.: Overcoming efficiency constraints on blind quantum computation. arXiv preprint (2014). arXiv:1411.4777
18. Giovannetti, V., Maccone, L., Morimae, T., Rudolph, T.G.: Efficient universal blind quantum computation. Phys. Rev. Lett. **111**(23), 230501 (2013)

19. Morimae, T., Fujii, K.: Blind quantum computation protocol in which alice only makes measurements. Phys. Rev. A **87**(5), 050301 (2013)
20. Sueki, T., Koshiba, T., Morimae, T.: Ancilla-driven universal blind quantum computation. Phys. Rev. A **87**(6), 060301 (2013)
21. Gheorghiu, A., Kashefi, E., Wallden, P.: Robustness and device independence of verifiable blind quantum computing. arXiv preprint (2015). arXiv:1502.02571
22. Hajdusek, M., Perez-Delgado, C.A., Fitzsimons, J.F.: Device-independent verifiable blind quantum computation. arXiv preprint (2015). arXiv:1502.02563
23. Reichardt, B.W., Unger, F., Vazirani, U.: Classical command of quantum systems. Nature **496**(7446), 456–460 (2013)
24. McKague, M.: Interactive proofs for BQP via self-tested graph states (2013). arXiv:1309.5675
25. Fitzsimons, J.F., Hajdušek, M.: Post hoc verification of quantum computation. arXiv preprint (2015). arXiv:1512.04375
26. Morimae, T., Fitzsimons, J.F.: Post hoc verification with a single prover. arXiv preprint (2016). arXiv:1603.06046
27. Hangleiter, D., Kliesch, M., Schwarz, M., Eisert, J.: Direct certification of a class of quantum simulations. arXiv preprint (2016). arXiv:1602.00703
28. Kapourniotis, T., Kashefi, E., Datta, A.: Blindness and verification of quantum computation with one pure qubit. In: 9th Conference on the Theory of Quantum Computation, Communication and Cryptography (TQC 2014), vol. 27, pp. 176–204 (2014)
29. Mills, D., Pappa, A., Kapourniotis, T., Kashefi, E.: Information theoretically secure hypothesis test for temporally unstructured quantum computation (2017)
30. Kapourniotis,T., Datta, A.: Nonadaptive fault-tolerant verification of quantum supremacy with noise (2017)
31. Shor, P.W.: Fault-tolerant quantum computation. In: 37th Annual Symposium on Foundations of Computer Science, Proceedings, pp. 56–65. IEEE (1996)

Graph Sketching and Streaming: New Approaches for Analyzing Massive Graphs

Andrew McGregor$^{(\boxtimes)}$

University of Massachusetts, Amherst, USA
mcgregor@cs.umass.edu

Abstract. In this invited talk, we will survey some of the recent work on designing algorithms for analyzing massive graphs. Such graphs may not fit in main memory, may be distributed across numerous machines, and may change over time. This has motivated a rich body of work on analyzing graphs in the data stream model and the development of general algorithmic techniques, such as graph sketching, that can help minimize the space and communication costs required to process these massive graphs.

1 Motivation and Definitions

If you pick up your favorite algorithms textbooks and turn to a random page, there is a reasonable chance that you will find an algorithm for solving a graph problem. This is perhaps unsurprising given that graphs are a natural abstraction whenever you have information about a set of basic entities and the relationships between these entities, e.g., people and their friendships, web-pages and hyperlinks; neurons and synapses; or IP addresses and network flows. However, many of the classical algorithms for analyzing graphs implicitly assume that the graphs are static and fit in the main memory of a single machine. Unfortunately, in a growing number of applications this is not the case and attention has turned to algorithms that can process streams of graph data and/or graph data that is distributed across numerous machines. In these scenarios, standard graph techniques and primitives such as constructing BFS or DFS trees, dynamic programming, and linear programming are no longer applicable and new techniques, such as graph sketching, are required.

In the accompanying talk, we will survey some of the recent work on these new approaches to analyzing massive graphs. In this document, we first present some of the relevant definitions and then collect together references for some of the main results that will be discussed.

Basic Definitions. The simplest version of the data stream model for processing graphs is the *insert-only model*. In this model, the input stream consists of a

---

This work was supported in part by NSF Awards CCF-0953754, IIS-1251110, CCF-1320719, CCF-1637536, and a Google Research Award.

P. Weil (Ed.): CSR 2017, LNCS 10304, pp. 20–24, 2017.
DOI: 10.1007/978-3-319-58747-9_4

sequence of unordered pairs $e = \{u, v\}$ where $u, v \in [n]$. Such a stream,

$$S = \langle e_1, e_2, \ldots, e_m \rangle$$

naturally defines an undirected graph $G = (V, E)$ where $V = [n]$ and $E = \{e_1, \ldots, e_m\}$. The goal is to design an algorithm that solves the required graph problem on the graph G while only accessing the input sequentially and using memory that is sublinear in m.

A natural extension of the model is the *insert-delete model* in which edges can be both inserted and deleted. In this case, the input is a sequence

$$S = \langle a_1, a_2, \ldots \rangle \quad \text{where} \quad a_i = (e_i, \Delta_i)$$

where e_i encodes an undirected edge as before and $\Delta_i \in \{-1, 1\}$. The multiplicity of an edge e is defined as $f_e = \sum_{i:e_i=e} \Delta_i$ and we typically restrict our attention to the case where $f_e \in \{0, 1\}$ for all edges e. Both the insert-only and insert-delete model can be extended to handle weighted graphs where the occurrence of each edge e in the input is replaced by the pair (e, w_e) where w_e indicates the weight of the edge.

An important algorithmic technique in the insert-delete model in particular, is that of *graph sketching*. A sketch is a random linear projection of a vector corresponding to the input data. In the context of graphs, this vector would be

$$\mathbf{f} \in \{0, 1\}^{\binom{n}{2}}$$

where entries correspond to the current f_e values and the sketch would be $M\mathbf{f} \subset \mathbb{R}^d$ where $d \ll n^2$ is the dimension of the sketch and M is a random matrix chosen according to an appropriate distribution, i.e., one from which the relevant properties of $\mathbf{f}$ can be inferred given $M\mathbf{f}$. Such a sketch can then be used as the basis for a data stream algorithm since $M\mathbf{f}$ can be computed incrementally: when (e, Δ) arrives in the stream we can update $M\mathbf{f}$ as follows:

$$M\mathbf{f} \leftarrow M\mathbf{f} + \Delta \cdot M^e$$

where M^e is the eth column of M. Hence, it suffices to store the current sketch and any random bits needed to compute the matrix M. The main challenge is therefore to design low-dimensional sketches as this results in small-space algorithms. Sketches are also useful for the purpose of reducing communication in various distributed models since if each machine communicates a sketch of their local input, then a sketch of the entire data set can be recovered simply by adding together the individual sketches.

2   Some Results and References

In this section, we briefly summarize some of the main results in the area. This is not intended to be an exhaustive survey and we focus on the most representative or most recent results on each problem. Further details of some of these algorithms can also be found in the survey [30] although many of the results postdate that survey.

Connectivity and Sparsification. One of the most basic graph problems is determining whether a graph is connected. This and many related problems can be solved relatively easily in the insert-only model using $O(n \text{ polylog } n)$ space, e.g., to test connectivity when there are no edge deletions, it suffices to keep track of the connected components of the graph. Furthermore, it can be shown that $\Omega(n \log n)$ space is necessary to solve this problem [37]. A more surprising result is that $O(n \text{ polylog } n)$ space also suffices in the insert-delete model [3]; this was one of the first applications of the graph sketching technique. Furthermore, the basic algorithm can be extended to testing k-edge connectivity [4] and approximate testing of k-node connectivity [20] using $O(kn \text{ polylog } n)$ space. Lastly, in $O(\epsilon^{-2}n \text{ polylog } n)$ space it is possible to construct combinatorial and spectral sparsifiers of the input graph [20,28]; these allow the size of all cuts to be approximated up to a $1 + \epsilon$ factor along with various properties related to the Laplacian of the graph.

Matching. Most of the work on approximating maximum matchings has focused on the insert-only model. The trivial greedy approach yields a 2-approximation using $O(n \log n)$ space in the unweighted case and after a long sequence of papers, a $(2+\epsilon)$-approximation algorithm using $O(\epsilon^{-1}n \log n)$ space in the weighted case is now known [19,35]. The best known lower bound is that no algorithm can beat a factor $e/(e-1) \approx 1.58$ while using only $O(n \text{ polylog } n)$ space [25] and closing the gap remains an open problem. Better approximation guarantees or lower space requirements are possible if the algorithm may take a small number of additional passes over the data stream [2,17,23,29] or if the edges of the graph are assumed to arrive in a random order [26,29]. Another line of work considers low arboricity graphs, e.g., the size of the maximum matching in a planar graph can be approximated up to a $(5 + \epsilon)$ factor using $O(\epsilon^{-2} \log n)$ space [15,32].

In the insert-delete model, it is known that $\Theta(n^2/\alpha^3 \cdot \text{polylog } n)$ space is necessary and sufficient to find a matching that is at least $1/\alpha$ times the size of the maximum matching [8,13]. This can be reduced to $O(n^2/\alpha^4 \cdot \text{polylog } n)$ space if we are only interested in estimating the size of the maximum matching [7]. Furthermore, if the size of the maximum matching is bounded by k, then $\Theta(k^2 \text{ polylog } n)$ space is necessary and sufficient to find a matching of maximum size [11,13].

And more... Other graph problems considered in the data stream model include finding the densest subgraph [10,18,31]; correlation clustering [1]; counting triangles [9,24,33], estimating the size of the maximum cut [27], finding large independent sets and cliques [14,21], and performing random walks [36]. Some of the above problems have also been considered in the *sliding window model*, a variant of the insert-only model in which the relevant graph is defined by only the most recent edges [16]. Another notable body of related work considers problems that can be described in terms of hypergraphs, i.e., every edge in the stream includes an arbitrary number of nodes rather than just two. Such problems include minimum set cover [5,6,12,22], maximum coverage [5,34], and minimum hitting set [13].

References

1. Ahn, K.J., Cormode, G., Guha, S., McGregor, A., Wirth, A.: Correlation clustering in data streams. In: International Conference on Machine Learning, pp. 2237–2246 (2015)
2. Ahn, K.J., Guha, S.: Linear programming in the semi-streaming model with application to the maximum matching problem. Inf. Comput. **222**, 59–79 (2013)
3. Ahn, K.J., Guha, S., McGregor, A.: Analyzing graph structure via linear measurements. In: ACM-SIAM Symposium on Discrete Algorithms, pp. 459–467 (2012)
4. Ahn, K.J., Guha, S., McGregor, A.: Graph sketches: sparsification, spanners, and subgraphs. In: ACM Symposium on Principles of Database Systems, pp. 5–14 (2012)
5. Assadi, S.: Tight Space-Approximation Tradeoff for the Multi-Pass Streaming Set Cover Problem. ArXiv e-prints, March 2017
6. Assadi, S., Khanna, S., Li, Y.: Tight bounds for single-pass streaming complexity of the set cover problem. In: ACM Symposium on Theory of Computing, pp. 698–711 (2016)
7. Assadi, S., Khanna, S., Li, Y.: On estimating maximum matching size in graph streams. In: ACM-SIAM Symposium on Discrete Algorithms, pp. 1723–1742 (2017)
8. Assadi, S., Khanna, S., Li, Y., Yaroslavtsev, G.: Maximum matchings in dynamic graph streams and the simultaneous communication model. In: ACM-SIAM Symposium on Discrete Algorithms, pp. 1345–1364 (2016)
9. Bera, S.K., Chakrabarti, A.: Towards tighter space bounds for counting triangles and other substructures in graph streams. In: Symposium on Theoretical Aspects of Computer Science, pp. 11:1–11:14 (2017)
10. Bhattacharya, S., Henzinger, M., Nanongkai, D., Tsourakakis, C.E.: Space- and time-efficient algorithm for maintaining dense subgraphs on one-pass dynamic streams. In: ACM Symposium on Theory of Computing, pp. 173–182 (2015)
11. Bury, M., Schwiegelshohn, C.: Sublinear estimation of weighted matchings in dynamic data streams. In: European Symposium on Algorithms, pp. 263–274 (2015)
12. Chakrabarti, A., Wirth, A.: Incidence geometries and the pass complexity of semi-streaming set cover. In: ACM-SIAM Symposium on Discrete Algorithms, pp. 1365–1373 (2016)
13. Chitnis, R., Cormode, G., Esfandiari, H., Hajiaghayi, M., McGregor, A., Monemizadeh, M., Vorotnikova, S.: Kernelization via sampling with applications to finding matchings and related problems in dynamic graph streams. In: ACM-SIAM Symposium on Discrete Algorithms, pp. 1326–1344 (2016)
14. Cormode, G., Dark, J., Konrad, C.: Independent set size approximation in graph streams. CoRR abs/1702.08299 (2017). http://arxiv.org/abs/1702.08299
15. Cormode, G., Jowhari, H., Monemizadeh, M., Muthukrishnan, S.: The sparse awakens: Streaming algorithms for matching size estimation in sparse graphs. CoRR abs/1608.03118 (2016). http://arxiv.org/abs/1608.03118
16. Crouch, M.S., McGregor, A., Stubbs, D.: Dynamic graphs in the sliding-window model. In: European Symposium on Algorithms, pp. 337–348 (2013)
17. Esfandiari, H., Hajiaghayi, M., Monemizadeh, M.: Finding large matchings in semi-streaming. In: IEEE International Conference on Data Mining Workshops, pp. 608–614 (2016)
18. Esfandiari, H., Hajiaghayi, M., Woodruff, D.P.: Brief announcement: applications of uniform sampling: Densest subgraph and beyond. In: ACM Symposium on Parallel Algorithms and Architectures, pp. 397–399 (2016)

19. Ghaffari, M.: Space-optimal semi-streaming for $(2 + \epsilon)$-approximate matching. CoRR abs/1701.03730 (2017). http://arxiv.org/abs/1701.03730
20. Guha, S., McGregor, A., Tench, D.: Vertex and hyperedge connectivity in dynamic graph streams. In: ACM Symposium on Principles of Database Systems, pp. 241–247 (2015)
21. Halldórsson, B.V., Halldórsson, M.M., Losievskaja, E., Szegedy, M.: Streaming algorithms for independent sets. In: International Colloquium on Automata, Languages and Programming, pp. 641–652 (2010)
22. Har-Peled, S., Indyk, P., Mahabadi, S., Vakilian, A.: Towards tight bounds for the streaming set cover problem. In: ACM Symposium on Principles of Database Systems, pp. 371–383 (2016)
23. Kale, S., Tirodkar, S., Vishwanathan, S.: Maximum matching in two, three, and a few more passes over graph streams. CoRR abs/1702.02559 (2017). http://arxiv.org/abs/1702.02559
24. Kallaugher, J., Price, E.: A hybrid sampling scheme for triangle counting. In: ACM-SIAM Symposium on Discrete Algorithms, pp. 1778–1797 (2017)
25. Kapralov, M.: Better bounds for matchings in the streaming model. In: ACM-SIAM Symposium on Discrete Algorithms, pp. 1679–1697 (2013)
26. Kapralov, M., Khanna, S., Sudan, M.: Approximating matching size from random streams. In: ACM-SIAM Symposium on Discrete Algorithms, pp. 734–751 (2014)
27. Kapralov, M., Khanna, S., Sudan, M., Velingker, A.: $1 + \Omega(1)$-approximation to MAX-CUT requires linear space. In: ACM-SIAM Symposium on Discrete Algorithms, pp. 1703–1722 (2017)
28. Kapralov, M., Lee, Y.T., Musco, C., Musco, C., Sidford, A.: Single pass spectral sparsification in dynamic streams. In: IEEE Symposium on Foundations of Computer Science, pp. 561–570 (2014)
29. Konrad, C., Magniez, F., Mathieu, C.: Maximum matching in semi-streaming with few passes. In: Gupta, A., Jansen, K., Rolim, J., Servedio, R. (eds.) APPROX/RANDOM -2012. LNCS, vol. 7408, pp. 231–242. Springer, Heidelberg (2012). doi:10.1007/978-3-642-32512-0_20
30. McGregor, A.: Graph stream algorithms: a survey. SIGMOD Record **43**(1), 9–20 (2014)
31. McGregor, A., Tench, D., Vorotnikova, S., Vu, H.T.: Densest subgraph in dynamic graph streams. In: Mathematical Foundations of Computer Science, pp. 472–482 (2015)
32. McGregor, A., Vorotnikova, S.: A note on logarithmic space stream algorithms for matchings in low arboricity graphs. CoRR abs/1612.02531 (2016). http://arxiv.org/abs/1612.02531
33. McGregor, A., Vorotnikova, S., Vu, H.T.: Better algorithms for counting triangles in data streams. In: ACM Symposium on Principles of Database Systems, pp. 401–411 (2016)
34. McGregor, A., Vu, H.T.: Better streaming algorithms for the maximum coverage problem. In: International Conference in Database Theory, pp. 22:1–22:18 (2017)
35. Paz, A., Schwartzman, G.: A $(2+\epsilon)$-approximation for maximum weight matching in the semi-streaming model. In: ACM-SIAM Symposium on Discrete Algorithms, pp. 2153–2161 (2017)
36. Sarma, A.D., Gollapudi, S., Panigrahy, R.: Estimating pagerank on graph streams. J. ACM **58**(3), 13 (2011)
37. Sun, X., Woodruff, D.P.: Tight bounds for graph problems in insertion streams. In: International Workshop on Approximation Algorithms for Combinatorial Optimization Problems, pp. 435–448 (2015)

Concatenation Hierarchies:
New Bottle, Old Wine

Thomas Place$^{(\boxtimes)}$ and Marc Zeitoun$^{(\boxtimes)}$

LaBRI, Bordeaux University, Bordeaux, France
{tplace,mz}@labri.fr

Abstract. We survey progress made in the understanding of concatenation hierarchies of regular languages during the last decades. This paper is an extended abstract meant to serve as a precursor of a forthcoming long version.

1 Historical Background and Motivations

Our objective in this extended abstract is to outline progress obtained during the last 50 years about *concatenation hierarchies of regular languages* over a fixed, finite alphabet A. Such hierarchies were considered in order to understand the interplay between two basic constructs used to build regular languages: Boolean operations and concatenation. The story started with Kleene's theorem [12], one of the core results in automata theory. It states that languages of finite words recognized by finite automata are exactly the ones that can be described by regular expressions, *i.e.*, are built from the singleton languages and the empty set using a finite number of times operations among three basic ones: union, concatenation, and iteration (also known as Kleene star).

As Kleene's theorem provides another syntax for regular languages, it makes it possible to classify them according to the hardness of describing a language by such an expression. The notion of star-height was designed for this purpose. The *star-height* of a regular expression is its maximum number of nested Kleene stars. The *star-height* of a regular language is the minimum among the star-heights of all regular expressions that define the language. Since there are languages of arbitrary star-height [7,8], this makes the notion an appropriate complexity measure, and justifies the question of computing the star-height of a regular language (it was raised by Eggan [8], see also Brzozowski [4]).

> *Given a regular language and a natural number n, is there an expression of star-height n defining the language?*

This question, called the *star-height problem*, is an instance of a *membership problem*. Given a class $\mathcal{C}$ of regular languages, the membership problem for $\mathcal{C}$ simply asks whether $\mathcal{C}$ is a decidable class, that is:

INPUT: A regular language L
OUTPUT: Does L belong to $\mathcal{C}$?

Funded by the DeLTA project (ANR-16-CE40-0007).

Thus, the star-height problem asks whether membership is decidable for each class $\mathcal{H}_n$, where $\mathcal{H}_n$ is the class of languages having star-height n. It was first solved by Hashiguchi [10], but it took several more years to obtain simpler proofs, see *e.g.*, [3,11].

Kleene's theorem also implies that adding *complement* to our set of basic operations does not make it possible to define more languages. Therefore, instead of just considering regular expressions, one may consider *generalized* regular expressions, where complement is allowed (in addition to union, concatenation and Kleene star). This yields the notion of *generalized star-height*, which is defined as the star-height, but replacing "regular expression" by "generalized regular expression". One may then ask the very same question: is there an algorithm to compute the *generalized star-height* of a regular language? Despite its very simple statement, this question, also raised by Brzozowski [4], is still open. Even more, one does not know whether there exists a regular language of generalized star-height greater than 1. In other words, membership is open for the class of languages of generalized star-height 1 (see [18] for a historical presentation).

This makes it relevant to already focus on languages of star height 0, *i.e.*, that can be described using only union, concatenation and Boolean operations (including complement), but *without* the Kleene star. Such languages are called *star-free*. Surprisingly, even this restricted problem turned out to be difficult. It was solved by Schützenberger [30] in a seminal paper.

Theorem 1 (Schützenberger [30]). *Membership is decidable for the class of star-free languages.*

The star-free languages rose to prominence due to their numerous characterizations, and in particular, the logical one, which is due to McNaughton and Papert [15]. Observe that one may describe languages with logical sentences. Indeed, any word may be viewed as a logical structure made of a linearly ordered sequence of positions, each one carrying a label. In first-order logic over words (denoted by FO($<$)), one may quantify these positions, compare them with a predicate "$<$" interpreted as the (strict) linear order, and check their labels (for any letter a, a unary predicate P_a selecting positions with label a is available). Each FO($<$) sentence states a property over words and defines the language of all words that satisfy it.

Theorem 2 (McNaughton-Papert [15]). *For any regular language L, the following properties are equivalent:*

- *L is star-free.*
- *L can be defined by an FO($<$) sentence.*

Let us point out that this connection is rather intuitive. Indeed, there is a clear correspondence between union, intersection and complement for star-free languages and the Boolean connectives in FO($<$) sentences. Moreover, concatenation corresponds to existential quantification.

1.1   The Dot-Depth and the Straubing-Thérien hierarchies

Just as the star-height measures how complex a regular language is, a natural complexity for star-free languages is the number of alternations between the concatenation product and the complement operation that are required to build a given language from basic star-free languages. This led Brzozowski and Cohen [5] to introduce in the 70s a hierarchy of classes of regular languages, called the *dot-depth hierarchy*. This hierarchy classifies all star-free languages into full levels, indexed by natural numbers: 0, 1, 2,..., and half-levels, indexed by half natural numbers: $\frac{1}{2}, \frac{3}{2}, \frac{5}{2}$, etc. Roughly speaking, level $n \in \mathbb{N}$ consists of all languages that can be expressed by a star-free expression having n alternations between concatenation and Boolean operations.

More formally, the hierarchy is built by using, alternately, two closure operations starting from level 0: *Boolean closure* and *polynomial closure*. Given a class of languages $\mathcal{C}$, its *Boolean closure* (denoted by $Bool(\mathcal{C})$) is the smallest Boolean algebra containing $\mathcal{C}$. Polynomial closure is slightly more complicated as it involves *marked concatenation*. Given two languages L_1 and L_2, a marked concatenation of L_1 with L_2 is a language of the form,

$$L_1 a L_2 \qquad \text{for some } a \in A.$$

We may now define the polynomial closure of $\mathcal{C}$ (denoted by $Pol(\mathcal{C})$) as the smallest class of languages containing $\mathcal{C}$ and closed under union, intersection and marked concatenation (*i.e.*, $L_1 a L_2 \in \mathcal{C}$ for any $L_1, L_2 \in \mathcal{C}$ and any $a \in A$).

The dot-depth hierarchy is now defined as follows:

- Level 0 is the class $\{\emptyset, \{\varepsilon\}, A^+, A^*\}$ (where A is the working alphabet).
- Each *half-level* is the *polynomial closure* of the previous full level: for any natural number $n \in \mathbb{N}$, level $n + \frac{1}{2}$ is the polynomial closure of level n.
- Each *full level* is the *Boolean closure* of the previous half-level: for any natural number $n \in \mathbb{N}$, level $n + 1$ is the Boolean closure of level $n + \frac{1}{2}$.

A side remark is that the above definitions are not the original ones. First, the historical definition of the dot-depth hierarchy started from another class of languages for level 0. However, both definitions coincide at level 1 and above. Next, the polynomial closure of a class $\mathcal{C}$ was historically defined as the smallest class containing $\mathcal{C}$ and closed under both union and marked concatenation. This original definition is intuitively weaker: it does not explicitly require $Pol(\mathcal{C})$ to be closed under intersection. However, it was shown by Arfi [1,2] that the two definitions are equivalent (provided that the class $\mathcal{C}$ satisfy some standard closure properties, which are always fulfilled for classes within concatenation hierarchies). This was also shown later by Pin [17]. We will present an alternative, elementary proof in the full version of this paper.

Clearly, the union of all levels in the dot-depth hierarchy is the whole class of star-free languages. Moreover, it was shown by Brzozowski and Knast that the dot-depth hierarchy is strict: any level contains strictly more languages than the previous one.

Theorem 3 (Brzozowski and Knast [6]). *The dot-depth hierarchy is strict when the alphabet contains at least two letters.*

This shows in particular that in general, Boolean closure does not preserve the property of being polynomially closed, and conversely. In other words, classes built using Boolean and polynomial closure do not satisfy the same closure properties: typically, when $\mathcal{C}$ is a class of languages, $Pol(\mathcal{C})$ is closed under marked concatenation but **not** under complement, while $Bool(\mathcal{C})$ is closed under complement but **not** under marked concatenation.

The fact that the hierarchy is strict motivates the investigation of the membership problem for all levels.

Problem 4 (Membership for the dot-depth hierarchy). For a fixed level in the dot-depth hierarchy, is the membership problem decidable for this level?

Using the framework developed by Schützenberger in his proof for deciding whether a language is star-free, Knast [13] established that level 1 has decidable membership, via a quite intricate proof from the combinatorial point of view.

Theorem 5 (Knast [13]). *Level 1 in the dot-depth hierarchy has decidable membership.*

The case of half levels required to adapt Schützenberger's methodology, since it was designed to deal with Boolean algebras only (recall that half-levels are *not* Boolean algebras, otherwise the hierarchy would collapse). This was achieved by Pin and Weil [21–23] and by Glaßer and Schmitz [9].

Theorem 6 (Pin and Weil [21–23], Glaßer and Schmitz [9]). *Levels $\frac{1}{2}$ and $\frac{3}{2}$ in the dot-depth hierarchy have decidable membership.*

One may now wonder why, in the definition of the dot-depth hierarchy, level 0 is $\{\emptyset, \{\varepsilon\}, A^+, A^*\}$. It would be natural to start from $\{\emptyset, A^*\}$, and to apply the very same construction for higher levels. This is exactly how the Straubing-Thérien hierarchy is defined. It was introduced independently by Straubing [33] and Thérien [35]. Its definition follows the same scheme as that of the dot-depth hierarchy, except that level 0 is $\{\emptyset, A^*\}$.

Like the dot-depth hierarchy, the Straubing-Thérien hierarchy is strict and spans the whole class of star-free languages. This can be shown by proving that level n in the dot-depth hierarchy sits between levels n and $n+1$ in the Straubing-Thérien hierarchy. This makes the membership problem again relevant for each level in this hierarchy.

Problem 7 (Membership for the Straubing-Thérien hierarchy). For a fixed level in the Straubing-Thérien hierarchy, is the membership problem decidable for this level?

Just as for the dot-depth hierarchy, level 1 in the Straubing-Thérien hierarchy was shown to be decidable (actually before the formal definition of the hierarchy itself), and the first half-levels were solved using the adaptation of the framework of Schützenberger to classes that are not closed under complement.

Theorem 8 (Simon [31,32]). *Level 1 in the Straubing-Thérien hierarchy has decidable membership.*

Theorem 9 (Arfi [1,2], Pin and Weil [21,22]). *Levels $\frac{1}{2}$ and $\frac{3}{2}$ in the Straubing-Thérien hierarchy have decidable membership.*

Both hierarchies are strongly related. First, as we already stated, they are interleaved. More importantly, Straubing established an effective reduction between the membership problems associated to their levels [34].

Theorem 10 (Straubing [34]). *Membership for level $n \in \mathbb{N}$ in the dot-depth hierarchy reduces to membership for level n in the Straubing-Thérien hierarchy.*

This theorem is crucial. Indeed, from a combinatorial view, membership is simpler to deal with for the Straubing-Thérien hierarchy rather than for the dot-depth. This is evidenced by all recent publications on the topic: most results for the dot-depth are indirect. They are corollaries of direct results for the Straubing-Thérien hierarchy via the above theorem. Thus, while the name "dot-depth" remains widely used, the Straubing-Thérien hierarchy is much more prominent.

1.2 Quantifier Alternation Hierarchies

Since star-free languages are exactly those that one can define in first-order logic, it is desirable to refine this correspondence level by level, in each of the hierarchies considered so far. A beautiful result of Thomas [36] establishes indeed such a correspondence, and it is very natural. To present it, we first need to slightly extend the standard signature used in first-order logic over words: we add four new predicates in addition to "$<$" and the unary predicates P_a for $a \in A$:

- The (binary) *successor*, interpreted as the successor between positions.
- The (unary) *minimum*, that selects the leftmost position of the word.
- The (unary) *maximum*, that selects the rightmost position of the word.
- The (nullary) *empty* predicate, which holds for the empty word only.

We denote by $\mathrm{FO}(<, +1, \min, \max, \varepsilon)$ the resulting logic. Notice that these predicates are all definable in $\mathrm{FO}(<)$. Therefore, adding them in the signature does not add to the overall expressive power of first-order logic. In other words, $\mathrm{FO}(<)$ and $\mathrm{FO}(<, +1, \min, \max, \varepsilon)$ are equally expressive. However, this enriched signature makes it possible to define fragments of first-order logic corresponding to levels of the dot-depth hierarchy.

To this end, we classify $\mathrm{FO}(<, +1, \min, \max, \varepsilon)$ sentences by counting their number of quantifier alternations. Given a natural number $n \in \mathbb{N}$, a sentence is said to be "$\Sigma_n(<, +1, \min, \max, \varepsilon)$" (resp. "$\Pi_n(<, +1, \min, \max, \varepsilon)$") when it is an $\mathrm{FO}(<, +1, \min, \max, \varepsilon)$-formula whose prenex normal form has either:

- *Exactly n* blocks of quantifiers, the leftmost being an "$\exists$" (resp. a "$\forall$") block, or
- *Strictly less* than n blocks of quantifiers.

For example, a formula over the signature $(<, +1, \min, \max, \varepsilon, (P_a)_{a \in A})$ whose prenex normal form is

$$\exists x_1 \exists x_2 \, \forall x_3 \, \exists x_4 \, \varphi(x_1, x_2, x_3, x_4) \quad (\varphi \text{ quantifier-free})$$

is $\Sigma_3(<, +1, \min, \max, \varepsilon)$. Observe that while $\mathrm{FO}(<)$ and $\mathrm{FO}(<, +1, \min, \max, \varepsilon)$ have the same expressiveness, the enriched signature increases the expressive power of each individual level.

Note also that the negation of a $\Sigma_n(<, +1, \min, \max, \varepsilon)$ sentence is not a $\Sigma_n(<, +1, \min, \max, \varepsilon)$ sentence in general (it is a $\Pi_n(<, +1, \min, \max, \varepsilon)$ sentence), and the corresponding classes of languages are not closed under complement. It is therefore meaningful to define $\mathcal{B}\Sigma_n(<, +1, \min, \max, \varepsilon)$ sentences as Boolean combinations of $\Sigma_n(<, +1, \min, \max, \varepsilon)$ and $\Pi_n(<, +1, \min, \max, \varepsilon)$ sentences. This gives a strict hierarchy of classes of languages depicted in Fig. 1, where, slightly abusing notation, each level denotes the class of languages defined by the corresponding set of formulas.

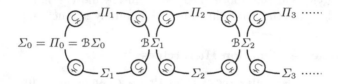

Fig. 1. Quantifier alternation hierarchy

The correspondence discovered by Thomas relates levels in the dot-depth hierarchy and levels in the quantifier alternation hierarchy of first-order logic, over the signature $(<, +1, \min, \max, \varepsilon, (P_a)_{a \in A})$.

Theorem 11 (Thomas [36]). *For any alphabet A, any $n \in \mathbb{N}$ and any language $L \subseteq A^*$, the two following properties hold:*

1. *L has dot-depth n iff L can be defined by a $\mathcal{B}\Sigma_n(<, +1, \min, \max, \varepsilon)$ sentence.*
2. *L has dot-depth $n + \frac{1}{2}$ iff L can be defined by a $\Sigma_{n+1}(<, +1, \min, \max, \varepsilon)$ sentence.*

Some years later, a similar correspondence was established between levels in the Straubing-Thérien hierarchy and levels in the quantifier alternation hierarchy over the original signature $(<, (P_a)_{a \in A})$. Such levels are defined analogously as for the enriched signature, and denoted by $\mathcal{B}\Sigma_n(<)$, $\Sigma_n(<)$, etc.

Theorem 12 (Perrin and Pin [16]). *For any alphabet A, any $n \in \mathbb{N}$ and any language $L \subseteq A^*$, the two following properties hold:*

1. *L has level n in the Straubing-Thérien hierarchy iff L can be defined by a $\mathcal{B}\Sigma_n(<)$ sentence.*
2. *L has level $n + \frac{1}{2}$ in the Straubing-Thérien hierarchy iff L can be defined by a $\Sigma_{n+1}(<)$ sentence.*

2   Generic Concatenation Hierarchies

Since the dot-depth and Straubing-Thérien hierarchies follow the very same construction scheme and enjoy similar properties, it is natural to generalize the definition. We will therefore define a generic notion of concatenation hierarchy. Such hierarchies should still classify languages according to the required number of alternations between concatenation and Boolean operations that are needed to define them. The only parameter in the construction is level 0, which is now any class of languages $\mathcal{C}$ satisfying some mild hypotheses (such as being a Boolean algebra). This parameter $\mathcal{C}$ is called the *basis* of the hierarchy. Once $\mathcal{C}$ is fixed, the construction process is uniform, exactly the same as for the two hierarchies we have already presented:

- Level 0 is the basis (*i.e.*, our parameter class $\mathcal{C}$).
- Each *half-level* is the *polynomial closure* of the previous full level: for any natural number $n \in \mathbb{N}$, level $n + \frac{1}{2}$ is the polynomial closure of level n.
- Each *full-level* is the *Boolean closure* of the previous half-level: for any natural number $n \in \mathbb{N}$, level $n + 1$ is the Boolean closure of level $n + \frac{1}{2}$.

For $q \subset \mathbb{N}$ or $q \in \frac{1}{2} + \mathbb{N}$, let $\mathcal{C}[q]$ denote level q of the concatenation hierarchy of basis $\mathcal{C}$. By definition, we have $\mathcal{C}[n] \subseteq \mathcal{C}[n + \frac{1}{2}] \subseteq \mathcal{C}[n + 1]$ for any $n \in \mathbb{N}$. However, note that these inclusions need not be strict. For instance, if the basis is closed under Boolean operations and marked concatenation (such as the class of star-free languages), the associated hierarchy collapses at level 0. Of course the interesting hierarchies are the strict ones. We give a graphical representation of the construction process of a concatenation hierarchy in Fig. 2 below.

$$0 \xrightarrow{Pol} \tfrac{1}{2} \xrightarrow[Bool]{} 1 \xrightarrow{Pol} \tfrac{3}{2} \xrightarrow[Bool]{} 2 \xrightarrow{Pol} \tfrac{5}{2} \xrightarrow[Bool]{} 3 \xrightarrow{Pol} \tfrac{7}{2} \cdots\cdots$$
$$\text{(basis)}$$

Fig. 2. A concatenation hierarchy

Notice that not all concatenations hierarchies are classifications of the star-free languages. Indeed, the generic definition now makes it possible to define hierarchies containing languages which are not star-free: it suffices to choose a basis containing such languages. The most famous one is the *group hierarchy* of Margolis and Pin [14], whose basis is the class of all regular languages recognized by an automaton in which every letter induces a permutation on the states.

The following result, which will be shown in the full version of this paper, generalizes Theorem 3 to any concatenation hierarchy whose basis is finite.

Theorem 13. *Let $\mathcal{C}$ be a finite Boolean algebra of regular languages over an alphabet of size at least 2. Then, the concatenation hierarchy of basis $\mathcal{C}$ is strict.*

Again, this theorem justifies the quest for algorithms deciding membership in levels of the hierarchy of basis $\mathcal{C}$.

Quantifier Alternation Hierarchies

The correspondence between star-free languages and first-order logic established by McNaughton and Papert in Theorem 2 can be lifted not only to the dot-depth and the Straubing-Thérien hierarchies (Theorems 11 and 12), but also to arbitrary concatenation hierarchies: for any basis $\mathcal{C}$, we associate a well-chosen first-order signature (also denoted by $\mathcal{C}$) such that the concatenation hierarchy of basis $\mathcal{C}$ and the quantifier alternation hierarchy within the variant $FO(\mathcal{C})$ of first-order logic equipped with this signature correspond. This signature contains all label predicates: for any $a \in A$, we have a unary predicate (denoted by "P_a") which is interpreted as the unary relation selecting all positions whose label is a. Moreover, for any language $L \in \mathcal{C}$, we add four predicates. To define them, we introduce the following notation: if $w = a_1 \cdots a_n$ is a word of length n, we denote by $w[i,j]$ its infix $a_i \cdots a_j$ (which is empty if $i > j$), and we let $w]i,j] = w[i+1,j]$, $w[i,j[= w[i,j-1]$ and $w]i,j[= w[i+1,j-1]$. We are now able to finish our description of the signature (associated to) $\mathcal{C}$. In addition to the strict order and the letter predicates, we add the following predicates for each language $L \in \mathcal{C}$:

- A binary predicate I_L. Its interpretation is as follows: given a word w and two positions i,j in w, $I_L(i,j)$ holds when $i < j$ and the infix $w]i,j[$ is in L.
- A unary predicate P_L. Its interpretation is as follows: given a word w and a position i in w, $P_L(i)$ holds when the prefix $w[1,i[$ is in L.
- A unary predicate S_L. Its interpretation is as follows: given a word w and a position i in w, $S_L(i)$ holds when the suffix $w]i,|w|]$ is in L.
- A nullary predicate N_L. Its interpretation is as follows: given a word w, N_L holds when w is in L.

Recall that we abuse notation and identify $\mathcal{C}$ with this signature. In other words, we denote by $FO(\mathcal{C})$ the associated variant of first-order logic.

We are now ready to state a generic correspondence between the concatenation hierarchy of basis $\mathcal{C}$ and the quantifier alternation hierarchy within $FO(\mathcal{C})$. We need an additional condition on $\mathcal{C}$: it should be closed under left and right quotients. That is, if L belongs to $\mathcal{C}$, then for any $a \in A$, so do its left and right quotients $a^{-1}L = \{w \in A^* \mid aw \in L\}$ and $La^{-1} = \{w \in A^* \mid wa \in L\}$.

Theorem 14. *Let $\mathcal{C}$ be a Boolean algebra of regular languages which is closed under left and right quotients. Then, for any finite alphabet A, any $n \in \mathbb{N}$ and any language $L \subseteq A^*$, the two following properties hold:*

1. *$L \in \mathcal{C}[n]$ if and only if L can be defined by a $\mathcal{B}\Sigma_n(\mathcal{C})$ sentence.*
2. *$L \in \mathcal{C}[n + \frac{1}{2}]$ if and only if L can be defined by a $\Sigma_{n+1}(\mathcal{C})$ sentence.*

3 Decision Problems

The membership problem for concatenation hierarchies is not well understood. For instance, although the dot-depth hierarchy has been given a lot of attention since 1971, obtaining membership algorithms for all of its levels remains one of the most famous open problems in automata theory. It has been under investigation for decades but progress is slow: as we explained above, the first known result is due to Knast [13] and yields an algorithm for dot-depth 1. Algorithms were later found for the half-levels $\frac{1}{2}$ in [21,22] and $\frac{3}{2}$ in [9,23]. However, it took more than thirty years to obtain an algorithm for the next full level: dot-depth 2 (see [25]). Furthermore, the problem is still open for dot-depth 3.

The result for level 2 is based on a new approach, which is the key idea we wish to convey in this survey. The approach relies on two main features:

1. It is generic to *all* concatenations hierarchies whose basis is *finite* (which is the case of the dot-depth and of the Straubing-Thérien hierarchies).
2. We consider decision problems which are *more general* than membership. While recent papers on the topic actually consider several such problems (see [28] for a global picture), we will focus on the simplest one: *separation*.

Let us define the separation problem. Consider a class of languages $\mathcal{C}$. Given two languages L_0 and L_1, we say that L_0 is $\mathcal{C}$-separable from L_1 if and only if there exists a third language $K \in \mathcal{C}$ such that $L_1 \subseteq K$ and $L_2 \cap K = \emptyset$. The separation problem for $\mathcal{C}$ is as follows:

> **INPUT:** Two regular languages L_0 and L_1
> **OUTPUT:** Is L_0 $\mathcal{C}$-separable from L_1?

The main reason why this problem is interesting is that solving it requires (and therefore, brings) much insight about the class $\mathcal{C}$. In particular, membership for $\mathcal{C}$ reduces to separation for $\mathcal{C}$. More interesting, if one has an algorithm in hand to decide *separation* for a given half-level in a concatenation hierarchy, then one can use it to obtain a new one deciding *membership* for the *next* half-level. This is what we formally state in the next theorem, which is essentially a result of [25] (note however that while the proof argument of [25] is generic to all hierarchies, the statement itself in [25] is specific to the Straubing-Thérien hierarchy).

Theorem 15. *Consider a basis $\mathcal{C}$ which is a Boolean algebra of regular languages closed under left and right quotients. Then, for any natural number $n \geq 1$, there exists an effective reduction from the membership problem for level $\mathcal{C}[n+\frac{1}{2}]$ to the separation problem for level $\mathcal{C}[n-\frac{1}{2}]$.*

This result is completed by the following theorem, which summarizes the recent results that have been obtained regarding the separation problem for low levels within concatenation hierarchies. The first two items are taken from [29] and the third one is an unpublished generalization of a result of [24] (which states that separation for level $\frac{5}{2}$ in the Straubing-Thérien hierarchy is decidable).

Theorem 16. *Consider an arbitrary finite Boolean algebra $\mathcal{C}$ which is closed under left and right quotients. Then the following results hold:*

1. *$Pol(\mathcal{C})$-separation is decidable.*
2. *$BPol(\mathcal{C})$-separation is decidable.*
3. *$Pol(BPol(\mathcal{C}))$-separation is decidable.*

Altogether, this yields that for any concatenation hierarchy whose basis is *finite*, levels $\frac{1}{2}$, 1 and $\frac{3}{2}$ have decidable separation. Moreover, this can be combined with Theorem 15 to obtain the decidability of membership for level $\frac{5}{2}$.

These results are generic to all concatenations hierarchies whose basis is finite. However, in the special case of the dot-depth and Straubing-Thérien hierarchies, one can do better and lift them to levels 2 and $\frac{5}{2}$ for separation (and thus to level $\frac{7}{2}$ for membership). These stronger results are based on a specific property of the Straubing-Thérien hierarchy: its level $\frac{3}{2}$ is also level $\frac{1}{2}$ in another concatenation hierarchy having a finite basis. Let us explain this statement in more details.

Back to the Dot-Depth and Straubing-Thérien Hierarchies

In this final part, we explain why one may lift all results one level higher in the dot-depth and Straubing-Thérien hierarchies. The argument relies on a theorem of Pin and Straubing [20], which implies that levels $\frac{3}{2}$ and above in the Straubing-Thérien hierarchy are also levels $\frac{1}{2}$ and above in the concatenation hierarchy whose basis is the *finite* class AT of alphabet testable languages, defined below. While simple, this result is crucial: it allows us to lift the separation results of Theorem 16 to levels 2 and $\frac{5}{2}$ of the Straubing-Thérien hierarchy.

Let us define the class AT of *alphabet testable languages*. It consists of all Boolean combinations of languages of the form,

$$A^*aA^* \quad \text{for some } a \in A.$$

Clearly AT is finite, and one may verify that it is a Boolean algebra closed under left and right quotients. It was proved by Pin and Straubing [20] that level $\frac{3}{2}$ in the Straubing-Thérien hierarchy[1] is also the class $Pol(\text{AT})$.

Note that the original formulation of this statement by Pin and Straubing is that level $\frac{3}{2}$ in the Straubing-Thérien hierarchy consists exactly of unions of languages of the form,

$$B_0^*a_1B_1^*a_2B_2^*\cdots a_nB_n^* \quad \text{with} \quad B_0,\ldots,B_n \subseteq A.$$

We reformulate this result in the following theorem.

Theorem 17 (Pin and Straubing [20]). *Level $\frac{3}{2}$ in the Straubing-Thérien hierarchy is exactly the class $Pol(\text{AT})$. In particular, any level $n \geq \frac{3}{2}$ (half or full) in the Straubing-Thérien hierarchy corresponds exactly to level $n - 1$ in the concatenation hierarchy of basis* AT.

The important point here is that while AT is more involved than $\{\emptyset, A^*\}$ as a basis, it remains *finite*. Therefore, Theorem 17 states that any level $n \geq \frac{3}{2}$ in the Straubing-Thérien hierarchy is also level $n - 1$ in another hierarchy whose basis is finite. This result is crucial. Indeed, this means that Theorem 16 does not only apply to levels $\frac{1}{2}$, 1 and $\frac{3}{2}$ of the Straubing-Thérien hierarchy but also to levels 2 and $\frac{5}{2}$. Altogether, we get the following corollary.

---

[1] In fact, the original formulation of Pin and Straubing considers level 2 in the Straubing-Thérien hierarchy and not level $\frac{3}{2}$.

Corollary 18. *The separation problem is decidable for levels* 2 *and* $\frac{5}{2}$ *in the Straubing-Thérien hierarchy. Moreover, the membership problem is decidable for level* $\frac{7}{2}$.

Finally, these results can be lifted to the dot-depth hierarchy using an approach which is similar to the one used by Straubing in Theorem 10. Indeed, recall from Theorem 10 that the Straubing-Thérien hierarchy can be viewed as "more fundamental" than the dot-depth. It turns out that the reduction provided by Straubing can actually be lifted to half-levels [23] and to separation [26].

Theorem 19. *For any level* n *in the dot-depth hierarchy, the following two properties hold:*

- *If membership is decidable for level* n *in the Straubing-Thérien, then it is decidable for level* n *in the dot-depth hierarchy as well.*
- *If separation is decidable for level* n *in the Straubing-Thérien, then it is decidable for level* n *in the dot-depth hierarchy as well.*

Corollary 20. *The separation problem for levels* 2 *and* $\frac{5}{2}$ *in the dot-depth hierarchy are decidable. Moreover, the membership problem is decidable for level* $\frac{7}{2}$.

4   Conclusion

In this extended abstract, we outlined part of the (slow) progress that occurred during the last decades regarding concatenation hierarchies. We refer the reader to the full version of the paper for details, and to [18,19,27,37] for surveys on this fascinating subject.

References

1. Arfi, M.: Polynomial operations on rational languages. In: Brandenburg, F.J., Vidal-Naquet, G., Wirsing, M. (eds.) STACS 1987. LNCS, vol. 247, pp. 198–206. Springer, Heidelberg (1987). doi:10.1007/BFb0039607
2. Arfi, M.: Opérations Polynomiales et Hiérarchies de Concaténation. Theoret. Comput. Sci. **91**(1), 71–84 (1991)
3. Bojanczyk, M.: Star height via games. In: 2015 30th Annual ACM/IEEE Symposium on Logic in Computer Science. IEEE Computer Society 2015, pp. 214–219 (2015)
4. Brzozowski, J.A.: Developments in the Theory of regular Languages. In: IFIP Congress, pp. 29–40 (1980)
5. Brzozowski, J.A., Cohen, R.S.: Dot-depth of star-free events. J. Comput. Syst. Sci. **5**(1), 1–16 (1971)
6. Brzozowski, J.A., Knast, R.: The Dot-depth hierarchy of star-free languages is infinite. J. Comput. Syst. Sci. **16**(1), 37–55 (1978)
7. Dejean, F., Schützenberger, M.P.: On a question of Eggan. Inf. Control **9**(1), 23–25 (1966)
8. Eggan, L.C.: Transition graphs and the star-height of regular events. Michigan Math. J. **10**(4), 385–397 (1963)
9. Glaßer, C., Schmitz, H.: Languages of dot-depth 3/2. Theory Comput. Syst. **42**(2), 256–286 (2007)

10. Hashiguchi, K.: Algorithms for determining relative star height and star height. Inf. Comput. **78**(2), 124–169 (1988)
11. Kirsten, D.: Distance desert automata and the star height problem. RAIRO-Theor. Inf. Appl. **39**(3), 455–509 (2005)
12. Kleene, S.C.: Representation of events in nerve nets and finite automata. In: Shannon, C., McCarthy, J. (eds.) Annals of Mathematics Studies 34, pp. 3–41. Princeton University Press, New Jersey (1956)
13. Knast, R.: A semigroup characterization of dot-depth one languages. RAIRO - Theor. Inform. Appl. **17**(4), 321–330 (1983)
14. Margolis, S.W., Pin, J.E.: Products of group languages. In: Budach, L. (ed.) FCT 1985. LNCS, vol. 199, pp. 285–299. Springer, Heidelberg (1985). doi:10.1007/BFb0028813
15. McNaughton, R., Papert, S.A.: Counter-Free Automata. MIT Press, Cambridge (1971)
16. Perrin, D., Pin, J.É.: First-order logic and star-free sets. J. Comput. Syst. Sci. **32**(3), 393–406 (1986)
17. Pin, J.É.: An explicit formula for the intersection of two polynomials of regular languages. In: Béal, M.-P., Carton, O. (eds.) DLT 2013. LNCS, vol. 7907, pp. 31–45. Springer, Heidelberg (2013). doi:10.1007/978-3-642-38771-5_5
18. Pin, J.-É.: Open problems about regular languages, 35 years later. In: The Role of Theory in Computer Science. Essays Dedicated to Janusz Brzozowski. World Scientific (2017)
19. Pin, J.-É.: The dot-depth hierarchy, 45 years later. In: The Role of Theory in Computer Science. Essays Dedicated to Janusz Brzozowski. World Scientific (2017)
20. Pin, J.-É., Straubing, H.: Monoids of upper triangular boolean matrices. In: Semigroups. Structure and Universal Algebraic Problems, vol. 39, pp. 259–272. North-Holland (1985)
21. Pin, J.-E., Weil, P.: Polynomial closure and unambiguous product. In: Fülöp, Z., Gécseg, F. (eds.) ICALP 1995. LNCS, vol. 944, pp. 348–359. Springer, Heidelberg (1995). doi:10.1007/3-540-60084-1_87
22. Pin, J.É., Weil, P.: Polynomial closure and unambiguous product. Theory Comput. Syst. **30**(4), 383–422 (1997)
23. Pin, J.É., Weil, P.: The wreath product principle for ordered semigroups. Commun. Algebra **30**(12), 5677–5713 (2002)
24. Place, T.: Separating regular languages with two quantifier alternations. In: 30th Annual ACM/IEEE Symposium on Logic in Computer Science, LICS 2015, pp. 202–213 (2015)
25. Place, T., Zeitoun, M.: Going higher in the first-order quantifier alternation hierarchy on words. In: Esparza, J., Fraigniaud, P., Husfeldt, T., Koutsoupias, E. (eds.) ICALP 2014. LNCS, vol. 8573, pp. 342–353. Springer, Heidelberg (2014). doi:10.1007/978-3-662-43951-7_29
26. Place, T., Zeitoun, M.: Separation and the successor relation. In: 32nd International Symposium on Theoretical Aspects of Computer Science, STACS 2015. Dagstuhl, Germany: Schloss Dagstuhl-Leibniz-Zentrum fuer Informatik, pp. 662–675 (2015)
27. Place, T., Zeitoun, M.: The tale of the quantifier alternation hierarchy of first-order logic over words. SIGLOG news **2**(3), 4–17 (2015)
28. Place, T., Zeitoun, M.: The covering problem: a unified approach for investigating the expressive power of logics. In: Proceedings of the 41st International Symposium on Mathematical Foundations of Computer Science, MFCS 2016, pp. 77:1–77:15 (2016)

29. Place, T., Zeitoun, M.: Separation for dot-depth two. In: 32th Annual ACM/IEEE Symposium on Logic in Computer Science, LICS 2017 (2017)
30. Schützenberger, M.P.: On finite monoids having only trivial subgroups. Inf. Control **8**(2), 190–194 (1965)
31. Simon, I.: Hierarchies of Events of Dot-Depth One. Ph.D. thesis. Waterloo, Ontario, Canada: University of Waterloo, Department of Applied Analysis and Computer Science (1972)
32. Simon, I.: Piecewise testable events. In: Brakhage, H. (ed.) GI-Fachtagung 1975. LNCS, vol. 33, pp. 214–222. Springer, Heidelberg (1975). doi:10.1007/3-540-07407-4_23
33. Straubing, H.: A generalization of the schützenberger product of finite monoids. Theoret. Comput. Sci. **13**(2), 137–150 (1981)
34. Straubing, H.: Finite semigroup varieties of the form V * D. J. Pure Appl. Algebra **36**, 53–94 (1985)
35. Thérien, D.: Classification of finite monoids: the language approach. Theoret. Comput. Sci. **14**(2), 195–208 (1981)
36. Thomas, W.: Classifying regular events in symbolic logic. J. Comput. Syst. Sci. **25**(3), 360–376 (1982)
37. Weil, P.: Concatenation product: a survey. In: Pin, J.E. (ed.) LITP 1988. LNCS, vol. 386, pp. 120–137. Springer, Heidelberg (1989). doi:10.1007/BFb0013116

Can We Locally Compute Sparse Connected Subgraphs?

Ronitt Rubinfeld[1,2(✉)]

[1] MIT, Cambridge, USA
`ronitt@csail.mit.edu`
[2] Tel Aviv University, Tel Aviv, Israel

1 Introduction

How can we solve optimization problems on data that is so large, that we cannot hope to view more than a miniscule fraction of it? When attempting to solve optimization problems on big data, we are presented with a double catastrophe, as *both the inputs to and the outputs from the computation are large*. One ray of hope is that often, the portion of the output that is needed by the user is, in fact, of a more manageable size. In such a situation, it would be useful if one could find very fast ways of computing only the portion of the output that is required by the user.

One approach to such settings is the study of sub-linear time algorithms. For the most part, sub-linear time algorithms have been studied in settings where the goal is to approximate only the *value* of an optimal solution, or to determine whether or not the input has a specified property. For sparse graphs and optimization problems, sublinear time approximations have been given for quantities such as the minimum spanning tree, optimal vertex cover, maximal matching, maximum matching, dominating set, sparse set cover, sparse packing and cover problems [15, 18, 20, 34, 46, 55, 59, 60, 62, 74].

In contrast, here we concern ourselves with the ability to provide access to the full description of a near-optimal solution (typically referred to as the "search" problem). For problems whose outputs are too large to view in their entirety, we consider algorithms within the context of *local computation algorithms* (LCAs), a framework introduced in [2, 65] to capture the subtle aspects of providing the user with sublinear time access to the parts of the output solution that they require.

Arguably, one of the most basic graph problems, for which the local version is first considered in [49], is the following: Suppose one wants to provide fast query access to a spanning tree of a given input graph – that is, design an oracle that has a particular spanning tree "in mind", and, given an edge query from the input graph, the oracle should quickly reply whether the edge is part of its spanning tree or not. One might ask whether the oracle could always say "yes" when asked about any edge in the graph, since every edge is part of *some*

R. Rubinfeld—Supported by ISF grant 1147/09 and NSF grant CCF-1650733.

P. Weil (Ed.): CSR 2017, LNCS 10304, pp. 38–47, 2017.
DOI: 10.1007/978-3-319-58747-9_6

spanning tree? The difficulty is that we would like the oracle's answers to be *consistent* with a single spanning tree.

While at first thought, it might seem plausible that such an easy task could be accomplished, it turns out that there is no sublinear time oracle that can provide such "spanning tree" answers! This can be seen by considering the behaviors of the oracle on the graph that is a path of length n and the graph that is a cycle of length n. On the path, the oracle must answer "yes" for every edge, while on the cycle, the oracle must answer "no" for a single edge, and "yes" for every other edge. Unfortunately, given an edge, determining whether it is in the cycle graph or the path graph requires linear time. Note that though the weight of a minimum spanning tree can be approximated in sublinear time for bounded weight graphs [15, 18, 20], this argument also shows that there is no sublinear time local computation algorithm which provides query access to a minimum spanning tree!

Given the impossibility of providing sublinear time query access to a spanning tree, in the rest of this survey, we describe work on a simpler question suggested in [49]: Is it possible to provide sublinear time query access to a sparse connected subgraph? This question in its full generality remains wide open. In the next section, we describe the local computation algorithms model, some context and related work. In the third section, we give a more detailed definition of the sparse connected subgraph question, as well as describe results that give partial answers.

2　Local Computation Algorithms

Locality in computation, or the ability to compute a function of the input by viewing few locations in the input, is an ubiquitous notion and is of central importance to parallel and distributed computing, sub-linear time algorithms, dynamic algorithms and data structures. LCAs can be used to model non-communicating entities computing in parallel on the same data, such as in cloud computations, wireless networks and game theory. Though the model of LCAs was introduced relatively recently, earlier works implicitly construct LCAs in a variety of settings. Some examples include locally (list)-decodable codes (e.g., [6, 11, 24, 25, 28, 30, 33, 36, 40, 41, 52, 72, 73]), local decompression (e.g., [26, 31, 57, 66]), local reconstruction and filters for monotone and Lipschitz functions [1, 7, 13, 19, 38, 67], and local reconstruction of graph properties [39]. The study of LCAs has been very active in the past few years, with recent results that include constructing maximal independent sets [2, 10, 23, 29, 51, 65], approximate maximum matchings [53, 54], satisfying assignments for k-CNF [2, 65], local computation mechanism design [35], local decompression [21], and local reconstruction of graph properties [14].

On input $x = (x_1, \ldots, x_n)$, a *local computation algorithm* (LCA) [2, 65], supports user queries to a correct output $y = (y_1, \ldots, y_m)$ of the computational problem on input x, such that after each *query* by a user to a specified location i of the output, the LCA computes and outputs y_i. If there is more than one legal output for input x, then it is required that all outputs be consistent with the same legal output y. The LCA is given *probe* access to bits of x.[1] The LCA does not have knowledge of future user queries. Maintaining consistency presents a challenge when the computation has several legal outputs. Though independent copies of the LCA cannot interact during the query-answering process, we may assume that they have access to a common initial random bit sequence.

For a given problem, the hope is that the query time and space complexity of an LCA is nearly proportional to the amount of the solution that is requested by the user. Even when such ambitious goals are not achieved, LCAs are of importance when the time and space complexity is sub-linear in the size of the input and output to the problem, and do not require storing the history of previous queries and answers.

Local distributed algorithms have received much attention in the distributed computing literature, where the model is such that the number of rounds is bounded by a constant, but the computation is performed by *all of the processors* in the distributed network [56,58]. Naor and Stockmeyer [58] and Mayer, Naor and Stockmeyer [56] investigate the question of what can be computed under these constraints, and show that there are nontrivial problems which can be solved via such algorithms. Several more recent works investigate local algorithms for various problems, including coloring, maximal independent set, dominating set (some examples are in [8,9,27,42–47,69,70]).

Although all of these algorithms are distributed algorithms, we note that those that use constant rounds yield (sequential) LCAs via a reduction of Parnas and Ron [62]. The cross-fertilization between the areas of distributed algorithms and LCAs has led to exciting recent results improving algorithms for maximal independent set in both models (see for example [2,10,23,29,51,65] and the references therein).

The input access allowed in the distributed settings is more restrictive in that distributed algorithms may only receive information from neighboring nodes. It is not clear how much the restriction affects the complexity of a problem – in particular, the recent work of [32] shows that for many graph problems, this restriction does not significantly increase the power of LCAs.

Local Algorithms for Massive Graphs. Local algorithms (which do not necessarily fall under the LCA model) have been demonstrated to be applicable for computations on the web graph. In [3,4,12,37,68], local algorithms are given which, for a given vertex v in the web graph, computes an approximation to v's personalized PageRank vector and computes the vertices that contribute significantly to v's PageRank. In these algorithms, evaluations are made only to the nearby neighborhood of v, so that the running time depends on the accuracy parameters input to the algorithm, but there is no running time dependence on the size of the webgraph. Local graph partitioning algorithms have been presented in several works (some early examples include [4,5,61,71,75]) to find subsets of

---

[1] We use the word *probe* to refer to the LCAs views of locations in the input and *query* to refer to the user requests to the LCA.

vertices whose internal connections are significantly richer than their external connections. The running time of these algorithms depends on the size of the cluster that is output, which can be much smaller than the size of the entire graph. However, even when the size of the cluster is guaranteed to be small, it is not obvious how to use these algorithms in the local computation setting where the cluster decompositions must be consistent among queries to all vertices.

3   Designing LCAs for Sparse Approximating Subgraphs

A somewhat more general set of questions than those described in the introduction are: Is it possible to provide fast random access to a sparsified version of a given input graph, that approximates various properties of the original input graph? Such algorithms have applications to motion planning and multiple entity coordination algorithms. For example, in a distributed and geometric setting, it has been shown that such algorithms can be used to design motion planning algorithms for multi-robot systems, and in particular, a provably correct flocking algorithm [16,17]. Such algorithms have also been used to design algorithms for a number of other basic problems such as minimum spanning graphs and sparse spanners, where spanner graphs are required to maintain low distortion of pairwise distances in the spanner as compared to the pairwise distances in the original graph. Progress in this line of problems has led to some answers and even more questions than were apparent at first sight!

The Sparse Spanning Graph Problem. As introduced in [49], LCAs for *sparse spanning subgraphs* consider the following scenario: Given a connected graph $G = (V, E)$, a basic task is to provide random access to a sparse graph $\tilde{G} = (V, \tilde{E})$ that is a connected spanning *subgraph* of G, where the sparsity is controlled by an input parameter ϵ such that the number of edges in $\tilde{E}$ is at most $(1 + \epsilon)n$. That is, given an edge (u, v) in G, the LCA decides whether (u, v) belongs to $\tilde{G}$, where the answers to each edge (u, v) must be consistent with a single connected sparse graph $\tilde{G}$ that is a subgraph of the input graph G. To achieve this, the LCA may probe the incidence relations of the graph G. There may be many possible $\tilde{G}$ – the LCA must answer consistently according to a single one.

In [49], it is shown that for general bounded-degree graphs, the probe complexity of any such algorithm must be $\Omega(\sqrt{n})$. The open question that remains is the following:

Problem 1. Are there LCAs for membership in sparse spanning subgraphs with probe complexity $o(n)$ for all bounded degree graphs?

More specifically, is $\theta(\sqrt{n})$ the right complexity for general bounded degree graphs? Partial progress in this direction has been made: In a sequence of works [48–50], several algorithms are developed and analyzed in terms of various graph parameters and properties.

- For graphs in which all vertices expand at a similar rate, and in particular, for good expanders, an algorithm of complexity $\tilde{\theta}(\sqrt{n/\epsilon})$ is given [49].

This algorithm is based on choosing $\theta(\sqrt{\epsilon n})$ centers at random and partitioning the vertices of the graph according to the nearest center. The edges of a breadth-first-search tree from the center to each of the vertices in its partition are included in the sparse spanning graph. In addition, a carefully chosen set of additional edges are included in the sparse spanning graph which connect partitions. By showing how to locally determine which partition each vertex is in, as well as which edges between partitions are included, an LCA that achieves the desired complexity is given.

- For the class of minor-free graphs, algorithms are given whose runtime does not depend on n, and has polynomial dependence on $1/\epsilon$ [50]. This algorithm also constructs a first phase partition by choosing centers, though a much larger number of centers than in the previous algorithm for expanders. After the first phase, there are many parts in the partition that are too large, so a method of refining the partition in a local manner must be designed.
- For the class of hyperfinite graphs, which generalize minor-free graphs, an algorithm is given, based on Kruskal's algorithm, whose complexity is independent of n, but the dependence on $1/\epsilon$ is exponential [49].
- For graphs in which every t-vertex subgraph has expansion $1/(\log t)^{1+o(1)}$ then there is an algorithm whose complexity is independent of n (but depends triply exponentially on $1/\epsilon$). On the other hand, there are graphs in which every t-vertex subgraph has expansion $1/(\log t)^{1-o(1)}$ for which dependence on n is necessary [48].

The previous describes various technical ideas that are appropriate for very different classes of bounded degree graphs. It remains to be seen whether one can design an LCA that runs in sub-linear time for *all* bounded degree graphs.

Low Weight Spanning Subgraphs. As we have seen in the introduction, it is not possible to give a fast LCA for the minimum spanning tree (MST) problem, but we consider the following weakening of the problem: An LCA for low weight sparse spanning graphs would, on query an edge e in the graph, tell the user whether the edge e is in E', such that E' is a low weight sparse spanning subgraph of G. Do sublinear time LCAs exist finding sparse low-weight spanning subgraphs?

Problem 2. Are there families of weighted graphs for which there are $o(n)$-time LCAs for low weight sparse spanning subgraphs?

For minor-free graphs, [50] give an LCA which provides query access to a low weight sparse spanning subgraph, such that the runtime has polynomial dependence on ϵ and d. The output spanning subgraph has the additional property of being a spanner. Graph spanners are subgraphs which give minimal distortion in terms of pairwise distances of the original input graph. More formally, given a (weighted) graph G, a subgraph G' of G is a k-spanner if for every pair of vertices u, v, the distance between u and v in G' is at most k times the distance between u and v in G (for various definitions of spanners, see e.g., [22,63,64]).

k is referred to as the *stretch* or *dilation*. The goal is to find a G' with a minimum number of edges or weight. Such subgraphs are important in distributed computation and in studying geometric network optimization.

4 Final Words

For such a seemingly basic problem, relatively little is known. Sublinear time LCAs have been designed for providing query access to sparse spanning subgraphs in two extreme cases, when the input graph is a very good expander, and when the input graph is highly nonexpanding. Several techniques have been developed to attack this problem. Nevertheless, the question of whether sublinear time LCAs are possible for general bounded degree graphs remains a mystery. Furthermore, nothing is known when there is no bound on the maximum degree of the input graph.

References

1. Ailon, N., Chazelle, B., Comandur, S., Liu, D.: Property-preserving data reconstruction. Algorithmica **51**(2), 160–182 (2008)
2. Alon, N., Rubinfeld, R., Vardi, S., Xie, N.: Space-efficient local computation algorithms. In: 23rd Annual ACM-SIAM Symposium on Discrete Algorithms (SODA 2012), Kyoto, January 2012
3. Andersen, R., Borgs, C., Chayes, J., Hopcroft, J., Mirrokni, V., Teng, S.: Local computation of pagerank contributions. Internet Math. **5**(1–2), 23–45 (2008)
4. Andersen, R., Chung, F., Lang, K.: Local graph partitioning using pagerank vectors. In: Proceedings of the 47th Annual IEEE Symposium on Foundations of Computer Science, pp. 475–486 (2006)
5. Andersen, R., Peres, Y.: Finding sparse cuts locally using evolving sets. In: Proceedings of the 41st Annual ACM Symposium on the Theory of Computing, pp. 235–244 (2009)
6. Arora, S., Sudan, M.: Improved low-degree testing and its applications. Combinatorica **23**(3), 365–426 (2003)
7. Awasthi, P., Jha, M., Molinaro, M., Raskhodnikova, S.: Limitations of local filters of Lipschitz and monotone functions. In: Gupta, A., Jansen, K., Rolim, J., Servedio, R. (eds.) APPROX/RANDOM-2012. LNCS, vol. 7408, pp. 374–386. Springer, Heidelberg (2012). doi:10.1007/978-3-642-32512-0_32
8. Barenboim, L., Elkin, M.: Distributed $(\Delta + 1)$-coloring in linear (in Δ) time. In: Proceedings of the 41st Annual ACM Symposium on the Theory of Computing, pp. 111–120 (2009)
9. Barenboim, L., Elkin, M.: Deterministic distributed vertex coloring in polylogarithmic time. In: Proceedings of the 29th ACM Symposium on Principles of Distributed Computing, pp. 410–419 (2010)
10. Barenboim, L., Elkin, M., Pettie, S., Schneider, J.: The locality of distributed symmetry breaking. J. ACM **63**(3), 20 (2016)
11. Ben-Aroya, A., Efremenko, K., Ta-Shma, A.: Local list-decoding with a constant number of queries, Technical report TR10-047, Electronic Colloquium on Computational Complexity, April 2010

12. Berkhin, P.: Bookmark-coloring algorithm for personalized pagerank computing. Internet Mathematics **3**(1), 41–62 (2006)
13. Bhattacharyya, A., Grigorescu, E., Jha, M., Jung, K., Raskhodnikova, S., Woodruff, D.P.: Lower bounds for local monotonicity reconstruction from transitive-closure spanners. SIAM J. Discrete Math. **26**(2), 618–646 (2012)
14. Campagna, A., Guo, A., Rubinfeld, R.: Local reconstructors and tolerant testers for connectivity and diameter. In: Raghavendra, P., Raskhodnikova, S., Jansen, K., Rolim, J.D.P. (eds.) APPROX/RANDOM-2013. LNCS, vol. 8096, pp. 411–424. Springer, Heidelberg (2013). doi:10.1007/978-3-642-40328-6_29
15. Chazelle, B., Rubinfeld, R., Trevisan, L.: Approximating the minimum spanning tree weight in sublinear time. SIAM J. Comput. **34**(6), 1370–1379 (2005)
16. Cornejo, A.: Local distributed algorithms for multi-robot systems. Ph.D. thesis, MIT (2012)
17. Cornejo, A., Kuhn, F., Ley-Wild, R., Lynch, N.: Keeping mobile robot swarms connected. In: Keidar, I. (ed.) DISC 2009. LNCS, vol. 5805, pp. 496–511. Springer, Heidelberg (2009). doi:10.1007/978-3-642-04355-0_50
18. Czumaj, A., Ergun, F., Fortnow, L., Magen, A., Newman, I., Rubinfeld, R., Sohler, C.: Sublinear-time approximation of euclidean minimum spanning tree. SIAM J. Comput. **35**(1), 91–109 (2005)
19. Czumaj, A., Sohler, C.: Sublinear-time algorithms. Bull. Eur. Assoc. Theor. Comput. Sci. **89**, 23–47 (2006)
20. Czumaj, A., Sohler, C.: Estimating the weight of metric minimum spanning trees in sublinear time. SIAM J. Comput. **39**(3), 904–922 (2009)
21. Dutta, A., Levi, R., Ron, D., Rubinfeld, R.: A simple online competitive adaptation of Lempel-Ziv compression with efficient random access support. In: Proceedings of the Data Compression Conference (DCC), pp. 113–122 (2013)
22. Elkin, M., Peleg, D.: The hardness of approximating spanner problems. Theor. Comput. Syst. **41**(4), 691–729 (2007)
23. Even, G., Medina, M., Ron, D.: Best of two local models: local centralized and local distributed algorithms. CoRR abs/1402.3796 (2014). http://arxiv.org/abs/1402.3796
24. Feigenbaum, D.B.J.: Hiding instances in multi-oracle queries. In: Proceedings of the 7th Annual STACS Conference, pp. 34–48 (1990)
25. Feigenbaum, J., Fortnow, L.: Random self-reducibility of complete sets. SIAM J. Comput. **22**, 994–1005 (1993)
26. Ferragina, P., Venturini, R.: A simple storage scheme for strings achieving entropy bounds. In: ACM-SIAM Symposium on Discrete Algorithms, pp. 690–696 (2007)
27. Fraigniaud, P., Korman, A., Peleg, D.: Local distributed decision. CoRR abs/1011.2152 (2010)
28. Gemmell, P., Lipton, R., Rubinfeld, R., Sudan, M., Wigderson, A.: Self-testing/correcting for polynomials and for approximate functions. In: Proceedings of the 23rd Annual ACM Symposium on the Theory of Computing, pp. 32–42 (1991)
29. Ghaffari, M.: An improved distributed algorithm for maximal independent set. In: Proceedings of the Twenty-Seventh Annual ACM-SIAM Symposium on Discrete Algorithms, SODA 2016, Arlington, 10–12 January 2016, pp. 270–277 (2016). http://dx.doi.org/10.1137/1.9781611974331.ch20
30. Goldreich, O., Rubinfeld, R., Sudan, M.: Learning polynomials with queries: the highly noisy case. SIAM J. Discrete Math. **13**(4), 535–570 (2000)

31. González, R., Navarro, G.: Statistical encoding of succinct data structures. In: Lewenstein, M., Valiente, G. (eds.) CPM 2006. LNCS, vol. 4009, pp. 294–306. Springer, Heidelberg (2006). doi:10.1007/11780441_27
32. Göös, M., Hirvonen, J., Levi, R., Medina, M., Suomela, J.: Non-local probes do not help with many graph problems. In: Gavoille, C., Ilcinkas, D. (eds.) DISC 2016. LNCS, vol. 9888, pp. 201–214. Springer, Heidelberg (2016). doi:10.1007/978-3-662-53426-7_15
33. Gopalan, P., Klivans, A.R., Zuckerman, D.: List-decoding Reed Muller codes over small fields. In: Proceedings of the 40th Annual ACM Symposium on Theory of Computing, pp. 265–274 (2008)
34. Hasidim, A., Kelner, J., Nguyen, H.N., Onak, K.: Local graph partitions for approximation and testing. In: Proceedings of the 50th Annual IEEE Symposium on Foundations of Computer Science, pp. 22–31 (2009)
35. Hassidim, A., Mansour, Y., Vardi, S.: Local computation mechanism design. CoRR abs/1311.3939 (2013)
36. Impagliazzo, R., Wigderson, A.: P = BPP if E requires exponential circuits: derandomizing the XOR lemma. In: Proceedings of the 29th Annual ACM Symposium on the Theory of Computing, pp. 220–229 (1997)
37. Jeh, G., Widom, J.: Scaling personalized web search. In: Proceedings of the 12th International Conference on World Wide Web, pp. 271–279 (2003)
38. Jha, M., Raskhodnikova, S.: Testing and reconstruction of Lipschitz functions with applications to data privacy. SIAM J. Comput. **42**(2), 700–731 (2013)
39. Kale, S., Peres, Y., Seshadhri, C.: Noise tolerance of expanders and sublinear expansion reconstruction. SIAM J. Comput. **42**(1), 305–323 (2013)
40. Katz, J., Trevisan, L.: On the efficiency of local decoding procedures for error-correcting codes. In: Proceedings of the 32nd Annual ACM Symposium on the Theory of Computing, pp. 80–86 (2000)
41. Kopparty, S., Saraf, S.: Local list-decoding and testing of sparse random linear codes from high-error, Technical report 115, Electronic Colloquium on Computational Complexity (ECCC) (2009)
42. Kuhn, F.: Local multicoloring algorithms: computing a nearly-optimal TDMA schedule in constant time. In: Proceedings of the 26th International Symposium on Theoretical Aspects of Computer Science, pp. 613–624 (2009)
43. Kuhn, F., Moscibroda, T.: Distributed approximation of capacitated dominating sets. In: Proceedings of the 19th Annual ACM Symposium on Parallelism in Algorithms and Architectures, pp. 161–170 (2007)
44. Kuhn, F., Moscibroda, T., Nieberg, T., Wattenhofer, R.: Fast deterministic distributed maximal independent set computation on growth-bounded graphs. In: Fraigniaud, P. (ed.) DISC 2005. LNCS, vol. 3724, pp. 273–287. Springer, Heidelberg (2005). doi:10.1007/11561927_21
45. Kuhn, F., Moscibroda, T., Wattenhofer, R.: What cannot be computed locally! In: Proceedings of the 23rd ACM Symposium on Principles of Distributed Computing, pp. 300–309 (2004)
46. Kuhn, F., Moscibroda, T., Wattenhofer, R.: The price of being near-sighted. In: Proceedings of the 17th ACM-SIAM Symposium on Discrete Algorithms, pp. 980–989 (2006)
47. Kuhn, F., Wattenhofer, R.: On the complexity of distributed graph coloring. In: Proceedings of the 25th ACM Symposium on Principles of Distributed Computing, pp. 7–15 (2006)
48. Levi, R., Moshkovitz, G., Ron, D., Rubinfeld, R., Shapira, A.: Constructing near spanning trees with few local inspections. Random Struct. Algorithms (2016)

49. Levi, R., Ron, D., Rubinfeld, R.: Local algorithms for sparse spanning graphs. In: Approximation, Randomization, and Combinatorial Optimization. Algorithms and Techniques, APPROX/RANDOM 2014, 4–6 September 2014, Barcelona, pp. 826–842 (2014)

50. Levi, R., Ron, D., Rubinfeld, R.: A local algorithm for constructing spanners in minor-free graphs. In: Approximation, Randomization, and Combinatorial Optimization. Algorithms and Techniques, APPROX/RANDOM 2016, 7–9 September 2016, Paris, pp. 38:1–38:15 (2016)

51. Levi, R., Rubinfeld, R., Yodpinyanee, A.: Brief announcement: local computation algorithms for graphs of non-constant degrees. In: Proceedings of the 27th ACM on Symposium on Parallelism in Algorithms and Architectures, SPAA 2015, Portland, 13–15 June 2015, pp. 59–61 (2015). http://doi.acm.org/10.1145/2755573.2755615

52. Lipton, R.: New directions in testing. In: Proceedings of the DIMACS Workshop on Distributed Computing and Cryptography (1989)

53. Mansour, Y., Rubinstein, A., Vardi, S., Xie, N.: Converting online algorithms to local computation algorithms. In: Czumaj, A., Mehlhorn, K., Pitts, A., Wattenhofer, R. (eds.) ICALP 2012. LNCS, vol. 7391, pp. 653–664. Springer, Heidelberg (2012). doi:10.1007/978-3-642-31594-7_55

54. Mansour, Y., Vardi, S.: A local computation approximation scheme to maximum matching. In: Raghavendra, P., Raskhodnikova, S., Jansen, K., Rolim, J.D.P. (eds.) APPROX/RANDOM-2013. LNCS, vol. 8096, pp. 260–273. Springer, Heidelberg (2013). doi:10.1007/978-3-642-40328-6_19

55. Marko, S., Ron, D.: Distance approximation in bounded-degree and general sparse graphs. In: Díaz, J., Jansen, K., Rolim, J.D.P., Zwick, U. (eds.) APPROX/RANDOM-2006. LNCS, vol. 4110, pp. 475–486. Springer, Heidelberg (2006). doi:10.1007/11830924_43

56. Mayer, A., Naor, S., Stockmeyer, L.: Local computations on static and dynamic graphs. In: Proceedings of the 3rd Israel Symposium on Theory and Computing Systems (1995)

57. Muthukrishnan, S., Strauss, M., Zheng, X.: Workload-optimal histograms on streams. Technical report 2005-19, DIMACS Technical Report (2005)

58. Naor, M., Stockmeyer, L.: What can be computed locally? SIAM J. Comput. **24**(6), 1259–1277 (1995)

59. Nguyen, H.N., Onak, K.: Constant-time approximation algorithms via local improvements. In: Proceedings of the 49th Annual IEEE Symposium on Foundations of Computer Science, pp. 327–336 (2008)

60. Onak, K., Ron, D., Rosen, M., Rubinfeld, R.: A near-optimal sublinear-time algorithm for approximating the minimum vertex cover size. In: 23rd Annual ACM-SIAM Symposium on Discrete Algorithms (SODA 2012), Kyoto, January 2012

61. Orecchia, L., Allen Zhu, Z.: Flow-based algorithms for local graph clustering. In: Proceedings of the Twenty-Fifth Annual ACM-SIAM Symposium on Discrete Algorithms, SODA 2014, Portland, 5–7 January 2014, pp. 1267–1286 (2014)

62. Parnas, M., Ron, D.: Approximating the minimum vertex cover in sublinear time and a connection to distributed algorithms. Theor. Comput. Sci. **381**(1–3), 183–196 (2007)

63. Peleg, D., Ullman, J.D.: An optimal synchronizer for the hypercube. In: PODC, pp. 77–85 (1987)

64. Pettie, S.: Distributed algorithms for ultrasparse spanners and linear size skeletons. Distrib. Comput. **22**(3), 147–166 (2010)

65. Rubinfeld, R., Tamir, G., Vardi, S., Xie, N.: Fast local computation algorithms. In: Proceedings of the Innovations in Computer Science Conference (2011)

66. Sadakane, K., Grossi, R.: Squeezing succinct data structures into entropy bounds. In: ACM-SIAM Symposium on Discrete Algorithms, pp. 1230–1239 (2006)
67. Saks, M.E., Seshadhri, C.: Local monotonicity reconstruction. SIAM J. Comput. **39**(7), 2897–2926 (2010)
68. Sarlos, T., Benczur, A., Csalogany, K., Fogaras, D., Racz, B.: To randomize or not to randomize: space optimal summaries for hyperlink analysis. In: Proceedings of the 15th International Conference on World Wide Web, pp. 297–306 (2006)
69. Schneider, J., Wattenhofer, R.: A log-star distributed maximal independent set algorithm for growth-bounded graphs. In: Proceedings of the 27th ACM Symposium on Principles of Distributed Computing, pp. 35–44 (2008)
70. Schneider, J., Wattenhofer, R.: A new technique for distributed symmetry breaking. In: Proceedings of the 29th ACM Symposium on Principles of Distributed Computing, pp. 257–266 (2010)
71. Spielman, D., Teng, S.: Nearly-linear time algorithms for graph partitioning, graph sparsification, and solving linear systems. In: Proceedings of the 36th Annual ACM Symposium on the Theory of Computing, pp. 81–90 (2004)
72. Sudan, M., Trevisan, L., Vadhan, S.: Pseudorandom generators without the XOR lemma. J. Comput. Syst. Sci. **62**(2), 236–266 (2001)
73. Yekhanin, S.: Private information retrieval. Commun. ACM **53**(4), 68–73 (2010)
74. Yoshida, Y., Yamamoto, Y., Ito, H.: An improved constant-time approximation algorithm for maximum matchings. In: Proceedings of the 41st Annual ACM Symposium on the Theory of Computing, pp. 225–234 (2009)
75. Zhu, Z.A., Lattanzi, S., Mirrokni, V.: A local algorithm for finding well-connected clusters. In: Proceedings of the Thirtieth International Conference on Machine Learning (2013)

Palindromic Decompositions
with Gaps and Errors

Michał Adamczyk[1], Mai Alzamel[2], Panagiotis Charalampopoulos[2],
Costas S. Iliopoulos[2], and Jakub Radoszewski[1,2(✉)]

[1] Faculty of Mathematics, Informatics and Mechanics,
University of Warsaw, Warsaw, Poland
{michal.adamczyk,jrad}@mimuw.edu.pl
[2] Department of Informatics, King's College London, London, UK
{mai.alzamel,panagiotis.charalampopoulos,costas.iliopoulos}@kcl.ac.uk

Abstract. Identifying palindromes in sequences has been an interesting line of research in combinatorics on words and also in computational biology, after the discovery of the relation of palindromes in the DNA sequence with the HIV virus. Efficient algorithms for the factorization of sequences into palindromes and maximal palindromes have been devised in recent years. We extend these studies by allowing gaps in decompositions and errors in palindromes, and also imposing a lower bound to the length of acceptable palindromes.

We first present an algorithm for obtaining a palindromic decomposition of a string of length n with the minimal total gap length in time $\mathcal{O}(n \log n \cdot g)$ and space $\mathcal{O}(n \cdot g)$, where g is the number of allowed gaps in the decomposition. We then consider a decomposition of the string in maximal δ-palindromes (i.e. palindromes with δ errors under the edit or Hamming distance) and g allowed gaps. We present an algorithm to obtain such a decomposition with the minimal total gap length in time $\mathcal{O}(n \cdot (g + \delta))$ and space $\mathcal{O}(n \cdot g)$.

1 Introduction

A palindrome is a symmetric word that reads the same backward and forward. The detection of palindromes is a classical and well-studied problem in computer science, language theory and algorithm design with a lot of variants arising out of different practical scenarios. String and sequence algorithms related to palindromes have long drawn the attention of stringology researchers [3,11,17]. Interestingly, in the seminal Knuth-Morris-Pratt paper presenting the well-known string matching algorithm [16], a problem related to palindrome recognition was

M. Alzamel is supported by the Saudi Ministry of Higher Education.

P. Charalampopoulos is supported by the Graduate Teaching Scholarship scheme of the Department of Informatics at King's College London.

J. Radoszewski is a Newton International Fellow and is supported by the Polish Ministry of Science and Higher Education under the 'Iuventus Plus' program grant no. 0392/IP3/2015/73.

© Springer International Publishing AG 2017
P. Weil (Ed.): CSR 2017, LNCS 10304, pp. 48–61, 2017.
DOI: 10.1007/978-3-319-58747-9_7

also considered. In combinatorics on words, for example, studies have investigated the inhabitation of palindromes in Fibonacci words or Sturmian words in general [6, 7]. There is also an interesting conjecture related to periodicity of infinite strings whose every factor can be decomposed into a bounded number of palindromes [9].

In computational biology, palindromes play an important role in regulation of gene activity and other cell processes because these are often observed near promoters, introns and specific untranslated regions. Hairpins (also called complemented palindromes) in the HIV virus are strings of the form $x\bar{x}^R$, where $\bar{x}^R$ is the reverse complement of x, while (even) palindromes are strings of the form xx^R. Algorithms for detecting palindromes can often be adapted to compute hairpins as well. Hence, we can identify palindromes in the DNA sequence and then align the part of the DNA sequence that contains them with the HIV virus.

In the beginnings of algorithmic study of palindromes, Manacher discovered an on-line sequential algorithm that finds all initial palindromes in a string [19]. A string $S[1 .. n]$ is said to have an initial palindrome of length k if $S[1 .. k]$ is a palindrome. Later Apostolico et al. observed that the algorithm given by [19] is able to find all maximal palindromic factors in the string in $\mathcal{O}(n)$ time [2]. Gusfield gave another linear-time algorithm to find all maximal palindromes in a string [14]. He also discussed the relation between biological sequences and gapped (separated) palindromes (i.e. strings of the form $xv\bar{x}^R$). Gupta et al. [13] presented an $\mathcal{O}(n)$-time algorithm to compute specific classes—based on length constraints—of such palindromes. Algorithms for finding the so-called gapped palindromes were also considered in [10, 17]. (In our study, we consider gaps *between* palindromes, not inside them.)

A problem that gained significant attention recently was decomposing a string into a minimal number of palindromes; any such decomposition is called a palindromic factorization. Fici et al. [8] presented an on-line $\mathcal{O}(n \log n)$-time algorithm for computing a palindromic factorization of a string of length n. A similar on-line algorithm was presented by I et al. [15] as well as an on-line algorithm with the same time complexity to factorize a string into maximal palindromes. Alatabbi et al. gave an off-line $\mathcal{O}(n)$-time algorithm for the latter problem [1]. In addition, Rubinchik and Shur [20] devised an $\mathcal{O}(n)$-sized data structure that helps locate palindromes in a string; they also show how it can be used to compute the palindromic factorization of a string in $\mathcal{O}(n \log n)$ time.

A similar problem, first studied by Galil and Seiferas in [12], asked whether a given string can be decomposed into k palindromes. Galil and Seiferas [12] presented an on-line $\mathcal{O}(n)$-time algorithm for $k = 1, 2$ and an off-line $\mathcal{O}(n)$-time algorithm for $k = 3, 4$. In 2014, Kosolobov et al. presented an on-line $\mathcal{O}(kn)$-time algorithm to decide this for arbitrary k [18].

Our work is a continuation of this line of research, motivated by possible errors and inconsistencies in the biological data. We extend the previous work by introducing a constraint on the length of the palindromes and allowing gaps and errors in the decompositions. By *gaps* we mean regions of the string that are not decomposed into palindromes of sufficient length. We allow *errors* in the palindromes, so that a *palindrome with errors* is a string having a small Hamming or edit distance from an ideal palindrome. We present two approaches for decomposing a string into sufficiently long palindromes; one allowing only gaps in the decomposition and the other allowing both gaps in the decomposition and errors in the palindromes. We first present an algorithm that computes a palindromic decomposition with the minimal total gap length of a string of length n in time $\mathcal{O}(n \log n \cdot g)$ and space $\mathcal{O}(n \cdot g)$, where g is the number of allowed gaps. Secondly, we present an $\mathcal{O}(n \cdot (g + \delta))$-time and $\mathcal{O}(n \cdot g)$-space algorithm for the decomposition of a string of length n into maximal palindromes with at most δ errors each, under the Hamming or edit distance, and g allowed gaps. The algorithms can be applied for both standard and complemented palindromes.

2   Notation and Terminology

Let $S = S[1]S[2] \cdots S[n]$ be a *string* of *length* $|S| = n$ over an alphabet Σ. We consider the case of an integer alphabet; in this case each letter can be replaced by its rank so that the resulting string consists of integers in the range $\{1, \ldots, n\}$. For two positions i and j, where $1 \leq i \leq j \leq n$, in S, we denote the *factor* $S[i]S[i+1] \cdots S[j]$ of S by $S[i..j]$. We denote the reverse string of S by S^R, i.e. $S^R = S[n]S[n-1] \cdots S[1]$. The empty string (denoted by ε) is the unique string over Σ of length 0. A string S is said to be a *palindrome* if and only if $S = S^R$. If $S[i..j]$ is a palindrome, the number $\frac{i+j}{2}$ is called the center of $S[i..j]$. Let $S[i..j]$, where $1 \leq i \leq j \leq n$, be a palindromic factor in S. It is said to be a *maximal palindrome* if there is no longer palindrome in S with center $\frac{i+j}{2}$. Note that a maximal palindrome can be a factor of another palindrome.

Definition 1. *We say that $S = p_1 p_2 \cdots p_\ell$ is a (maximal) palindromic decomposition of S if all the strings p_i are (maximal) palindromes.*

Definition 2. *A (maximal) palindromic decomposition of S such that the number of (maximal) palindromes is minimal is called a (maximal) palindromic factorization of S.*

Note that any single letter is a palindrome and, hence, every string can always be decomposed into palindromes. However, not every string can be decomposed into maximal palindromes; e.g. consider $S = $ abaca [1].

Let f be an *involution* on the alphabet Σ, i.e., a function such that $f^2 = $ id. We extend f into a morphism on strings over Σ. We say that a string x is a *generalized palindrome* if $x = f(x^R)$. Two known notions fit this definition:

- If $f = $ id, then a generalized palindrome is a standard palindrome.
- If $\Sigma = \{$A, C, G, T$\}$ and $f($A$) = $ T, $f($C$) = $ G, $f($G$) = $ C, $f($T$) = $ A, then a generalized palindrome corresponds to a so-called complemented palindrome [14].

Example 3. The string A G T A C T T C A T G A is a standard palindrome and the string T A G T C G A C T A is a complemented palindrome.

We also consider (generalized) palindromes with errors. Let us recall two well-known metrics on strings. Let u and v be two strings. If $|u| = |v|$, then the *Hamming distance* between u and v is the number of positions where u and v do not match. The *edit (or Levenshtein) distance* between u and v is the minimum number of edit operations (insertions, deletions, substitutions) needed to transform u into v. We say that x is a *generalized δ-palindrome* under the Hamming distance (or the edit distance) if the minimum Hamming distance (edit distance, respectively) from x to any generalized palindrome is *at most δ*.

A generalized palindrome $S[i \mathinner{..} j]$ is called *maximal* if there is no longer generalized palindrome with the same center. Similarly, a generalized δ-palindrome $S[i \mathinner{..} j]$ under the Hamming/edit distance is called *maximal* if there is no longer generalized δ-palindrome under the same distance measure with the same center.

Example 4. All maximal 0-palindromes/1-palindromes in GTATCG (for $f = $ id) under the Hamming and under the edit distance are as follows:

Center	1	1.5	2	2.5	3	3.5	4	4.5	5	5.5	6
0	G	ε	T	ε	TAT	ε	T	ε	C	ε	G
1 under Hamming	G	GT	GTA	TA	GTATC	AT	ATC	TC	TCG	CG	G
1 under edit	G	GT	GTA	GTAT	GTATC	GTATCG	ATC	TC	TCG	CG	G

For instance, the whole string GTATCG is a 1-palindrome under the edit distance, as deleting its fifth letter yields a palindrome GTATG.

The computational problems we study can be formally stated as follows.

GENERALIZED PALINDROMIC DECOMPOSITION WITH GAPS
Input: A string S of length n, an involution f, and integers $g, m \geq 1$
Output: A decomposition of S into generalized palindromes with the minimal possible total length of gaps, $\sum_i^q |g_i|$, such that:

- There are at most g gaps, i.e. $q \leq g$
- Each palindrome is of length at least m

GENERALIZED MAXIMAL δ-PALINDROMIC DECOMPOSITION WITH GAPS
Input: A string S of length n, an involution f, and integers $g, m, \delta \geq 1$
Output: A decomposition of S into maximal generalized δ-palindromes with the minimal possible total length of gaps, $\sum_i^q |g_i|$, such that:

- There are at most g gaps, i.e. $q \leq g$
- Each generalized δ-palindrome is of length at least m

We apply several instances of dynamic programming. For simplicity of presentation, we only show how to compute the minimal total length of gaps and omit describing the retrieval of the decomposition itself. To compute the latter, in each of the dynamic programming matrices we would store a pointer to the cell that gave us the minimum value so that we could actually compute the decomposition with the minimal total length of the gaps by backtracing.

3   Palindromic Decomposition with Gaps

In this section we develop an efficient solution to the GENERALIZED PALIN-DROMIC DECOMPOSITION WITH GAPS problem. It is based on several transformations of the algorithm for computing a palindromic factorization by Fici et al. [8]. For a string S of length n this algorithm works in $\mathcal{O}(n \log n)$ time. The algorithm consists of two steps:

1. Let P_j be the sorted list of starting positions of all palindromes ending at position j in S. This list may have size $\mathcal{O}(j)$. However, it follows from combinatorial properties of palindromes that the sequence of consecutive differences in P_j is non-increasing and contains at most $\mathcal{O}(\log j)$ distinct values. Let $P_{j,\Delta}$ be the maximal sublist of P_j containing elements whose predecessor in P_j is smaller by exactly Δ. Then there are $\mathcal{O}(\log j)$ such sublists in P_j. Hence, P_j can be represented by a set G_j of size $\mathcal{O}(\log j)$ which consists of triples of the form (i, Δ, k) that represent $P_{j,\Delta} = \{i, i + \Delta, \dots, i + (k-1)\Delta\}$. The triples are sorted according to decreasing values of Δ and all starting positions in each triple are greater than in the previous one. Fici et al. show that G_j can be computed from G_{j-1} in $\mathcal{O}(\log j)$ time.
2. Let $PL[j]$ denote the number of palindromes in a palindromic factorization of $S[1 .. j]$. Fici et al. show that it can be computed via a dynamic programming approach, using all palindromes from G_j in $\mathcal{O}(\log j)$ time. Their algorithm works as follows. Let $PL_\Delta[j]$ be the minimum number of palindromes we can decompose $S[1 .. j]$ in, provided that we use a palindrome from $(i, \Delta, k) \in G_j$. Then $PL_\Delta[j]$ can be computed in constant time using $PL_\Delta[j - \Delta]$ based on the fact that if $(i, \Delta, k) \in G_j$ and $k \geq 2$, then $(i, \Delta, k-1) \in G_{j-\Delta}$. Exploiting this fact, $PL_\Delta[j]$ can be computed by only considering $PL_\Delta[j - \Delta]$ and the shortest palindrome in (i, Δ, k). Finally, we compute $PL[j]$ from all such $PL_\Delta[j]$ values.

In Appendix A we show for completeness that the same approach works for generalized palindromes for any involution f.

To solve the GENERALIZED PALINDROMIC DECOMPOSITION WITH GAPS problem, we first need to modify each of the triples in G_j to reflect the length constraint (m). More precisely, due to the length constraint, in each G_j some triples will disappear completely, and at most one triple will get *trimmed* (i.e. the parameter k will be decreased).

Our algorithm then computes an array $MG[1 .. n][0 .. g]$ such that $MG[j][q]$ is the minimum possible total length of gaps in a palindromic decomposition of $S[1 .. j]$, provided that there are at most q gaps. Simultaneously, our algorithm computes an auxiliary array $MG'[1 .. n][0 .. g]$ such that $MG'[j][q]$ is the minimum possible total length of gaps up to position j provided that this position belongs to a gap: at most the q-th one.

For $j > 0$ and $q \geq 0$ we have the following formula:

$$MG[j][q] = \min(MG'[j][q], \min_{\Delta}\{MG_{\Delta}[j][q]\})$$

where $MG_{\Delta}[j][q]$ is the partial minimum computed only using generalized palindromes from $(i, \Delta, k) \in G_j$. The formula means: either we have a gap at position j, or we use a generalized palindrome ending at position j. We also set $MG[0][q] = 0$ for any $q \geq 0$.

We compute $MG_{\Delta}[j][q]$ for $(i, \Delta, k) \in G_j$ using the same approach as Fici et al. [8] used for PL_{Δ}, ignoring the triples that disappear due to the length constraint. If there is a triple that got trimmed, then the corresponding triple at position $j - \Delta$ (from which we reuse the values in the dynamic programming) must have got trimmed as well. More precisely, if the triple (i, Δ, k) is trimmed to (i, Δ, k') at position j, then at position $j - \Delta$ there is a triple $(i, \Delta, k - 1)$ which is trimmed to $(i, \Delta, k' - 1)$; that is, by the same number of generalized palindromes. Consequently, to compute $MG_{\Delta}[j][q]$ from $MG_{\Delta}[j-\Delta][q]$, we need to include one additional generalized palindrome (the shortest one in the triple) just as in Fici et al.'s approach.

Example 5. Consider the string AACCAACCAACCAACCAA, $f = \mathrm{id}$, and let $m = 7$.

A	A	C	C	A	A	C	C	A	A	C	C	A	A	C	C	A	A
1	2	3	4	5	6	7	8	9	10	11	12	13	14	15	16	17	18

Then $G_{18} = \{(1, \infty, 1), (5, 4, 4), (18, 1, 1)\}$, where:

- $(1, \infty, 1)$ represents the whole string,
- $(5, 4, 4)$ represents {AACCAACCAACCAA, AACCAACCAA, AACCAA, AA} which will get trimmed by 2 palindromes due to the length constraint, becoming $(5, 4, 2)$,
- $(18, 1, 1)$ represents {A} and disappears.

Now looking at position $j - \Delta = 18 - 4 = 14$ for the trimmed group, we had $(5, 4, 3) \in G_{14}$ representing {AACCAACCAA, AACCAA, AA}, and this also gets trimmed by 2 palindromes, becoming $(5, 4, 1)$.

Finally, for $j > 0$ and $q > 0$ we compute MG' using the following formula:

$$MG'[j][q] = \min(MG'[j - 1][q], MG[j - 1][q - 1]) + 1.$$

The first case corresponds to continuing the gap from position j, whereas the second to using a generalized palindrome finishing at position $j - 1$ or a gap finishing at position $j - 1$ (the latter will be suboptimal). Here the border cases are $MG'[j][0] = \infty$ for $j \geq 0$ and $MG'[0][q] = \infty$ for $q > 0$.

Thus we arrive at the complete solution to the problem.

Theorem 6. *The* GENERALIZED PALINDROMIC DECOMPOSITION WITH GAPS *problem can be solved in $\mathcal{O}(n \log n \cdot g)$ time and $\mathcal{O}(n \cdot g)$ space.*

4   Computing Maximal Palindromes with Errors

Recall that all maximal (standard) palindromes in a string can be computed in $\mathcal{O}(n)$ time by Manacher's [5,19] and Gusfield's [14] algorithms. These algorithms perform different computations for odd- and for even-length palindromes. Recall that we defined the centers of odd-length palindromes as integers and the centers of even-length palindromes as odd multiples of $\frac{1}{2}$.

Gusfield's algorithm [14] applies Longest Common Extension (LCE) Queries in the string $T = S\$S^R$, where $\$ \notin \Sigma$ is a sentinel character. An $LCE(i,j)$ query returns the length of the longest common prefix of the suffixes $T[i \mathinner{.\,.} |T|]$ and $T[j \mathinner{.\,.} |T|]$. For example, to compute the length of the maximal even-length palindrome centered between positions i and $i + 1$, the algorithm computes $LCE(i+1, 2n+2-i)$ in T. Recall that LCE queries in a string (over an integer alphabet) can be answered in $\mathcal{O}(1)$ time after linear-time preprocessing [4].

Gusfield's approach can be easily adapted to generalized palindromes: it suffices to apply LCE-queries on $T = S\$f(S^R)$. To further simplify the description of this approach, we introduce the Longest Gapped Palindrome (LGPal) Queries, such that $LGPal(i,j)$ is the maximum k such that $f(S[i-k+1 \mathinner{.\,.} i]^R) = S[j \mathinner{.\,.} j+k-1]$; see Fig. 1. As we have already noticed, LGPal-queries are equivalent to LCE-queries in $T = S\$f(S^R)$.

Fig. 1. To find the longest complemented 1-palindrome under the Hamming distance centered at position 7.5 in $S =$ GACATTCGAACGT, it suffices to ask two LGPal-queries: $LGPal(7,8) = 3$ finds the first mismatch, and $LGPal(3,12)$ extends the 1-palindrome after the mismatch. Note that each of these LGPal-queries is equivalent to an appropriate LCP-query in $S\$f(S^R)$.

It is known (see [14]) that all maximal generalized δ-palindromes under the Hamming distance can be computed in $O(n \cdot \delta)$ time via at most δ applications of the LGPal-query for each possible center position. Below we show how to compute maximal generalized δ-palindromes under the edit distance within the same time complexity.

Recall that if u is a generalized δ-palindrome under the edit distance, then there exists a generalized palindrome v such that the minimal number of edit operations (insertion, deletion, substitution) required to transform u to v is at most δ. The following simple observation shows that we can restrict ourselves to deletions and substitutions only, which we call in what follows the *restricted edit operations*. Intuitively, instead of inserting at position i a character to match the character at position $|u|-i+1$, we can delete the character at position $|u|-i+1$.

Observation 7. *Let u be a generalized δ-palindrome and v a generalized palindrome such that the edit distance between u and v is minimal. Then there exists a generalized palindrome v′ such that the number of restricted edit operations needed to transform u to v′ is equal to the edit distance between u and v.*

We can extend a maximal generalized δ-palindrome $S[i \mathinner{\ldotp\ldotp} j]$ to a maximal generalized $(δ+1)$-palindrome in three ways; either ignore the letter $S[i-1]$ and then perform an LGPal-query, or ignore the letter $S[j+1]$ and then perform an LGPal-query, or ignore both and then perform the LGPal-query. More formally:

Definition 8. *Assume that $S[i \mathinner{\ldotp\ldotp} j]$ is a generalized δ-palindrome. Then we say that each of the factors $S[i' \mathinner{\ldotp\ldotp} j']$ for:*

- *$i' = i - 1 - d$, $j' = j + d$, where $d = LGPal(i - 2, j + 1)$*
- *$i' = i - d$, $j' = j + 1 + d$, where $d = LGPal(i - 1, j + 2)$*
- *$i' = i - 1 - d$, $j' = j + 1 + d$, where $d = LGPal(i - 2, j + 2)$*

is an extension of $S[i \mathinner{\ldotp\ldotp} j]$. If the index i' is smaller than 1 or the index j' is greater than $|S|$, the corresponding extension is not possible. We also say that $S[i \mathinner{\ldotp\ldotp} j]$ can be extended to any of the three strings $S[i' \mathinner{\ldotp\ldotp} j']$.

Clearly, the extensions of a generalized δ-palindrome are always generalized $(δ+1)$-palindromes.

To facilitate the case of δ-palindromes being prefixes or suffixes of the text, we also introduce the following *border-reductions* for $S[i \mathinner{\ldotp\ldotp} j]$ being a generalized δ-palindrome:

- If $i = 1$, a border reduction leads to $S[1 \mathinner{\ldotp\ldotp} j - 1]$.
- If $j = n$, a border reduction leads to $S[i + 1 \mathinner{\ldotp\ldotp} n]$.

If any of the reductions is possible, we also say that $S[i \mathinner{\ldotp\ldotp} j]$ can be border-reduced to the corresponding strings. As previously, border-reductions of a generalized δ-palindrome are always generalized $(δ + 1)$-palindromes.

Lemma 9. *Given a maximal generalized δ-palindrome $S[i' \mathinner{\ldotp\ldotp} j']$ with $δ > 0$, there exists a maximal generalized $(δ - 1)$-palindrome $S[i \mathinner{\ldotp\ldotp} j]$ which can be extended or border-reduced to $S[i' \mathinner{\ldotp\ldotp} j']$.*

Proof. Consider a shortest sequence of restricted edit operations that transforms $u = S[i' \mathinner{\ldotp\ldotp} j']$ into a generalized palindrome v. Let us consider the position where we perform a restricted edit operation that is closest to i' or j'. Assume w.l.o.g. that this position—denote it by e—is not further to i' than to j'.

Assume first that this edit operation is a substitution. Then $S[i \mathinner{\ldotp\ldotp} j]$, for $i = e + 1$ and $j = j' - (e + 1 - i')$, is a generalized $(δ - 1)$-palindrome (the witness generalized palindrome is the corresponding factor of v); see Fig. 2. Moreover, it is a maximal generalized $(δ - 1)$-palindrome, as otherwise $S[e] = S[i - 1]$ would be equal to $f(S[j + 1])$, which means that the substitution at the position e would not be necessary. This completes the proof in this case.

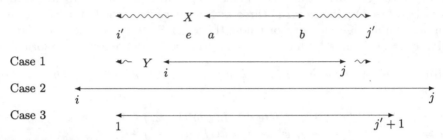

Fig. 2. If the outermost restricted edit operation on $S[i'\mathinner{\ldotp\ldotp}j']$ is a substitution (from letter X to letter Y), then $S[i'\mathinner{\ldotp\ldotp}j']$ is an extension of the third type of the maximal generalized $(\delta-1)$-palindrome $S[i\mathinner{\ldotp\ldotp}j]$.

Case 1

Case 2

Case 3

Fig. 3. Three cases resulting when the outermost edit operation on $S[i'\mathinner{\ldotp\ldotp}j']$ is a deletion of a character X.

Now assume that the edit operation at the position e was a deletion. Let $a = e+1$ and $b = j' - (e - i')$. Again, we see that clearly $S[a\mathinner{\ldotp\ldotp}b]$ is a generalized $(\delta-1)$-palindrome. If it is maximal, then we are done. Otherwise, consider the maximal generalized $(\delta-1)$-palindrome $S[i\mathinner{\ldotp\ldotp}j]$ centered at the same position as $S[a\mathinner{\ldotp\ldotp}b]$ ($a - i = j - b > 0$). Now we have three cases; see Fig. 3.

1. If $j \le j'$, then we can obtain $S[i'\mathinner{\ldotp\ldotp}j']$ by an extension (of the first type) of $S[i\mathinner{\ldotp\ldotp}j]$; i.e. ignoring the letter $S[i-1]$.
2. If $j > j'$, then we have that $S[i'\mathinner{\ldotp\ldotp}j'+1]$ is a generalized $(\delta-1)$-palindrome. If, additionally, $i' > 1$, then $S[i'-1\mathinner{\ldotp\ldotp}j'+1]$ is a generalized δ-palindrome, which contradicts the maximality of $S[i'\mathinner{\ldotp\ldotp}j']$.
3. Finally, if $j > j'$ and $i' = 1$, then $i = 1$, $j = j' + 1$. Hence, $S[i'\mathinner{\ldotp\ldotp}j']$ obtained from $S[i\mathinner{\ldotp\ldotp}j]$ by a border-reduction.

This completes the proof of the lemma.   □

The combinatorial characterization of Lemma 9 yields an algorithm for generating all maximal generalized d-palindromes, for all centers and subsequent $d = 0, \ldots, \delta$. Maximal generalized 0-palindromes are computed using Gusfield's approach (LGPal-queries). For a given $d < \delta$, we consider all the maximal generalized d-palindromes and try to extend each of them in all three possible ways (and border-reduce, if possible). This way we obtain a number of generalized $(d+1)$-palindromes amongst which, by Lemma 9, are all maximal generalized $(d+1)$-palindromes. To exclude the non-maximal ones, we group the generalized $(d+1)$-palindromes by their centers (in $\mathcal{O}(n)$ time via bucket sort) and retain only the longest one for each center. We arrive at the following intermediate result.

Lemma 10. *Under the edit distance, all maximal generalized δ-palindromes in a string of length n can be computed in $\mathcal{O}(n \cdot \delta)$ time and $\mathcal{O}(n)$ space.*

5   Maximal Palindromic Decomposition with Gaps and Errors

Let $\mathcal{F}$ be a set of factors of the text $S[1 \, .. \, n]$. In this section we develop a general framework that allows to decompose S into factors from $\mathcal{F}$, allowing at most g gaps. We call such a factorization a $(g, \mathcal{F})$-factorization of S. Our goal is to find a $(g, \mathcal{F})$-factorization of S that minimizes the total length of gaps. We aim at the time complexity $\mathcal{O}((n + |\mathcal{F}|) \cdot g)$ and space complexity $\mathcal{O}(n \cdot g + |\mathcal{F}|)$.

In our solution we use dynamic programming to compute two arrays, similar to the ones used in Sect. 3:

$MG[1 \, .. \, n][0 \, .. \, g]$: $MG[j][q]$ is the minimum total length of gaps in a $(q, \mathcal{F})$-factorization of $S[1 \, .. \, j]$.

$MG'[1 \, .. \, n][0 \, .. \, g]$: $MG'[j][q]$ is the minimum total length of gaps in a $(q, \mathcal{F})$-factorization of $S[1 \, .. \, j]$ for which the position j belongs to a gap.

We use the following formulas, for $j > 0$ and $q > 0$:

$$MG[j][q] = \min(MG'[j][q], \min_{S[a..j] \in \mathcal{F}} MG[a - 1][q])$$

$$MG'[j][q] = \min(MG[j - 1][q - 1], MG'[j - 1][q]) + 1$$

The border cases are exactly the same as in Sect. 3.

Clearly, the space complexity of this solution is $\mathcal{O}(n \cdot g + |\mathcal{F}|)$. Let us analyse its time complexity. Fix $q \in \{0, \dots, g\}$. The number of transitions using the factors from $\mathcal{F}$ in the dynamic programming is $|\mathcal{F}|$ in total, as each factor is used only for the position j where it ends. Hence, the formulas for $MG[j][q]$ take $\mathcal{O}(n \cdot g + |\mathcal{F}| \cdot g)$ time to evaluate. Computing the $MG'[j][q]$ values takes $\mathcal{O}(n \cdot g)$ time. Thus we arrive at the desired time complexity of $\mathcal{O}((n + |\mathcal{F}|) \cdot g)$.

We apply this approach to maximal generalized δ-palindromes in each of the considered metrics (see the classic result from [14] for the Hamming distance and Lemma 10 for the edit distance) to obtain the following result.

Theorem 11. *The* GENERALIZED MAXIMAL δ-PALINDROMIC DECOMPOSITION WITH GAPS *problem under the Hamming distance or the edit distance can be solved in* $\mathcal{O}(n \cdot (g + \delta))$ *time and* $\mathcal{O}(n \cdot g)$ *space.*

Example 12. Consider the following string[1] of length 92:

GGACTCGGCTTGCTGAGGTGCACACAGCAAGAGGCGAGAGCGGCGACTGGTGAGTACGCCAAATTT
TGACTAGCGGAGGCTAGAAGGAGAGA

We have used our implementation of the algorithm from Theorem 11 to compute the decomposition of the string into maximal complemented 3-palindromes of length at least 14 under the **edit distance** with at most 4 gaps ($g = 4$, $\delta = 3$, $m = 14$) with the minimal total gap length:

---

[1] See http://www.cesshiv1.org/disview.php?accession=AB220944.

[GGACTCG] G̲CTTGCTG*A*GGTGCACA̲CAGCAAGA̲ [GGCGAGAGC] GGCGACTG̲G̲T̲GAGT*A*CGCC
[AAATTTTG] A̲CTAGCG̲G̲AGGCTAGA̲ [AGGAGAGA]

The gaps are given in square brackets. Edit operations are underlined, with deletes additionally given in italics. The gaps have total length 32.

In comparison, the optimal decomposition of this string under the **Hamming distance** with the same parameters ($g = 4$, $\delta = 3$, $m = 14$) uses four gaps of total length 46.

6   Conclusions

We have presented two algorithms for finding palindromic decompositions: one allowing gaps and the other allowing both gaps in the decomposition and errors in palindromes. The first algorithm shows that (somewhat surprisingly) Fici et al.'s algorithm [8] for finding an exact palindromic factorization can be extended to handle gaps, a constraint on the palindromes length, and complements in palindromes as well. In the second algorithm we decompose a string into maximal palindromes with errors; the most involved part here was computing all such maximal palindromes under the edit distance.

In the problems that were defined in the beginning, the objective was to minimize the total length of gaps, allowing a certain number of gaps. However, the approaches that were presented in this paper can be used to solve different variants of the problems, like minimizing only the total number of gaps or maximizing the total length of palindromes, regardless of the number of gaps.

An open question is to efficiently compute decompositions into palindromes that may contain errors and are not necessarily maximal. This problem seems to be hard, as δ-palindromes do not have such a strong combinatorial structure as palindromes without errors.

A   Appendix

Generalized Palindromic Factorization

In this section we show that the approach of Fici et al. [8] works for generalized palindromes for any involution f. The following auxiliary lemma extends the combinatorial properties of standard palindromes used in [8] (see Lemmas 1–3 therein) to generalized palindromes. Recall that a string y is called a *border* of a string x if it is both a prefix and a suffix of x. A number p is called a *period* of x if $x[i] = x[i + p]$ for all $i = 1, \ldots, |x| - p$. It is well known that x has a period p iff it has a border of length $|x| - p$; see [4,5].

Lemma 13. *(a) Let y be a suffix of a generalized palindrome x. Then y is a border of x iff y is a generalized palindrome.*

(b) Let x be a string with a border y such that $|x| \leq 2|y|$. Then x is a generalized palindrome iff y is a generalized palindrome.

(c) Let y be a proper suffix of a generalized palindrome x. Then $|x| - |y|$ is a period of x iff y is a generalized palindrome. In particular, $|x| - |y|$ is the smallest period of x iff y is the longest generalized palindromic proper suffix of x.

Proof. (a) Let y' be the prefix of x of length $|y|$. As x is a generalized palindrome, $y' = f(y^R)$. ($\Rightarrow$) If y is a border of x, then $y = y' = f(y^R)$, so y is a generalized palindrome. ($\Leftarrow$) If y is a generalized palindrome, then $y' = f(y^R) = y$, so y is a border of x.

(b) ($\Rightarrow$) From (a), if x is a generalized palindrome and y is its border, then y is a generalized palindrome. ($\Leftarrow$) If y is a generalized palindrome, $f(x^R)$ has a border $f(y^R) = y$. This border covers the whole string $f(x^R)$ and is the same as the border of x, so $x = f(x^R)$ and x indeed is a generalized palindrome.

(c) This is a consequence of part (a) and the relation between borders and periods of a string.   □

The crucial combinatorial property of standard palindromes used in Step 1 of the algorithm in Sect. 3 is that the sequence of consecutive differences in P_j is non-increasing and contains at most $\mathcal{O}(\log j)$ distinct values. We show that the same observation holds for generalized palindromes; this follows from the next lemma, parts (1) and (2). The proof of Lemma 14 follows exactly the lines of the proof of the corresponding Lemma 4 in [8]; due to space constraints, we refer the reader to Fig. 3 illustrating the proof in [8].

Lemma 14. *Let x be a generalized palindrome, y the longest generalized palindromic proper suffix of x, and z the longest generalized palindromic proper suffix of y. Let u and v be strings such that $x = uy$ and $y = vz$. Then:*

(1) $|u| \geq |v|$;
(2) if $|u| > |v|$ then $|u| > |z|$;
(3) if $|u| = |v|$ then $u = v$.

Proof. (1) By Lemma 13(c), $|u| = |x| - |y|$ is the smallest period of x, and $|v| = |y| - |z|$ is the smallest period of y. Since y is a factor of x, either $|u| > |y| > |v|$ or $|u|$ is a period of y too, and thus it cannot be smaller than $|v|$.

(2) By Lemma 13(a), y is a border of x and thus v is a prefix of x. Let w be a string such that $x = vw$. Then z is a border of w and $|w| = |zu|$. Since we assume $|u| > |v|$, we must have $|w| > |y|$. Suppose to the contrary that $|u| \leq |z|$. Then $|w| = |zu| \leq 2|z|$, and by Lemma 13(b), w is a generalized palindrome. But this contradicts y being the longest generalized palindromic proper suffix of x.

(3) In the proof of (2) we saw that v is a prefix of x, and so is u by definition. Thus $u = v$ if $|u| = |v|$.   □

We have thus shown that, also in case of generalized palindromes, the set P_j can be compactly represented by a set G_j, as described in Sect. 3. To complete Step 1 of the algorithm, we need to show that G_j can be computed from G_{j-1} in $\mathcal{O}(\log j)$ time. For this, just as in [8], we show that each triple $(i, \Delta, k) \in G_{j-1}$

will be either eliminated or replaced by $(i-1, \Delta, k)$ in G_j. The proof exploits part (3) of Lemma 14.

Lemma 15. *Let p_i and p_{i+1} be two consecutive elements of $P_{j-1,\Delta}$. Then $p_i - 1 \in P_j$ iff $p_{i+1} - 1 \in P_j$.*

Proof. By definition, $p_{i+1} - p_i = \Delta$, and the predecessor of p_i in P_j is $p_{i-1} = p_i - \Delta$. The strings $x = S[p_{i-1} .. j - 1]$, $y = S[p_i .. j - 1]$, and $z = S[p_{i+1} .. j - 1]$ form the situation of Lemma 14(3). Hence, $S[p_i - 1] = S[p_{i+1} - 1] = c$. Thus, $p_i - 1 \in P_j$ iff $S[j] = f(c)$ iff $p_{i+1} - 1 \in P_j$. □

After this transformation, one might need to update pairs of adjacent triples in G_j because the gaps between them might have changed. This simple process is explained in detail in [8] and takes only $\mathcal{O}(\log j)$ additional time.

As for Step 2 of the algorithm, it suffices to show that the following combinatorial observation holds for generalized palindromes. Again we follow the lines of the proof from [8] (cf. Fig. 5 in that paper).

Lemma 16. *If $(i, \Delta, k) \in G_j$ and $k \geq 2$, then $(i, \Delta, k - 1) \in G_{j-\Delta}$.*

Proof. By definition, $(i, \Delta, k) \in G_j$ is equivalent to saying that $P_{j,\Delta} = \{i, i + \Delta, \ldots, i + (k-1)\Delta\}$, and we need to show that $P_{j-\Delta,\Delta} = \{i, i + \Delta, \ldots, i + (k-2)\Delta\}$. We will show first that $P_{j-\Delta,\Delta} \cap [i - \Delta + 1 .. j - \Delta] = \{i, i + \Delta, \ldots, i + (k-2)\Delta\}$ and then that $P_{j-\Delta,\Delta} \cap [1 .. i - \Delta] = \emptyset$.

Since $y = S[i .. j]$ and $x = S[i - \Delta .. j]$ are generalized palindromes and y is the longest proper border of x (by Lemma 13(a)), $S[i - \Delta .. j - \Delta] = y = S[i .. j]$. Thus for all $\ell \in [i .. j]$, $\ell \in P_j$ iff $\ell - \Delta \in P_{j-\Delta}$. In particular, the consecutive differences in both cases are the same and for all $\ell \in [i + 1 .. j]$, $\ell \in P_{j,\Delta}$ iff $\ell - \Delta \in P_{j-\Delta,\Delta}$. Thus $P_{j-\Delta,\Delta} \cap [i - \Delta + 1 .. j - \Delta] = \{i, i + \Delta, \ldots, i + (k-2)\Delta\}$.

We still need to show that $P_{j-\Delta,\Delta} \cap [1 .. i - \Delta] = \emptyset$, which is true if and only if $i - 2\Delta \notin P_{j-\Delta}$. Suppose to the contrary that $S[i - 2\Delta .. j - \Delta]$ is a generalized palindrome and let $w = S[i - 2\Delta .. i - \Delta - 1]$. Then $S[j - 2\Delta + 1 .. j - \Delta] = f(w^R)$. Since $z = S[i - \Delta .. j - \Delta]$ and $S[i - \Delta .. j]$ are generalized palindromes too, we have that $S[i - \Delta .. i - 1] = w$ and $S[j - \Delta + 1 .. j] = f(w^R)$. Finally, since z is a generalized palindrome, $S[i - 2\Delta .. j] = wzf(w^R)$ is a generalized palindrome. This implies that $i - 2\Delta \in P_j$ and thus $i - \Delta \in P_{j,\Delta}$, which is a contradiction. □

References

1. Alatabbi, A., Iliopoulos, C.S., Rahman, M.S.: Maximal palindromic factorization. In: Stringology, pp. 70–77 (2013)
2. Apostolico, A., Breslauer, D., Galil, Z.: Parallel detection of all palindromes in a string. Theor. Comput. Sci. **141**(1), 163–173 (1995). http://dx.doi.org/10.1016/0304-3975(94)00083-U
3. Breslauer, D., Galil, Z.: Finding all periods and initial palindromes of a string in parallel. Algorithmica **14**(4), 355–366 (1995). http://dx.doi.org/10.1007/BF01294132

4. Crochemore, M., Hancart, C., Lecroq, T.: Algorithms on Strings. Cambridge University Press, Cambridge (2007)
5. Crochemore, M., Rytter, W.: Jewels of Stringology. World Scientific, Singapore (2003)
6. Droubay, X.: Palindromes in the Fibonacci word. Inf. Process. Lett. **55**(4), 217–221 (1995). http://dx.doi.org/10.1016/0020-0190(95)00080-V
7. Droubay, X., Pirillo, G.: Palindromes and Sturmian words. Theor. Comput. Sci. **223**(1–2), 73–85 (1999). http://dx.doi.org/10.1016/S0304-3975(97)00188-6
8. Fici, G., Gagie, T., Kärkkäinen, J., Kempa, D.: A subquadratic algorithm for minimum palindromic factorization. J. Discret. Algorithms **28**(C), 41–48 (2014). http://dx.doi.org/10.1016/j.jda.2014.08.001
9. Frid, A., Puzynina, S., Zamboni, L.: On palindromic factorization of words. Adv. Appl. Math. **50**(5), 737–748 (2013). http://dx.doi.org/10.1016/j.aam.2013.01.002
10. Fujishige, Y., Nakamura, M., Inenaga, S., Bannai, H., Takeda, M.: Finding gapped palindromes online. In: Mäkinen, V., Puglisi, S.J., Salmela, L. (eds.) IWOCA 2016. LNCS, vol. 9843, pp. 191–202. Springer, Cham (2016). doi:10.1007/978-3-319-44543-4_15
11. Galil, Z.: Real-time algorithms for string-matching and palindrome recognition. In: Proceedings of the Eighth Annual ACM Symposium on Theory of Computing, pp. 161–173. ACM (1976). http://doi.acm.org/10.1145/800113.803644
12. Galil, Z., Seiferas, J.: A linear-time on-line recognition algorithm for "palstar". J. ACM **25**(1), 102–111 (1978). http://doi.acm.org/10.1145/322047.322056
13. Gupta, S., Prasad, R., Yadav, S.: Searching gapped palindromes in DNA sequences using dynamic suffix array. Indian J. Sci. Technol. **8**(23), 1 (2015)
14. Gusfield, D.: Algorithms on Strings, Trees, and Sequences: Computer Science and Computational Biology. Cambridge University Press, New York (1997)
15. I, T., Sugimoto, S., Inenaga, S., Bannai, H., Takeda, M.: Computing palindromic factorizations and palindromic covers on-line. In: Kulikov, A.S., Kuznetsov, S.O., Pevzner, P. (eds.) CPM 2014. LNCS, vol. 8486, pp. 150–161. Springer, Cham (2014). doi:10.1007/978-3-319-07566-2_16
16. Knuth, D.E., Morris Jr., J.H., Pratt, V.R.: Fast pattern matching in strings. SIAM J. Comput. **6**(2), 323–350 (1977)
17. Kolpakov, R., Kucherov, G.: Searching for gapped palindromes. Theor. Comput. Sci. **410**(51), 5365–5373 (2009). http://dx.doi.org/10.1016/j.tcs.2009.09.013
18. Kosolobov, D., Rubinchik, M., Shur, A.M.: Pal^k is linear recognizable online. In: Italiano, G.F., Margaria-Steffen, T., Pokorný, J., Quisquater, J.-J., Wattenhofer, R. (eds.) SOFSEM 2015. LNCS, vol. 8939, pp. 289–301. Springer, Heidelberg (2015). doi:10.1007/978-3-662-46078-8_24
19. Manacher, G.: A new linear-time "on-line" algorithm for finding the smallest initial palindrome of a string. J. ACM (JACM) **22**(3), 346–351 (1975)
20. Rubinchik, M., Shur, A.M.: EERTREE: an efficient data structure for processing palindromes in strings. In: Lipták, Z., Smyth, W.F. (eds.) IWOCA 2015. LNCS, vol. 9538, pp. 321–333. Springer, Cham (2016). doi:10.1007/978-3-319-29516-9_27

Cascade Heap: Towards Time-Optimal Extractions

Maxim Babenko[1,2], Ignat Kolesnichenko[2,3($\boxtimes$)], and Ivan Smirnov[3]

[1] National Research University Higher School of Economics, Moscow, Russia
maxim.babenko@gmail.com
[2] Yandex LLC, Moscow, Russia
[3] Moscow Institute of Physics and Technology, Moscow, Russia
ignat1990@gmail.com, ifsmirnov@yandex.ru

Abstract. *Heaps* are well-studied fundamental data structures, having myriads of applications, both theoretical and practical.

We consider the problem of designing a heap with an "optimal" EXTRACT-MIN operation. Assuming an arbitrary linear ordering of keys, a heap with n elements typically takes $O(\log n)$ time to extract the minimum. Extracting *all* elements faster is impossible as this would violate the $\Omega(n \log n)$ bound for comparison-based sorting. It is known, however, that is takes only $O(n + k \log k)$ time to sort just k smallest elements out of n given, which prompts that there might be a faster heap, whose EXTRACT-MIN performance depends on the number of elements extracted so far.

In this paper we show that is indeed the case. We present a version of heap that performs INSERT in $O(1)$ time and takes only $O(\log^* n + \log k)$ time to carry out the k-th extraction (where $\log^*$ denotes the iterated logarithm). All the above bounds are worst-case.

1 Introduction

Heap is a data structure consisting of elements with some assigned *keys*. Typically heaps enable inserting new elements (INSERT), extracting the element with the minimum key (EXTRACT-MIN), and decreasing keys of existing elements (DECREASE-KEY). Multitude of heap flavors are known, and their design and possible applications have been studied quite extensively [1, Chap. 6.5].

Let n denote the number of elements in the heap. Historically, the first version of heap was suggested by Williams [2]. This *binary* heap is able to perform INSERT, EXTRACT-MIN, and DECREASE-KEY in $O(\log n)$ time. A *Fibonacci heap* introduced by Fredman and Tarjan [3] improves upon these bounds and carries INSERT and DECREASE-KEY in $O(1)$ time but still takes $O(\log n)$ time to perform EXTRACT-MIN (these bounds are amortized). Another quite sophisticated data structure is suggested by Brodal [4]. Similarly to Fibonacci heaps, it requires $O(1)$ time per INSERT and DECREASE-KEY and $O(\log n)$ time per EXTRACT-MIN. In contrast to Fibonacci heaps, all these bounds are worst-case rather than amortized. Much effort [5,6] has been devoted to study heaps in RAM model, where keys are restricted to be integers in a certain range.

© Springer International Publishing AG 2017
P. Weil (Ed.): CSR 2017, LNCS 10304, pp. 62–70, 2017.
DOI: 10.1007/978-3-319-58747-9_8

Let us focus on just INSERT and EXTRACT-MIN operations. For any comparison-based algorithm a sequence of n insertions and n extractions cannot be performed faster than in $\Omega(n \log n)$ time, as it follows from the well-known sorting bound. The latter bound, in fact, shows that EXTRACT-MIN cannot be performed faster that $O(\log n)$ *on average*. Still it is possible to *partial-sort* an array (i.e. find and sort $k \le n$ smallest keys out of n given) in just $O(n + k \log k)$ time using, e.g. linear-time selection of the k-th order statistic [7]. This indicates that EXTRACT-MIN could run faster if the number of extractions is small. See [8] for some practical randomized algorithms.

As it is widely known, $O(n + k \log k)$ bound for partial-sorting keys is optimal in comparison-based model. Hence one would expect some heap data structure with $O(1)$ time per INSERT and $O(\log k)$ time per EXTRACT-MIN to exist (where k stands for the total number of elements extracted so far).

Hereinafter we use the following notation: $\log^{(r)}$ indicates the r-th iteration of logarithm, i.e. $\log^{(0)} x = x$, $\log^{(k+1)} x = \log \log^{(k)} x$; $\log^*$ indicates the *iterated logarithm*, i.e. $\log^* x$ is the minimum k such that $\log^{(k)} x \le 1$.

In this paper we make some progress towards obtaining a data structure with the optimal time complexity by presenting *Cascade Heaps* with $O(1)$ insertion and $O(\log^* n + \log k)$ extraction times. These bounds are worst-case.

The rest of the paper is organized as follows. In Sect. 2 we introduce some notation and auxiliary subroutines to be used later. In Sect. 3 we give a high-level overview of Cascade Heaps and present an amortized version. Section 4 deamortizes the data structure. In Sect. 5 we conclude.

2   Preliminaries

2.1   Binary Heaps

Let K be a linear-ordered universe of possible keys. Our algorithms are comparison-based, i.e. it is assumed that any pair of keys can be compared in $O(1)$ time. By $H(L)$ we denote the family of all binary heaps with keys in some linear ordered set L.

Recall the classical *binary heap* data structure [2]. It consists of an almost complete binary tree T with keys assigned to its nodes. This tree satisfies the following *heap-like* property: the key at each node is larger than or equal to the key at its parent (if any). For the sake of simplicity, we assume all keys in our heaps to be distinct. To insert an element into a binary heap we place it to the end of the last level of the tree (forming a new one if the current last level is full) and then run the standard SIFT-UP procedure, restoring the heap-like property. The minimum key can always be found in the root of such a tree. To extract the minimum, we move the last element from the last tree level to the root and then invoke SIFT-DOWN procedure at the root. A collection of n keys can be turned into a binary heap in $O(n)$ time using MAKE-HEAP routine. For more details, please refer to [1, Chap. 6].

2.2   Recursive Heaps

Let us introduce the notion of a *recursive heap*, which will be the corner-stone of all our data structures. In what follows, we will need to be able to compare not only individual keys but also keys with heaps and heaps themselves. To this aim, for the sake of comparison each heap is identified with its minimum key. This way, we can introduce a linear order on, e.g., $H(K) \cup K$. Now denote K by $H^{(0)}(K)$ and for $m \geq 1$ define $H^{(m)}(K)$ to be $H(H^{(m-1)}(K)) \cup K$. In other words, $H^{(0)}(K)$ is the set of keys, $H^{(1)}(K)$ is the family of keys and heaps of keys, $H^{(2)}(K)$ is the family consisting of keys and heaps whose elements are keys and heaps of keys, etc. We call heaps in $H^{(m)}(K)$ *recursive*.

To make comparisons cheap, for each recursive heap and all its contained subheaps we explicitly maintain their minimum keys.

For a recursive heap α, we call its *rank* the smallest m such that $\alpha \in H^{(m)}(K)$. Also let top-size(α) denote the number of elements in the almost complete binary tree (called the *top-tree*) representing α. Please keep in mind that elements of this tree could be trees themselves, so top-size(α) is *not* the same as the number of keys stored in α.

Given a recursive heap $\alpha \in H^{(m)}(K)$, a new element β (either a key or a heap from $H^{(m-1)}(K)$) can be inserted in $O(\text{top-size}(\alpha))$ time by attaching β as a new leaf of the top-tree of α and applying SIFT-UP.

Also the minimum key in α can be extracted without increasing the rank of α and increasing top-size(α) on the constant. This will be done by the following RECURSIVE-EXTRACT-MIN procedure. Denote the root of α by root(α). Note that the root of a recursive heap may be either a key, or a heap itself. If root(α) is a key then we perform the usual binary heap extraction, which takes $O(\log(\text{top-size}(\alpha)))$ time and clearly cannot increase the rank of α and decreases top-size(α).

Otherwise let root(α) be a tree. We split α into root *root*(α) and its left and right subtrees, denoted by left(α) and right(α) respectively and proceed to extracting the minimum in *root*(α). After at most $m-1$ such iterations we obtain a tree where the root node is just a key. We split this tree as earlier. It remains to re-assemble α from the left and right subtrees: left(α), right(α), left(root(α)), right(root(α)), etc. Note that, the ranks of the left and right subtrees (viewed as recursive heaps) do not exceed the rank of the tree itself. Hence we can use heaps left(root(root(α))), right(root(root(α))), ... (of rank at most $m-2$) as elements to construct a new heap β with rank at most $m-1$ in $O(m)$ time using MAKE-HEAP routine. The remaining job is to merge left(α) and right(α) (of rank at most m) with heaps β, left(root(α)) and right(root(α)) (of rank at most $m-1$). First, we place β into the root and attach subtrees left(α) and right(α) to β as its children. Then we invoke SIFT-DOWN to restore the heap-like property for the constructed tree α'. After that we insert left(root(α)) and right(root(α)) as elements into α' using INSERT procedure for binary heap. Altogether this takes $O(\log(\text{top-size}(\alpha)) + m)$ time and increases top-size(α) by at most two.

We summarize as follows:

Theorem 1. *Inserting an element (either a key or a heap) into a recursive heap α takes $O(\log top\text{-}size(\alpha))$ time and increases top-size(α) by one. Extracting the minimum key from a recursive heap α of rank m takes $O(m + \log top\text{-}size(\alpha))$ time, increases top-size(α) by at most 2, and does not increase the rank of α.*

3 Cascade Heaps

3.1 Separating Insertions and Extractions

We start by describing a heap, which runs INSERT in $O(\log n)$ worst-case time and EXTRACT-MIN in $O(\log k)$ worst-case time. This shows the core techniques that will help us to get closer to the $O(\log k)$ worst-case time bound for EXTRACT-MIN.

Our heap consists of two parts: the *insertion heap* IH from $H(K)$ and the *extraction heap* EH from $H^{(2)}(K)$. Let us describe the insertion and extraction routines. The insertion routine works as a simple heap insertion to IH. The extraction is done as follows. First we insert IH (as an element) to EH and set the new IH to be the empty heap. Second we run RECURSIVE-EXTRACT-MIN to extract the minimum key from EH as described in Sect. 2. Note that the top-size of EH increases by at most three (inserting IH into EH increases the top-size by one and RECURSIVE-EXTRACT-MIN increases it by at most two). Therefore the top-size of EH is $O(k)$, where k is the number of extractions performed so far.

The above data structure is very simple and elegant. Its main drawback is that insertion requires $O(\log n)$ time both in the average and in the worst cases. In the rest of this section we show how to reduce the insertion time to $O(1)$ in amortized sense.

3.2 Amortized `CascadeHeap`

Let us generalize by adding more intermediate levels to our data structure. We first describe its amortized version. We maintain a collection of recursive heaps $IH_1, \ldots, IH_{\log^* n}$, and, as earlier, a recursive heap EH.

The following *tetration* notation will be useful:

$$^{n}a = \underbrace{a^{a^{\cdot^{\cdot^{\cdot^{a}}}}}}_{n}.$$

For each i, the rank of IH will be at most i and its top-size will remain less than $^{i}2$. The rank of EH will be at most $\log^* n + 2$ and its top-size will be bounded by $O(k)$.

INSERT for `CascadeHeap` works as follows. First, we insert the element into IH_1. If this heap reaches its top-size limit of 2 (we call this situation an *overflow*), we insert IH_1 (as an element) into IH_2 and set IH_1 to be empty. If IH_2 also reaches its top-size limit, then we insert IH_2 into IH_3, etc. This maintains the top-size and rank invariants.

Clearly, if at most n elements are being inserted in `CascadeHeap`, then $\log^* n$ levels of IH_i are enough. As n grows, $\log^* n$ may increase by 1 thus forming a new level.

Consider EXTRACT-MIN for `CascadeHeap`. We take all $IH_1, \ldots, IH_{\log^* n}$, construct a heap (of rank at most $\log^* n + 1$) out of them, and insert the resulting element into `EH` (recall that the latter is of rank at most $\log^* n + 2$). Finally we extract the minimum key from `EH`. Altogether top-size(`EH`) increases by at most three (by one during insertion and by two during extraction), hence it remains $O(k)$. The top-size and the rank invariants are obviously maintained.

Time complexities of the above INSERT and EXTRACT-MIN are as follows. Constructing a heap out of IH_i takes $O(\log^* n)$ time, inserting the latter into `EH` takes $O(\log k)$ time, extracting the minimum from `EH` takes $O(\log^* n + \log k)$, which is $O(\log^* n + \log k)$ in total.

Analyzing insertions is a bit more tricky, so instead of proving amortized bounds we compute the *average* complexity (we will present a formal deamortized $O(1)$ worst-case version in the next section anyway). Consider a sequence of N insertions. Then IH_i overflows at most

$$\frac{N}{{}^1 2 \cdot {}^2 2 \cdot \ldots \cdot {}^i 2}$$

times. Each overflow involves inserting IH_i into IH_{i+1}, which costs $O(\log({}^{i+1} 2)) = O({}^i 2)$ time since top-size(IH_{i+1}) $< {}^{i+1} 2$. Therefore, the total cost of handling overflows at IH_i is

$$\frac{N}{{}^1 2 \cdot {}^2 2 \cdot \ldots \cdot {}^{i-1} 2} = O(N \cdot 2^{-i}).$$

Summing over all i, one gets $O(N)$ in total or $O(1)$ per inserted element on average, as promised.

4   Deamortization

4.1   Delayed Insertions

In this chapter we describe how to deamortize `CascadeHeap`. We start by introducing a *heap with delayed insertions*. Informally this is just the usual binary heap where each insertion is split into $\log n$ separate elementary steps. When an insertion starts the only possible operations are to continue the insertion or to cancel it. Once the final step of the insertion is complete, we obtain the heap which is the same as if we would have performed the insertion using SIFT-UP routine.

Let us describe this data structure formally. It can be in one of two states: a *regular* state or an *insertion* state. In the regular state the data structure is represented by a regular binary heap and supports INSERT and EXTRACT-MIN operations. EXTRACT-MIN extracts the root element and splits the tree into two subtrees. INSERT operation receives and memorizes an element x and changes the state of the heap to insertion. In the insertion state the data structure supports two operations: CONTINUE and CANCEL. CONTINUE operation performs some insertion steps. If the insertion completes after CONTINUE the heap goes to the regular state and appears to be changed to the heap with additional element x, otherwise no visible changes occur. If CANCEL is called then the insertion is interrupted; it returns the element x being inserted. All described operations take $O(1)$ worst-case time. The full insertion requires $O(\log n)$ steps (calls to CONTINUE) to finish.

The insertion executes SIFT-UP routine gradually, which adds x as a new leaf and then lifts x up to the root step-by-step. At each step x is compared against its parent v, if the key of x is less than the key of v then we swap v and x, otherwise we finish SIFT-UP. However, we must ensure that no changes to the tree are visible before the insertion is finished. To this aim, we employ a variant of *copy-on-write* technique as follows. For readers, the heap is represented by a pointer to its root node. In INSERT we initially find a spot for x in the tree (at the end of the last level or at the beginning of the new level if the last one is full) and set v to be the parent of x. Child pointers of v are not updated, though, so this child-parent relation between x and v is "virtual". On each step we compare the keys of x and v. If the key of x is larger then x is attached to v, and INSERT finishes. Otherwise we need to "swap" x and v. To do this without affecting the readers, we make a (shallow) copy v' of v and attach x to v', swap x and v', and reset v to its parent. The invariant is that if, at some point, we do actually attach x to v then what we get is essentially the tree as it would have appeared at this point of SIFT-UP. Also note that x could be lifted to the very root of the tree, i.e. v could become null. In this case INSERT finishes by replacing the pointer to the root of the tree. At the end, one prunes the nodes that are no longer used (these could be detected by, e.g., a simple reference counting). Figure 1 provides an example.

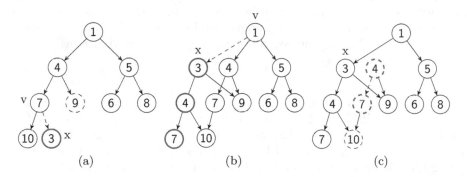

(a)   (b)   (c)

Fig. 1. Delayed insertion of x into the heap. (a) Just after the insertion started; (b) After SIFT-UP; (c) After the final step. Blue bold indicates new nodes created during insertion. Red dashed bold indicates nodes that are pruned after insertion. (Color figure online)

4.2 Deamortizing CascadeHeap

Now we use the above heaps with delayed insertions to implement the top-trees for each of IH_i. INSERT for CascadeHeap works as follows: we first execute up to γ (which is a constant to be chosen latter) calls to CONTINUE for those IH_i that are in the insertion state, starting from the smallest i. If after these calls some of IH_i is in the regular state and reaches its top-size limit of $^i2$, then we start inserting IH_i into IH_{i+1}. As we show later, at this point IH_{i+1} will be in the regular state, i.e. insertions will not interfere.

EXTRACT-MIN cancels all insertions and thus receives certain un-inserted recursive heaps $\{\alpha_1, \ldots, \alpha_t\}$, $t = O(\log^* n)$. Then it constructs a heap out of all α_i and IH_i and inserts the result into EH. Clearly this takes $O(\log^* n + \log k)$, as in the amortized version.

To facilitate locating appropriate IH_i during INSERT, we maintain a sorted linked list L of the form $i_1 < i_2 < \ldots < i_t$ consisting of all i such that IH_i is in the insertion state. During INSERT, this list is scanned from left to right. For each $i \in L$, we invoke CONTINUE for IH_i. We stop either when L is fully scanned or after γ calls to CONTINUE. If after some CONTINUE call the insertion into IH_i completes, we remove IH_i from the list. In the latter case it is also possible that IH_i overflows, i.e. reaches the top-size limit of $^i2$. We collect all such i thus forming a new list F (initially empty). Finally, once all (up to γ) CONTINUE calls complete, we handle overflows: scan F is the *reverse* order (from largest indexes to smallest) and for each $i \in F$ start inserting IH_i into IH_{i+1} (and thus clear IH_i), remove i from F and add $i+1$ to beginning of L (note that at this point L does not contain elements that less than or equal to $i+1$, therefore this operation preserves sorted order of elements in list L). Obviously the total overhead of maintaining L and F is $O(\gamma)$, which does not affect the time complexity.

It remains to prove that the insertions in IH_i do not interfere (and choose the appropriate value of γ). Note that it suffices to assume that no extractions are being performed since an extraction immediately cancels all ongoing activities. Fix $k \in \{1, \ldots, \log^* n\}$ and consider IH_k. It gets overflowed after each

$$N = {}^12 \cdot {}^22 \cdot \ldots \cdot {}^k2$$

insertions into CascadeHeap. Consider a moment when IH_k overflows. We need at most $\delta \cdot \log(^{k+1}2) = \delta \cdot {}^k2$ calls to CONTINUE to insert IH_k into IH_{k+1}. (Here $\delta > 0$ is some constant depending on the implementation details of heaps with delayed insertion.) The next overflow will happen after exactly N insertions. During these insertions the algorithm is ready to spend time performing up to γN CONTINUE calls.

Recall that these calls first address smaller insertion heaps before moving to larger ones. Let us estimate the total number of CONTINUE calls needed to handle insertions in $IH_1, \ldots, IH_k$. Fix $i \in \{1, \ldots, k-1\}$, then IH_i overflows

$$\frac{N}{^12 \cdot {}^22 \cdot \ldots \cdot {}^i2} = {}^{i+1}2 \cdot {}^{i+2}2 \cdot \ldots \cdot {}^k2$$

times during N insertions into `CascadeHeap`. Handling each overflow takes at most $\delta \cdot {}^{i}2$ calls to CONTINUE for $\text{IH}_1, \ldots, \text{IH}_k$. Hence at least

$$\gamma N - \delta \sum_{i=1}^{k-1} {}^{i}2 \cdot {}^{i+1}2 \cdot \ldots \cdot {}^{k}2 \tag{1}$$

of calls to CONTINUE remain "spare" and thus are offered to IH_{k+1}, IH_{k+2} , etc. It remains to prove that (1) is at least $\delta \cdot {}^{k}2$. Indeed, set $\gamma := 3\delta$, then

$$\gamma N - \delta \sum_{i=1}^{k-1} {}^{i}2 \cdot {}^{i+1}2 \cdot \ldots \cdot {}^{k}2 =$$

$$N \cdot \left(\gamma - \delta \sum_{i=1}^{k-1} \frac{1}{{}^{1}2 \cdot {}^{2}2 \cdot \ldots \cdot {}^{i-1}2} \right) \geq$$

$$N \cdot \left(\gamma - \delta \sum_{i=1}^{k-1} \frac{1}{2^{i-1}} \right) \geq N \cdot (\gamma - 2\delta) = \delta N.$$

Therefore we have at least δN spare CONTINUE calls but only need $\delta \cdot {}^{k}2 < \delta N$. The proof follows.

We summarize:

Theorem 2. *For CascadeHeap,* INSERT *takes* $O(1)$ *worst-case time and* EXTRACT-MIN *takes* $O(\log^* n + \log k)$ *worst-case time, where n denotes the current number of keys in the heap and k denotes the number of* EXTRACT-MIN *calls performed so far.*

5 Conclusions

We have presented a new `CascadeHeap` data structure, which is almost optimal in sense of insertion and extraction times. In our construction, we only rely on binary heaps and simple recursion ideas, which suggests that our approach could be quite practical.

One could try to extend `CascadeHeap` with operations like UNITE or DECREASE-KEY. This seems doable but the exact implementation details and complexity bounds are yet to be examined.

The major remaining open problem obviously is: can we obtain a truly-optimal heap, i.e. a heap with $O(1)$ worst-case time INSERT $O(\log k)$ worst-case time EXTRACT-MIN? The additional $O(\log^* n)$ term in `CascadeHeap` is minute, but could turn out to be notoriously difficult to eliminate.

References

1. Cormen, T.H., Leiserson, C.E., Rivest, R.L., Stein, C.: Introduction to Algorithms, 3rd edn. The MIT Press, Cambridge (2009)
2. Williams, J.W.J.: Heapsort. Commun. ACM **7**, 347–348 (1964)
3. Fredman, M.L., Tarjan, R.E.: Fibonacci heaps and their uses in improved network optimization algorithms. J. ACM **34**(3), 596–615 (1987)
4. Brodal, G.S.: Worst-case efficient priority queues. In: Proceedings of the Seventh Annual ACM-SIAM Symposium on Discrete Algorithms (SODA 1996), pp. 52–58, Philadelphia, PA, USA. Society for Industrial and Applied Mathematics (1996)
5. van Emde Boas, P.: Preserving order in a forest in less than logarithmic time. In: Proceedings of the 16th Annual Symposium on Foundations of Computer Science (SFCS 1975), pp. 75–84, Washington, DC, USA. IEEE Computer Society (1975)
6. Thorup, M.: On ram priority queues. SIAM J. Comput. **30**(1), 86–109 (2000)
7. Blum, M., Floyd, R.W., Pratt, V., Rivest, R.L., Tarjan, R.E.: Time bounds for selection. J. Comput. Syst. Sci. **7**(4), 448–461 (1973)
8. Navarro, G., Paredes, R.: On sorting, heaps, and minimum spanning trees. Algorithmica **57**(4), 585–620 (2010)

Entropic Uniform Sampling of Linear Extensions in Series-Parallel Posets

Olivier Bodini[1], Matthieu Dien[2(✉)], Antoine Genitrini[2],
and Frédéric Peschanski[2]

[1] Laboratoire d'Informatique de Paris-Nord, CNRS UMR 7030 - Institut
Galilée - Université Paris-Nord, 99, avenue Jean-Baptiste Clément,
93430 Villetaneuse, France
Olivier.Bodini@lipn.univ-paris13.fr
[2] Sorbonne Universités, UPMC Univ Paris 06, CNRS, LIP6 UMR 7606,
4 place Jussieu, 75005 Paris, France
{Matthieu.Dien,Antoine.Genitrini,Frederic.Peschanski}@lip6.fr

Abstract. In this paper, we introduce a uniform random sampler for
linear extensions of Series-Parallel posets. The algorithms we present
ensure an essential property of random generation algorithms: entropy.
They are in a sense optimal in their consumption of random bits.

1 Introduction

The state-space of a concurrent program is interpreted, in what is called the
interleaving semantics, as the linearization of partially ordered sets (posets).
This linearization process is highly combinatorial, a topic we studied thoroughly
in previous papers. The uniform random generation of linear extensions provides a stochastic approach for the linearization process, hopefully avoiding the
so-called "combinatorial explosion". For instance, in [1] we develop an efficient
algorithm to draw linear extensions of tree-shaped posets. The algorithm has
worst-case time complexity $O(n \log n)$ (counting arithmetic operations) with n
the size of the poset (more precisely, the number of nodes of its covering tree).
A uniform random sampler for posets of dimension 2 is introduced in [2]. A
perfect sampling algorithm for arbitrary posets is presented in [4]. These are
polynomial algorithms but in the order $\tilde{O}(n^3)$ hence not usable on large posets.
Our goal is to identify subclasses of posets for which more efficient algorithms
can be proposed. In this paper, we introduce a uniform random sampler for linear extensions of *Series-Parallel (SP) posets*. This represents a very important
class of posets that can be found in many practical settings. Generating linear
extensions uniformly at random for this subclass can be done in a relatively
straightforward way. However, such a naive algorithm fails an essential property
of random generation algorithms: *entropy*. When studying random generation,
the consumption of random bits is a very important measure. An entropic algorithm minimizes this consumption, which has a major impact on efficiency. The
algorithms we describe in the paper are *entropic*, i.e. they are in a sense optimal
in their consumption of random bits.

© Springer International Publishing AG 2017
P. Weil (Ed.): CSR 2017, LNCS 10304, pp. 71–84, 2017.
DOI: 10.1007/978-3-319-58747-9_9

The outline of the paper is as follows. First, in Sect. 2 we define a canonical representation of Series-Parallel posets. Based on this representation we develop our random generation algorithms in Sect. 3. We propose two variants of the random sampler: a bottom-up and a top-down variant. In Sect. 4 we describe the common stochastic core of both algorithms. This is where we discuss, and prove, the property of *entropy*. In this extended abstract, we only provide outlines for the correctness and entropy proofs.

2   Canonical Representation of Series-Parallel Posets

The Series-Parallel posets are not easily handled from an algorithmic point of view. The ground set, the set of relations, and the inductive structure of a Series-Parallel poset must be adapted in order tu use them automatically through algorithms, see for example [11]. So, to work with Series-Parallel posets we introduce a canonical representation of such posets, based on their *covering* directed acyclic graph (DAG). Then we design an effective algorithm that can be precisely analyzed, thanks to the canonicity of our representation. In this section we detail such a DAG representation whose main objective is to preserve the necessary informations such that the uniformity property of the sampling process is preserved. We first recall the classical construction of Series-Parallel posets [10] (SP posets). This is based on the two basic composition rules below.

Definition 1 (Poset compositions). *Let P and Q be two independent partial orders, i.e. posets whose ground sets are disjoint.*

– *The parallel composition R of P and Q, denoted by $R = P \parallel Q$, is the partial order obtained by the disjoint union of the ground sets of P and Q and such that*

$$\forall E \in \{P, Q\}, \forall x, y \in E, x \prec_R y \ iff \ x \prec_E y.$$

– *The series composition S of P and Q, denoted by $S = P.Q$, is the partial order obtained by the disjoint union of the ground sets of P and Q and such that*

$$\forall x, y \in P \cup Q, x \prec_S y \ iff \ \begin{cases} x \in P \ and \ y \in Q \\ x, y \in P \ and \ x \prec_P y \\ x, y \in Q \ and \ x \prec_Q y \end{cases}.$$

We are now ready for the definitions of SP posets.

Definition 2 (Series-Parallel partial orders). *The class of Series-Parallel orders is the smallest class containing the atomic order and closed under series and parallel composition.*

In the mathematical definition, a partial order is a set of relations between points of the ground set. We must adapt this context to obtain an efficient representation in computers. So, a common way to handle such a poset for an algorithmic need, is to exploit their *covering* DAG.

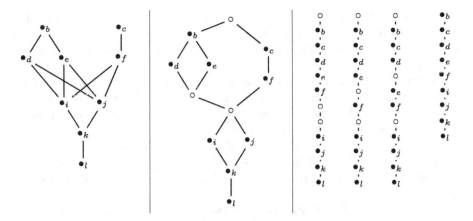

Fig. 1. (Left to Right) The covering of a Poset P; A SP DAG for P; The topological sorts of a given equivalence class and the associated linear extension.

Definition 3. *Let P be a partial order. The covering G of P is the directed acyclic graph (DAG) whose nodes are the points of the ground set V of P and whose set of directed edges is such that*

$$\{(x,y) \in V^2 \mid (x \prec_P y) \wedge \neg(\exists z \in V, \ x \prec_P z \prec_P y)\}.$$

Note that the combinatorial class of coverings is the one of intransitive DAG.

The leftmost part of Fig. 1 represents the covering of a Series-Parallel poset. It represents the poset with ground set $\{b, c, d, e, f, i, j, k, l\}$. The set of order relation is $\{b < d, b < e, c < f, d < i, d < j, e < i, e < j, f < i, f < j, i < k, j < k, k < l\}$ and all transitive relations. We will use the common representation of coverings as Hasse diagrams in which edges are directed from top to bottom.

Definition 4. *Let P be a poset. A linear extension of P is a total order of the points of the ground set of P and satisfying the relations of P.*

The rightmost chain represented in Fig. 1, is the following linear extension of P: $b < c < d < e < f < i < j < k < l$. For the rest of the paper, the points of P will be called nodes. The following is *folklore*.

Fact 1. *Let P be a poset. Each linear extension of P corresponds to a topological sort of the nodes of the covering of P.*

We now introduce our special flavor of coverings, with the objective of getting effective and elegant algorithms presented in the next sections.

Let P a poset, and G its covering. First we want to distinguish the different children of a node in G. The classical way for this consists in a *combinatorial embedding* of G in the plane (cf. [5, Chap. 3])[1]. In our context, we choose an

---

[1] The combinatorial embedding allows to distinguish the two successors of a node: the left one and the right one.

arbitrary embedding, e.g. in Fig. 1, we have chosen that b is on the left of c. Note that this arbitrary choice only impacts the representation and not the poset itself. Thus, we will identify the covering of a poset with the chosen combinatorial embedding.

A second simplification is that we only consider unary-binary DAGs instead of ones with nodes of arbitrary in-degree and out-degree. This is important otherwise an extra level of loop would be required in the algorithms. To reach this goal, we use the *left-leaning* principle [8] directly on the combinatorial embedding: a poset composed of several posets in parallel, associated to a combinatorial embedding of its covering, $P_1 \parallel P_2 \parallel \cdots \parallel P_n$ is seen as a poset with a binary parallel composition (relying on its associativity property): $(\ldots (P_1 \parallel P_2) \parallel \ldots) \parallel P_n$.

To encode the covering G with only unary-binary nodes, we need to introduce some "silent" nodes that we call *white nodes* in the rest of the paper, among the original *black nodes* of G (the points of the ground set of P). In the following, we explain the construction of this new structure. We also show how to recover the linear extensions of P from this representation.

Definition 5. *Let Ψ be the following function, from the set of combinatorial embeddings of Series Parallel poset coverings to the set of bicolored unary-binary (combinatorially embedded) DAGs. It is inductively defined as:*

$$
\begin{cases}
\Psi(\varnothing) = \circ \qquad \Psi(\bullet) = \bullet \qquad \Psi\left(\begin{matrix} \bullet \\ \vdots \\ \bullet \end{matrix}\right) = \begin{matrix} \bullet \\ \vdots \\ \bullet \end{matrix} \\[2em]
\Psi(P.(Q \parallel R).S) = \begin{matrix} \Psi(P) \\ \swarrow \quad \searrow \\ \Psi(Q) \qquad \Psi(R) \\ \searrow \quad \swarrow \\ \Psi(S) \end{matrix} ,
\end{cases}
$$

where P, Q, R and S are arbitrary coverings of Series-Parallel posets. P and S may be empty. Q and R must not be empty and R must verify $\neg(\exists R_1, R_2, \ R = R_1 \parallel R_2)$.

The Series-Parallel DAG (SP DAG) of a poset P is the image of an arbitrary combinatorial embedding of the covering of P by the function Ψ.

When the poset looks like $P_1 \parallel P_2 \parallel \cdots \parallel P_n$, the last pattern condition means $R = P_n$. Note that the unique constraint for P and S is that they are SP posets, eventually empty. In particular, they can contain substructure like $(A \parallel B)$. Thus the application of the last rule is not deterministic, in the sense that we can apply it sometimes successively, but the order of application is arbitrary. However the rules are trivially confluent.

In Fig. 1, on the left, a poset is represented by its covering. In the middle, it is the SP DAG we associate to this covering. A SP DAG contains bicolored nodes. The black nodes are the nodes of the initial covering, while the white nodes have been added to fulfill the unary-binary arity constraint.

The SP DAGs are by essence recursively decomposable, and thus we use the classical notation from Analytic Combinatorics [3] for their characterization.

Proposition 1. *The class $\mathcal{D}$ of SP DAGs is unambiguously specified by:*

$$
\left\{
\begin{array}{l}
\mathcal{D} \;=\; \bullet \;+\; \begin{array}{c} \bullet \\ | \\ \mathcal{D}_t \end{array} \;+\; \begin{array}{c} \bullet + \circ \\ \diagup \quad \diagdown \\ \mathcal{D} \qquad \mathcal{D}_r \\ \diagdown \quad \diagup \\ \mathcal{D} + \circ \end{array} \\[3em]
\mathcal{D}_t \;=\; \bullet \;+\; \begin{array}{c} \bullet \\ | \\ \mathcal{D}_t \end{array} \;+\; \begin{array}{c} \bullet \\ \diagup \quad \diagdown \\ \mathcal{D} \qquad \mathcal{D}_r \\ \diagdown \quad \diagup \\ \mathcal{D} + \circ \end{array} \\[3em]
\mathcal{D}_r \;=\; \mathcal{D}_t \;+\; \begin{array}{c} \circ \\ \diagup \quad \diagdown \\ \mathcal{D} \qquad \mathcal{D}_r \\ \diagdown \quad \diagup \\ \mathcal{D} \end{array}
\end{array}
\right.
$$

Let us recall the basic notation from Analytic Combinatorics, by describing the first equation. A DAG in $\mathcal{D}$ is either a single node $\bullet$ or a root $\bullet$ followed by a DAG from the class $\mathcal{D}_t$ or a *top* root (either a black node $\bullet$ or a white node $\circ$) with a *left* substructure belonging to $\mathcal{D}$ and a *right* substructure from $\mathcal{D}_r$, both followed by a *bottom* substructure corresponding either to a DAG in $\mathcal{D}$ or to a white node $\circ$. In the rest of the paper we will use the terms top, left, right and bottom substructures.

Remark that the class $\mathcal{D}_t$ contains the connected Series-Parallel coverings with a single source (i.e. a node smaller that all other nodes), and $\mathcal{D}_r$ encodes the class of connected Series-Parallel coverings.

Theorem 2. *For each Series-Parallel poset, we choose an arbitrary combinatorial embedding of the covering. Let $\mathcal{E}$ be the set of the chosen embeddings of the Series-Parallel posets. Ψ is a bijection from $\mathcal{E}$ to the set of SP DAGs.*

The proof is direct by a structural induction.

By successively using the combinatorial embedding, the left-leaning principle and the transformation Ψ, we have an effective and canonical way for representing Series-Parallel posets.

Another central property is the correspondence between the linear extensions of the poset P and the topological sorts of $\Psi(P)$.

Let P be a poset, and $S = \Psi(P)$ be its associated SP DAG. We consider the set of topological sorts of S. Obviously each of them contain all nodes of S, the black ones but also the white ones, although the latter have no meaning for P.

Definition 6. *Let P be a poset, and $S = \Psi(P)$ be its associated SP DAG. Let ρ be the function, from the set of topological sorts of S to the set of linear extensions of P, such that, applied to a topological sort all white nodes are removed.*

Definition 7. *Let P be a Series-Parallel poset, and $S = \Psi(P)$ be its associated SP DAG. We define the following equivalence relation based on the function ρ: Two topological sorts s_1 and s_2 of S are equivalent if and only if $\rho(s_1) = \rho(s_2)$.*

Let us remark some fundamental constraint in the equivalence relation: for two distinct linear extensions of a poset, the numbers of sorts in the two corresponding equivalence classes are not necessary equal. In fact, in Fig. 1 (right-hand side), we have represented the three topological sorts corresponding to the same linear extension. But the following linear extension $c < f < b < d < e < i < j < k < l$, is given by a single topological sort, where both white nodes appear between e and i. As a consequence, we cannot directly sample uniformly a topological sort of S and then apply the transformation ρ to obtain *uniformly* a linear extension of P.

Definition 8. *Let P be a Series-Parallel poset, $S = \Psi(P)$ be its associated SP DAG. In an equivalence class of topological sorts of S, we define the representative to be the sort whose white nodes appear as soon as possible. If two white nodes are incomparable and appear successively in the representative we choose the leftmost one (in S) to appear first.*

In Fig. 1 (on the right side), among the three topological sorts, the rightmost one is considered to be the representative for the linear extension. The case where several white nodes are incomparable and successive in the topological sort is handled similarly.

Theorem 3. *Let P be a Series-Parallel poset, $S = \Psi(P)$ be its associated SP DAG. An uniform sampling of the linear extension of a Series-Parallel poset P is obtained by drawing uniformly at random a representative $\Psi(P)$ and by applying the function ρ to it.*

The proof of the theorem is a direct consequence of the previous results.

To conclude this section, we exhibit the computational complexity for the construction of the SP DAG resulting of a Series-Parallel poset.

Proposition 2. *The SP DAG corresponding to a Series-Parallel poset, whose ground set contains n nodes is built in $O(n)$ time complexity.*

The number of black nodes of the SP DAG is n and the number of white nodes is at most $2(n-1)$.

This last proposition guarantees that the complexity of building and using a SP DAG in place of its associated Series-Parallel poset, will be negligible in front of the complexity of the algorithms presented below.

3   Random Generation of Linear Extensions

Based on the SP DAG structures defined in the previous section, we now begin the presentation of the uniform random samplers. In this section, we give the outline of the algorithms, and in the next section we discuss their common stochastic core. We present two complementary generation schemes: a *bottom-up* scheme that recursively generates the linear extensions "from the inside-out", and a *top-down* variant that does the converse. Both algorithms have an interest. The bottom-up approach is arguably the simplest, in that it follows the recursive decomposition of the input DAG. The correctness proofs are much easier in this setting. Comparatively, the top-down approach seems more convoluted, and it is best seen as a transformation of the bottom-up scheme in the proofs. One advantage the top-down algorithm is that it can be implemented *in-place*.

Algorithm 1. Bottom-up variant of the uniform random generation of linear extensions

---

    function RandLinExt-BU(P)
        if $P = \circ$ **then return** [ ]
        else if $P = \bullet_x$ **then return** $[x]$
        else if $P = \bullet_x$. T **then return** cons(x, RandLinExt-BU(Q))
        else if $P = \square$. $(L \parallel R)$. T **then**
            $h :=$ Shuffle(RandLinExt-BU(L), RandLinExt-BU(R))
            $t :=$ RandLinExt-BU(T)
            if $\square = \bullet_x$ **then return** concat(cons(x, h), t)
            else return concat(h, t)

The bottom-up variant is Algorithm 1. We illustrate it on the poset example of Fig. 1. The root is an unlabeled white node, with two subposets with respective roots b and c. The join node is white and also the root of the poset comprising the labels $\{i, j, k, l\}$. In the case of such a fork/join structure, the algorithm works as follows. First, the algorithm is recursively applied on the two subposets in parallel: the one with labels $\{b, d, e\}$ and the one with labels $\{c, f\}$. For the latter, there is only one possibility: taking first the label c and then f, resulting in the partial linear extension $[c, f]$. For the left part, the label b is prepended to the *uniform shuffle* of the singleton linear extensions $[d]$ and $[e]$. The shuffle algorithm will be presented in detail, but in this situation it simply consists in taking either $[d, e]$ or $[e, d]$ both with a probability of $\frac{1}{2}$. Suppose we take the latter, we ultimately obtain the linear extension $[b, e, d]$. Note that nothing is appended at the end since the join node is white in this subposet. In the next step, the extensions $[c, f]$ and $[b, e, d]$ are shuffled uniformly, one possible outcome being $[c, b, e, d, f]$. This is then concatenated with a linear extension of the downward poset. For example $[c, b, e, d, f, i, j, k, l]$ is one such possibility, and is thus a possible output of the algorithm.

    The top-down variant is described by Algorithm 2. The main difference with the bottom-up algorithm is that it samples *positions* in an array instead of *labels* directly. The advantage is that most operations can then be performed in-place, at the price of having to deal with one level of indirection. The *rankings* structure is a mapping associating the node labels to a position in the sampled linear extension. The *positions* are organized as a stack structure, initially containing all the available positions from 1 to $|P|$ (the size of the poset in the number

Algorithm 2. Top-down variant of the uniform random generation of linear extensions

---

> **function** RANDLINEXT-TD(P)
>> **function** RECRANDLINEXT-TD(P, *rankings*, *positions*)
>>> **if** $P = \circ$ **then return** *rankings*
>>> **else if** $P = \bullet_x$ **then**
>>>> *rankings*[x] := pop(*positions*)
>>>> **return** *rankings*
>>> **else if** $P = \bullet_x \,.\, T$ **then**
>>>> *rankings*[x] := pop(*positions*)
>>>> **return** RECRANDLINEXT-TD(T, *rankings*, *positions*)
>>> **else if** $P = \square \,.\, (L \mid R) \,.\, T$ **then**
>>>> **if** $\square = \bullet_x$ **then** *rankings*[x] := pop(*positions*)
>>>> *upPositions* := *positions*[0 ... $|L| + |R| - 1$]
>>>> *botPositions* := *positions*[$|L| + |R|$ ... $|P| - 1$]
>>>> l, r := SPLIT(*upPositions*, $|L|$, $|R|$)
>>>> *rankings* := RECRANDLINEXT-TD(L, *rankings*, l)
>>>> *rankings* := RECRANDLINEXT-TD(R, *rankings*, r)
>>>> **return** RECRANDLINEXT-TD(T, *rankings*, *botPositions*)
>> *rankings* := an empty dictionary
>> *positions* := [1 ... $|P|$]
>> **return** RECRANDLINEXT-TD(P, *rankings*, *positions*)

---

of labels i.e. the black nodes). In our example poset, the initial contents of *positions* is $[1, 2, 3, 4, 5, 6, 7, 8, 9]$. The *rankings* map is empty. In the first step, the white root is simply skipped and the two sets of positions are computed: the *upPositions* taking the front part of the poset i.e. $[1, 2, 3, 4, 5]$ and *botPositions* what is remaining i.e. $[6, 7, 8, 9]$. The algorithm then performs an *uniform split* of the positions 1 to 5 e.g. in a subset $l = \{2, 3, 4\}$ for $\{b, d, e\}$ and $r = \{1, 5\}$ for $\{c, f\}$. The details about the splitting process are given below. The rankings of each subposet are computed recursively, and the result naturally interleaves since we work with disjoint sets of positions. Once again, we can ultimately obtain the linear extension $[c, b, e, d, f, i, j, k, l]$. We have to show that it is obtained with the same exact (uniform) probability as in the bottom-up case.

In fact, the two algorithms are dual. In the bottom-up case, randomness comes from the SHUFFLE function which is the dual of the SPLIT function in the sense of a coproduct. A shuffle takes two lists and mixes them into one, while the split takes one list and divides it into two. The key property comes from the fact that the shuffle (resp. split) of one (resp. two) list(s) is sampled uniformly: each shuffle (resp. splits) has the same probability to be drawn. For example, there is $\binom{5}{2} = 10$ possible shuffles between the sets $\{a, b, c\}$ and $\{d, e\}$. Equivalently there is 10 possibles splits of the set $\{a, b, c, d, e\}$ into two subsets, one of size 3 and the other of size 2.

Both algorithms operate in the same way: they draw a random combination of p elements among $p + q$, then shuffle or split using this combination. This will

---

Algorithm 3. Algorithm of uniform random splitting and shuffling

function SPLIT(S, p, q)
  ℓ, r := [ ], [ ]
  i := 0
  v := RandomCombination(p, q)
  for all $e \in v$ **do**
    if e **then**
      append $S[i]$ to ℓ
    else
      append $S[i]$ to r
  return ℓ, r

function SHUFFLE(ℓ, r)
  t := [ ]
  v := RandomCombination($|\ell|$, $|r|$)
  for all $e \in v$ **do**
    if e **then**
      append pop(ℓ) to t
    else
      append pop(r) to t
  return t

---

be discussed in the next section. Based on the assumption that the stochastic process is uniform, we obtain a first important result about the algorithms.

Theorem 4. *Algorithms 1 and 2 both generate a linear extension of a series-parallel poset P uniformly at random. Their worst-case time complexity is $\Theta(n^2)$ (by measuring the number of memory writes). The average time complexity is equivalent to $\frac{1}{4}\sqrt{\frac{\pi\, n^3}{3\sqrt{2}-4}}$.*

Fact 5 *(Möhring [7]). Let P be a SP poset and ℓ_P be its number of linear extensions. If $P = P_1 \times P_2$ is the series composition of P_1 and P_2, then $\ell_P = \ell_{P_1} \cdot \ell_{P_2}$. If $P = P_1 + P_2$ is the parallel composition of P_1 and P_2, then $\ell_P = \binom{n_1+n_2}{n_1} \cdot \ell_{P_1} \cdot \ell_{P_2}$, where n_1 (resp. n_2) is the size of P_1 (resp. P_2).*

The correctness of both algorithms is easily proved by a structural induction and based on the Fact 5.

To compute the average complexity, the idea is to find a recurrence equation for the number of SP DAGs of size n and another recurrence equation for counting the number of cumulated memory writes on SP DAGs of size n. Thus, using standard analytic combinatorics tools [3], we derive the asymptotic behaviors of the solutions of both recurrences. The result is obtained by dividing the asymptotic number of cumulated memory writes by the one of SP DAGs of size n.

4 Entropic Sampling Core

The bottom-up and top-down algorithms described in the previous section both depend upon the same *stochastic core*: namely the procedure we named RANDOMCOMBINATION. Random generation must adopt the point of view of probabilistic Turing machines, i.e. deterministic machines with a tape containing random bits. As a consequence, an important measure of complexity for such an algorithm is the entropy of the targeted distribution: the number of random bits consumed to produce a possible output. Our objective is to define *entropic* algorithms, according to the following definition.

Definition 9 (Entropic algorithm). *Let A be an algorithm sampling an element of a finite set S at random according to a probability distribution μ. We say that A is entropic if the average number of random bits n_e it uses to sample one element $e \in S$ is proportional to the entropy of μ, in the sense of Shannon entropy [9]:*

$$\exists K > 0, \forall e \in S, n_e \leqslant K \cdot \sum_{x \in S} -\mu(x) \log_2(\mu(x)).$$

The key idea in the following entropic algorithms is to show that Bernoulli random variable (r.v.) of small or big parameter has weak entropy and because we are unable to use fraction of bits we group it in packs to draw Bernoulli r.v. of entropy 1. The maximum entropy of a Bernoulli is reached when the parameter is $\frac{1}{2}$, in this case the Bernoulli r.v. is just a random bit.

Algorithm 4. Algorithm of uniform random generation of combination

---

function RANDOMCOMBINATION(p, q)
   $l := [\,]$
   ♯**True** is the number of **True** in l
   ♯**False** is the number of **False** in l
   $rndBits :=$ a stream of random booleans produced with k-BERNOULLI$\left(\frac{p}{p+q}\right)$
   if $p > \log(q)^2 \wedge q > \log(p)^2$ **then**
      while ♯**True** $<= p \wedge$ ♯**False** $<= q$ **do**
         if pop($rndBits$) **then** $l :=$ **cons**(**True**, l)
         else $l :=$ **cons**(**False**, l)
      $remaining := \neg$**pop**(l)
   else
      if $p < q$ **then**
         $l :=$ a list of q times **False**
         $remaining :=$ **True**
      else
         $l :=$ a list of p times **True**
         $remaining :=$ **False**
   for $i :=$ ♯**True** $+$ ♯**False** $- 1$ to $p + q - 1$ **do**
      $j :=$ **uniformRandomInt**$[0 \ldots i]$
      insert $remaining$ at position j in l
      return l

The core of the random samplers is presented in Algorithm 4. The objective is to draw, in an entropic way, a list l of booleans of size $p + q$ such that p cells contains a **True** value and the q remaining are set to **False**.

We give an example of the sampling process for $p = 6$ and $q = 2$. In the first step, the list l is filled with **True** and **False** values, with respective probability $\frac{p}{p+q}$ and $\frac{q}{p+q}$. For this we use a stream of Bernoulli random variables $rndBits$ that is produced by a function named k-BERNOULLI, which we explain below. The filling process stops when one of the boolean values is drawn once more than needed (e.g. reaching $p+1$ (resp. $q+1$) times True (resp. False) values). The last value is then discarded. For example, if $l = \text{F} :: \text{T} :: \text{T} :: \text{F} :: \text{T} :: \text{F} :: \text{T} :: [\,]$ then F is drawn 3 times although only 2 is needed. So the last F is discarded. In the second step of the algorithm, a number *remaining* of boolean values is needed to complete the list l. These are randomly inserted among the bits already drawn, by using uniform integer random variables. For example:

$$l = \text{T} :: \text{T} :: \text{F} :: \text{T} :: \text{F} :: \text{T} :: [\,]$$
$$\hookrightarrow l = \text{T} :: \text{T} :: \text{F} :: \underline{\text{T}} :: \text{T} :: \text{F} :: \text{T} :: [\,]$$
$$\hookrightarrow l = \text{T} :: \text{T} :: \text{F} :: \text{T} :: \text{T} :: \text{F} :: \text{T} :: \underline{\text{T}} :: [\,]$$

At the end, the list l contains the required number of booleans values, and as we justify below, it is drawn uniformly.

An important part of the algorithm is the first test $p > \log(q)^2 \wedge q > \log(p)^2$. This tests if p is largely smaller than q (or q largely smaller than p). In this case we skip the Bernoulli drawing step to directly insert the smallest number of booleans in the bigger one. This particular case is due to a change of rate in the distribution of the binomial coefficient.

Theorem 6. *The* RANDOMCOMBINATION *algorithm uniformly samples a list of p True and q False. It uses an entropic number of random bits.*

Proof (sketch). Let l be a drawn list of p True and q False. The correctness proof, is divided into two cases:

- l was drawn entirely during the first step: in this case the probability to draw l is directly the product of the probability to draw each boolean *i.e.* $\left(\frac{p}{p+q}\right)^p \left(\frac{q}{p+q}\right)^q$ and because this probability does not depend of l it is the same for each l
- l was drawn after $p + q - k$ Bernoulli samples: using the previous argument, this combination of $p + q - k$ values is uniformly drawn. Then each remaining boolean is uniformly inserted, and so, each combination of $p + q - k + 1$ to $p + q$ booleans is uniformly built from the previous one

Concerning the random bit efficiency of the algorithm, we assume that k-BERNOULLI is entropic, which we will establish later. Assuming this, the main idea of the proof is to analyze the number of consumed booleans in the stream *rndBits*.

To do this we let the random variable T to be the sum of $\sharp$True and $\sharp$False at the end of the first step (when the list l if filled initially). Thus, we have

$$\mathbb{P}(T = t) = \binom{t}{p} \left(\frac{p}{p+q}\right)^{p+1} \left(\frac{q}{p+q}\right)^{t-p} + \binom{t}{q} \left(\frac{p}{p+q}\right)^{q-t} \left(\frac{q}{p+q}\right)^{q+1}$$

Then, we compute the expected value T and get $\mathbb{E}[T] = (p+q) + o(p+q)$ when $p > \log(q)^2$ (resp. $q > \log(p)^2$). The latter condition justifies the first test of the algorithm. It remains to count the number of random bits used and to compare it to the entropy of the combinations of p among $p + q$ elements *i.e.* the entropy of $\binom{p}{p+q}$.

We recall that a uniform integer between 0 and n can be drawn with $\mathcal{O}(\log n)$ random bits. We let $B_{p,q}$ the number of random bits used to sample a Bernoulli random variable of parameter $\frac{p}{p+q}$. We recall that the entropy of such variable is $-p \log \frac{p}{p+q} - q \log \frac{q}{p+q}$.

Thus, the number of average random bits used is $\mathbb{E}[T]B_{p,q} + o(p+q)\mathcal{O}(\log n)$. The $o(p+q)\mathcal{O}(\log n)$ term come from the second step of the algorithm *i.e.* the uniform insertions of remaining booleans. To conclude, the average number of random bits used is asymptotically equal to $(p+q)B_{p,q}$ which is equivalent to the entropy of $\binom{p}{p+q}$.   □

Algorithm 5. Sampling of k Bernoulli random variables

function k-BERNOULLI(p)   ▷ p is less than 1
    function k-BERNOULLIAUX(p)
        $k := \left\lfloor \frac{\log \frac{1}{2}}{\log p} \right\rfloor$, $i := 0$
        $v :=$ a vector of k times **True**
        while ¬BERNOULLI($\sum_{\ell=0}^{i} \binom{k}{\ell} p^{k-\ell}(1-p)^{\ell}$) **do**
            $i := i + 1$
            $j :=$ **uniformRandomInt**($[0 \ldots k-1]$)
            $v[j] :=$ **False**
        return v
    if $p < \frac{1}{2}$ **then return negate**(k-BERNOULLIAUX($1-p$))
    else return k-BERNOULLIAUX(p)

The entropic property of the RANDOMCOMBINATION relies on the entropy of the k-BERNOULLI function, which is described by Algorithm 5. The key idea is to draw Bernoulli random variables of parameter p by packs, using the fact that a successful Bernoulli r.v. of parameter p^k corresponds to a sequence of k successes of a Bernoulli r.v. of parameter p. Thus, the parameter $k = \left\lfloor \frac{\log \frac{1}{2}}{\log p} \right\rfloor$ is such that p^k is close to $\frac{1}{2}$. This allows to draw Bernoulli r.v. of parameter close to $\frac{1}{2}$, for which BERNOULLI is entropic. Let us consider the following example of a call to k-BERNOULLI with the argument $\frac{2}{7}$. In that case we let $p = 1 - \frac{2}{7} = \frac{5}{7}$ and so $k = 2$. Then we present the different possible runs in the form of a decision tree. We draw a Bernoulli r.v. of parameter $\left(\frac{5}{7}\right)^2$ and:

- if the Bernoulli draw is successful then two successes of Bernoulli r.v. of parameter $\frac{2}{7}$ are returned
- else it is a fail, which means that at least one of the two Bernoulli r.v. of parameter $\frac{2}{7}$ is a fail, which has a probability $2 \cdot \frac{5}{7} \cdot \frac{2}{7}$ to happen, so we need to redraw a new r.v.
  - if it is successful, it means that only one variable is a fail and so we need decide which one

- else we need to draw one more r.v. of parameter 1, in other words a successful r.v., and we return two failed r.v.

The last brick of this framework is the BERNOULLI algorithm, already known in the literature (as explained by [6]).

Algorithm 6. Sampling of Bernoulli random variable

---

    function BERNOULLI(p)   ▷ p is less than 1
        function RECBERNOULLI(a, b, p)
            if RandomBit() $= 0$ **then**
                if $m > p$ **then return** False
                else RECBERNOULLI($\frac{a+b}{2}$, b, p)
            else
                if $m < p$ **then return** True
                else RECBERNOULLI(a, $\frac{a+b}{2}$, p)
        return RECBERNOULLI(0, 1, p)

Theorem 7. *The Algorithm 6 draws a Bernoulli random variable of parameter p using, in average, 2 random bits.*

Proof. The correction proof is direct. We just remark that if we note K the number of calls to the recursive RECBERNOULLI function, K is equal to the length K prefix of the binary writings of p.

The average number of random bits used is the expectation of K:

$$\mathbb{E}[K] = \sum_{k=1}^{\infty} k \cdot \frac{1}{2}^{k} = \frac{1}{2} \cdot \left(\frac{\mathrm{d}}{\mathrm{d}z} \frac{1}{1-z}\right)\Big|_{z=\frac{1}{2}} = 2$$

□

Theorem 8. *The Algorithm 5 draws $\left\lfloor \frac{\log\frac{1}{2}}{\log p} \right\rfloor$ Bernoulli random variable of parameter p entropically.*

Proof (sketch). Let N be the number of iteration of the **while** loop in the k-BERNOULLI algorithm, we get that the number of random bits used is upper bounded by $2N + N \log k$: the number of bits used to draw N Bernoulli plus the number of bits used for the N uniform (over $[0\ldots k]$) r.v. draws.

So, the expected number of random bits used is

$$\sum_{n=0}^{k} \binom{k}{n} p^{k-n}(1-p)^n (2(n+1) + n \log k) = 2 + (1-p)(2 + \log k)$$

We have to average it by the number of Bernoulli r.v. drawn this way *i.e.* k. So, the average number of random bits used to draw one Bernoulli r.v. of parameter p is $\frac{2}{k} + (1-p)(2 + \log k)$. The minimum of this function in k is reached when $k = \frac{2\log 2}{1-p}$, in other word when the average number of random bits used is greater or equal to 2. This corresponds to the case where $k = 1$ in the algorithm k-BERNOULLI: in this case we should directly use BERNOULLI. In the other case, we obtain that the average number of random bits used is entropic. □

References

1. Bodini, O., Genitrini, A., Peschanski, F.: A quantitative study of pure parallel processes. Electron. J. Comb. **23**(1), 39 (2016). P1.11
2. Felsner, S., Wernisch, L.: Markov chains for linear extensions, the two-dimensional case. In: SODA, pp. 239–247. Citeseer (1997)
3. Flajolet, P., Sedgewick, R.: Analytic Combinatorics. Cambridge University Press, Cambridge (2009)
4. Huber, M.: Fast perfect sampling from linear extensions. Discr. Math. **306**(4), 420–428 (2006)
5. Klein, P., Mozes, S.: Optimization Algorithms for Planar Graphs (to appear)
6. Lumbroso, J.: Optimal discrete uniform generation from coin flips, and applications. CoRR abs/1304.1916 (2013). http://arxiv.org/abs/1304.1916
7. Möhring, R.H.: Computationally Tractable Classes of Ordered Sets. Report, Institut für Ökonometrie und Operations Research (1987)
8. Sedgewick, R.: Left-leaning red-black trees. In: Dagstuhl Workshop on Data Structures. p. 17 (2008)
9. Shannon, C.E.: A mathematical theory of communication. ACM SIGMOBILE Mobile Comput. Commun. Rev. **5**(1), 3–55 (2001)
10. Stanley, R.P.: Enumerative Combinatorics, vol. 1, 2nd edn. Cambridge University Press, New York (2011)
11. Valdes, J., Tarjan, R.E., Lawler, E.L.: The recognition of series parallel digraphs. In: Proceedings of the Eleventh Annual ACM Symposium on Theory of Computing (STOC 1979), pp. 1–12. ACM, New York (1979). http://doi.acm.org/10.1145/800135.804393

Parameterized Counting of Trees, Forests and Matroid Bases

Cornelius Brand$^{(\boxtimes)}$ and Marc Roth

Saarland University and Cluster of Excellence (MMCI), Saarbrücken, Germany
{cbrand,mroth}@mmci.uni-saarland.de

Abstract. We prove #W[1]-hardness of counting (1) trees with k edges in a given graph, (2) forests with k edges in a given graph, and (3) bases of a given matroid of rank (or nullity) k representable over an arbitrary field of characteristic two, where k is the parameter.

1 Introduction

Parameterized counting complexity has produced results on the hardness of computing the number of paths, cliques and cycles with k edges in a given graph. One important step was the proof of #W[1]-hardness for computing the number of k-matchings in a simple graph [3]. This line of research culminated in a classification theorem of Curticapean and Marx [4] for the following problem: Given a graph H from a class of graphs $\mathcal{H}$ and an arbitrary graph G, compute the number of all subgraphs of G that are isomorphic to H, parameterized by $|V(H)|$. They proved that this problem is fixed-parameter tractable if the vertex cover number of all graphs in $\mathcal{H}$ is bounded by a constant[1], and #W[1]-hard otherwise.

This theorem does not cover the problem of counting all occurrences of *all* subgraphs of a certain size that are contained in a fixed class $\mathcal{H}$ of graphs. For example, using their theorem, we can classify the problem of counting all k-cliques in a graph as #W[1]-hard as follows: For the class $\mathcal{H}$, we take $\mathcal{H} = \{K_n \mid n \in \mathbb{N}\}$. As n goes to infinity, so does the vertex cover number of K_n, and the hardness follows. Of course, we might take $\mathcal{H}$ to be the set of all trees or all forests, but then the theorem speaks about the complexity of computing the number of *one specific* tree or forest in some given graph, instead of counting *all* trees or forests of a given size. It is the complexity of these two problems, namely counting all trees and counting all forests with a given number of edges in a given graph that we are concerned with in this paper.

Another problem that has yet escaped a parameterized analysis is the problem of counting bases in matroids. Matroids have been studied over decades and play a central role in numerous combinatorial applications (see e.g. [13]). Although they were treated in the parameterized world (see e.g. [5, Chap. 12] for

---

[1] That is, $\sup\{\tau(H) \mid H \in \mathcal{H}\} < \infty$, where $\tau(H)$ is the size of a minimum vertex cover of H.

© Springer International Publishing AG 2017
P. Weil (Ed.): CSR 2017, LNCS 10304, pp. 85–98, 2017.
DOI: 10.1007/978-3-319-58747-9_10

an overview), the problem of computing the number of bases was only addressed from the classical point of view so far [12,14,15]. It should be noted that this problem comprises also a generalization of counting forests in a graph, which gives the connection to the previously mentioned problems. Building on our results on counting forests, this gap in knowledge is one we address in the subsequent sections.

1.1 Related Work

It is known that computing the number of all (labeled) trees and computing the number of all forests are #P-hard problems, even on planar graphs [8,9,16]. A general theorem of Eppstein implies their being fixed-parameter tractable on planar graphs [6].

The problem of counting k-independent sets in a binary matroid is #P-hard. This follows from the well-known fact that the k-forests of a graph G correspond one-to-one to the k-independent sets of the binary matroid represented by the incidence matrix of G over $\mathbb{F}_2$ (see e.g. [14]). Also, counting the bases of a binary matroid is #P-hard (see e.g. [15]). On the other hand, the number of bases of a regular matroid can be computed in polynomial time [12].

2   Preliminaries

The integers are denoted by $\mathbb{Z}$. The polynomial ring over some ring R in the variables $x_1, \dots, x_n$ is written $R[x_1, \dots, x_n]$. The set of matrices with m rows, n columns and entries from a set X is $\mathrm{Mat}(n \times m, X)$. For an integer ℓ and a finite set X, we suggestively denote with $\binom{X}{\ell}$ the set of subsets of X of size ℓ, the number of which is $\binom{|X|}{\ell}$.

2.1   Parameterized Counting Complexity

We begin with basic definitions of parameterized counting complexity, following closely Chapt. 14 of the textbook [7], which we recommend to the interested reader for a more comprehensive overview of the topic. Our fundamental object of study is the following. A *parameterized counting problem* (F, k) consists of a function $F : \{0,1\}^* \to \mathbb{N}$ and a polynomial-time computable function $k : \{0,1\}^* \to \mathbb{N}$, called the *parameterization*.

A parameterized counting problem (F, k) is called *fixed-parameter tractable* if there is an algorithm A for computing F, a constant $c > 0$ and a computable function $f : \mathbb{N} \to \mathbb{N}$ such that A is running in time $f(k(x)) \cdot |x|^c$ for all $x \in \{0,1\}^*$. We say that such an algorithm runs in *fpt-time*. Let (F, k) and (F', k') be two parameterized counting problems. Then, a function $R : \{0,1\}^* \to \{0,1\}^*$ is called an *fpt parsimonious reduction from* (F, k) *to* (F', k') if

1. For all $x \in \{0,1\}^*$, $F(x) = F'(R(x))$.
2. R runs in fpt-time.

3. There is some computable $g : \mathbb{N} \to \mathbb{N}$ such that $k'(R(x)) \leq g(k(x))$ for all $x \in \{0,1\}^*$.

An algorithm A with oracle access to F' is called an *fpt Turing reduction from* (F,k) *to* (F',k') if

1. A computes F.
2. A runs in fpt-time.
3. There is some computable $g : \mathbb{N} \to \mathbb{N}$ such that for all $x \in \{0,1\}^*$ and for all instances y for which the oracle is queried during the execution of $A(x)$, $k'(y) \leq g(k(x))$.

The parameterized counting problem #k-CLIQUE is defined as follows, and is parameterized by k: Given a graph G and an integer k, compute the number of cliques of size k in G. The class #W[1] is defined as the set of all parameterized counting problems (F,k) such that there is an fpt parsimonious reduction from (F,k) to #k-Clique. A parameterized counting problem (F,k) is called #W[1]-*hard* if there is an fpt Turing reduction from #k-CLIQUE to (F,k).

2.2   Matroids

A *matroid* is a pair $M = (E,\mathcal{I})$ consisting of a finite ground set E and a family $\mathcal{I} \neq \emptyset$ of subsets of E that satisfies the following axioms:

1. $\mathcal{I}$ is downward closed, i.e. if $I \in \mathcal{I}$ and $I' \subset I$, then $I' \in \mathcal{I}$.
2. $\mathcal{I}$ has the exchange property, i.e. If $I_1, I_2 \in \mathcal{I}$ and $|I_1| < |I_2|$, then there is some $e \in I_2 - I_1$ such that $I_1 \cup \{e\} \in \mathcal{I}$.

Note that this entails $\emptyset \in \mathcal{I}$. The elements of $\mathcal{I}$ are called *independent sets*, and an inclusion-wise maximal element of $\mathcal{I}$ is called a *basis* of M. The exchange property warrants that all bases have the same cardinality, and we call this cardinality the *rank* of M, written as rk M. Furthermore we define $(|E| - \text{rk } M)$ as the *nullity* of M. The pair $M_k = (E, \mathcal{I}_k)$ with $\mathcal{I}_k = \{I \in \mathcal{I} \mid |I| \leq k\}$ is again a matroid, called the *k-truncation M_k* of M.

For a field F, a *representation of M over F* is a mapping $\rho : E \to V$, where V is a vector space over F, such that for all $A \subseteq E$, A is independent if and only if $\rho(A)$ is linearly independent. M is called *representable* if it is representable over some field. If there is such a representation, we call M *representable over F*, or *F-linear*, and it holds that $\text{rk}(\rho) = \text{rk}(M)$. Conversely, every matrix over a field F induces an F-linear matroid on its columns, where sets of columns are independent if and only if they are linearly independent. A *k-truncation* of a matrix is the matrix of a representation of the k-truncation of the corresponding linear matroid, possibly over a different field. Recently, Lokshtanov et al. proved that a k-truncation of a matrix can be computed in deterministic polynomial time (see [10], Theorem 3.23).[2] In the following we will write $\mathbb{F}_q$ for the field with q elements.

---

[2] A slightly weaker version of this result with a simpler proof that still suffices for our application seems to follow along the lines of Snook [14].

Given a matroid $M = (E, \mathcal{I})$, the *dual matroid* M^* of M is a matroid on the same ground set as M, and $B \subseteq E$ is a basis of M^* if and only if $E \backslash B$ is a basis of M. Given a representation of a matroid M, a representation of M^* in the same field can be found in polynomial time (see e.g. [11]).

In the following, *all* matroids will be assumed to be representable, and encoded using a representing matrix ρ and a suitable encoding for the ground field. Furthermore, we can, without loss of generality, always assume that ρ has $\mathrm{rk}(M)$ rows, because row operations (multiplying a row by a non-zero scalar, and adding such multiples to other rows) do not affect linear independence of the columns of ρ. Hence, any ρ' obtained from ρ through row operations is a representation of M. In particular, by Gaussian elimination, we may assume all but the first $\mathrm{rk}(M)$ rows of ρ to be zero.

2.3 Graphs and Matrices

We consider simple graphs without self-loops unless stated otherwise. Given a graph G we will write n for the number of vertices of G and m for the number of edges. A *k-forest* is an acyclic graph consisting of k edges and a *k-tree* is a connected k-forest. We say that two graphs $G_1 = (V_1, E_1)$ and $G_2 = (V_2, E_2)$ are *isomorphic* if there is a bijection $\varphi : V_1 \to V_2$ such that for all $u, v \in V_1$, $\{u, v\} \in E_1$ if and only if $\{\varphi(u), \varphi(v)\} \in E_2$. A *$k$-matching* of a graph $G = (V, E)$ is a subset of k edges such that no pair of edges has a common vertex. The following was established by Curticapean.

Theorem 1 ([3]). *Given a graph G and a parameter k, it is #W[1]-hard to compute the number of matchings of size k in G.*

Given a graph G with vertices $v_1, \ldots, v_n$ and edges $e_1, \ldots, e_m$ we define the (unoriented) *incidence matrix* $M[G] \in \mathrm{Mat}(n \times m, \mathbb{F}_2)$ of G by $M[G](i, j) := 1$ if $v_i \in e_j$ and 0 otherwise. A subset of columns of $M[G]$ is linearly independent (over $\mathbb{F}_2$) if and only if the corresponding edges form a k-forest in G.

3 Counting Trees

Definition 1. *Given a graph G and a natural number k, we denote the problem of counting all acyclic, connected subsets of edges of G of size k as #k-TREES, parameterized by k.*

In this section we will prove the hardness of counting trees. For the proof, we show hardness of an intermediate problem.

Definition 2 (Weighted k-trees). *Given a graph $G = (V, E)$ with edge weights $\{w_e\}_{e \in E}$ and $k \in \mathbb{N}$, #k-WTREES is defined as the problem of computing*

$$\mathrm{WT}_k(G) := \sum_{t \in T_k(G)} \prod_{e \in t} w_e,$$

where $T_k(G)$ is the set of all acyclic, connected edge sets of size k in G.

Note that this polynomial bears some similarity with the multivariate forest polynomial, to be defined in the next section. In fact, both polynomials have in common that they can be viewed as the multivariate generating functions of the respective structures in a graph.

Borrowing an idea from [1] in the context of counting forests, we first consider the above polynomial on a modified graph, namely after adding an apex, and show that this reveals information on the number of k-matchings in the original graph. More precisely, edges incident to the apex will be assigned a weight z which leads to the following intermediate problem.

Definition 3. *Let k be a natural number, $G = (V, E)$ be a graph with an apex $a \in V$, that is, a vertex that is adjacent to every other vertex, and edge weights $\{w_e\}_{e \in E}$ such that $w_e = 1$ for all edges e that are not adjacent to a and $w_e = z$ for a fixed $z \leq k$ otherwise. Then we denote the problem of computing $\mathrm{WT}_k(G)$ parameterized by k as $\#k$-WAPEXTREES.*

Lemma 1. *$\#k$-WAPEXTREES is $\#W[1]$-hard.*

Proof. First, we will outline the proof: The problem we are reducing from is the problem of counting k-matchings. That is, given a graph G we want to compute the number of k matchings by using an oracle for $\#k$-WAPEXTREES. To do so, we are going to add x isolated vertices to G and after that, add an apex a (that is adjacent to every vertex in G and to every isolated vertex). We call the resulting graph G_x. The first step in the reduction is to count the number of trees containing $2k$ edges such that exactly k edges are incident to a. This number can be computed by assigning weight z to every edge incident to a (yielding a graph we call $G_{x,z}$), computing the value of WT_{2k}, which is a polynomial in z, for different values of z and then interpolating the coefficient of z^k. Note that we can compute this number *for different values of x*. After that we show that those numbers induce a system of linear equations with a unique solution. Finally we prove that one entry of the solution vector is the number of $2k$-trees in G_0 such that (1) exactly k edges are incident to a and (2) for each of those edges $\{v_1, a\}, \ldots, \{v_k, a\}$ there exist $u_1, \ldots, u_k$ such that $\{u_1, v_1\}, \ldots, \{u_k, v_k\}$ are also contained in the tree. As trees do not contain cycles, we have that the u_i are pairwise different which implies that trees satisfying (1) and (2) correspond (up to a factor of 2^k) to the k-matchings in G.

As stated before, we reduce from the problem of counting k-matchings. Let $G = (V, E)$ be a graph with $n = |V|$. For x and z we construct the graph $G_{x,z}$ as follows:

- Add x isolated vertices to G.
- Add a vertex a to G and connect a to all other vertices, including the isolated ones (i.e., a is an apex).
- Assign weights to the edges as follows: If $a \notin e$ then set $w_e = 1$. Otherwise set $w_e = z$.

The unweighted version of $G_{x,z}$ is just denoted by G_x. Furthermore we denote the set of edges adjacent to a as E_x^a. Now fix x and compute for all $i \in \{0, \ldots, 2k\}$

the value $P_i(x) = \mathrm{WT}_{2k}(G_{x,i})$ with an oracle for #k-WApexTrees. This corresponds to evaluating the following polynomial in points $0, \dots, 2k$:

$$Q_x(z) = \sum_{t \in T_{2k}(G_x)} \prod_{e \in t} w_e = \sum_{t \in T_{2k}(G_x)} \prod_{\substack{e \in t \\ a \notin e}} 1 \cdot \prod_{\substack{e \in t \\ a \in e}} z = \sum_{t \in T_{2k}(G_x)} z^{|E_x^a \cap t|}$$

Therefore, we can interpolate all coefficients. In particular, we are interested in the coefficient of z^k, which is the number of trees of size $2k$ in G_x, such that exactly k edges of the tree are adjacent to the apex a. We call these trees *apex-fair*, and denote their set as F_x. Now, consider an edge $e = \{v, a\}$ of such an apex-fair tree t that is adjacent to a. We call e *apex-isolated* if v is not incident to any other edge of t. Otherwise we call e *apex-connected*, and we call the subtree t^c of t without apex-isolated edges *apex-connected* as well.[3] Note that t^c is indeed a tree, not only a forest. Furthermore, t^c is a tree in G_0 as edges that are connected to isolated vertices cannot be apex-connected.

We observe that for an apex-fair tree t in G_x, the subtree t^c in G_0 has exactly $s - k$ apex-connected edges, where s is the number of edges in t^c: Since t is apex-fair, we know that exactly k of the edges in t are not incident to a, and thus by definition, such an edge cannot be apex-isolated. Hence, these k edges are also in t^c, meaning that the remaining $s - k$ edges in t^c have to be incident to the apex, and indeed, they have to be apex-connected, since all apex-isolated edges were removed from t when constructing t^c.

We can partition the set F_x of apex-fair trees in G_x by the number s of edges that the corresponding apex-connected tree contains, say

$$F_x = \bigcup_{s=0}^{2k} B_x(s)$$

where $B_x(s)$ is the set of apex-fair trees t in G_x such that t^c is of size s. Clearly, this union is disjoint, and $B_x(i) = \emptyset$ for $0 \leq i \leq k$, since we consider apex-fair trees, meaning that $|F_x| = \sum_{s=k}^{2k} |B_x(s)|$. Now, consider a single apex-fair tree $t \in B_x(s)$ for some fixed s, which is hence of size $2k$. Then, as argued, t^c has exactly $s - k$ apex-connected edges, and k edges not incident to the apex. This leaves $2k - s$ edges of t to be chosen, and since there are already k edges present in t^c that are not incident to the apex, and t is apex-fair, all these $2k - s$ edges have be incident to the apex, and also have to be apex-isolated, since otherwise they would be in t^c. By the construction of G_x, there are $n + x$ possible choices for the apex-isolated edges in total, but s of these are already occupied by t^c, leaving $n + x - s$ choices for the $2k - s$ remaining edges. Hence, for every apex-connected tree t^c, there are exactly $\binom{n+x-s}{2k-s}$ fair trees t' such that $t'^c = t^c$. Letting β_s be the number of apex-connected trees of size s, this amounts to

$$|B_x(s)| = \beta_s \cdot \binom{n + x - s}{2k - s}$$

---

[3] This terminology stems from the fact that the removal of a leaves v isolated.

and thus

$$|F_x| = \sum_{s=k+1}^{2k} \beta_s \cdot \binom{n+x-s}{2k-s}.$$

For convenience, we perform an index shift on β, and find

$$|F_x| = \sum_{s=1}^{k} \beta_{s+k} \cdot \binom{n+x-(k+s)}{k-s}.$$

We can then evaluate $|F_x|$ for $x \in \{1, \ldots, k\}$ and solve for the β_s. To do so, we show that the corresponding matrix $A \in \mathbb{N}^{k \times k}$, defined by

$$A_{i,j} = \binom{i+n-k-j}{k-j}$$

has full rank.

Its j-th column is an evaluation vector of the polynomial

$$R_j(x) = \binom{x+n-k-j}{k-j}.$$

The degrees of the polynomials R_j are pairwise distinct and hence, the polynomials and the evaluation vectors are as well. It follows that A has full rank, i.e., we can indeed compute the coefficients β_s.

Finally, consider β_{2k}: This is the number of apex-connected trees in G_0 with k edges in G_0 and k apex-connected edges. It follows that these trees correspond to k-matchings in G, more precisely, for every k-matching in G, there are 2^k such apex-connected trees. This concludes the proof. $\square$

In the proof of Lemma 1 we exploited the fact that the weights of edges incident to the apex can be used to enforce some desired structure via interpolation. One might wonder why we explicitly enforced the edges with weight $\neq 1$ to be incident to the apex and the weights of those edges to be equal in the definition of $\#k$-WApexTrees, instead of just defining a canonical edge-weighted variant of the problem of counting trees. The reason for this is the fact that we need this very structure in order to get rid of the weights and reduce to $\#k$-Trees, at least in our reduction. Before we do this reduction, we prove another lemma.

Lemma 2. *Let $G = (V, E)$ be a simple graph and $A \subseteq E$ a subset of edges of size z. Then, the problem of computing the number of $(k + z)$-trees whose edges contain A can be solved in time $O(2^z) \cdot \text{poly}(|V|)$ if access to an oracle for $\#k$-Trees is provided.*

Proof. Let S be a subset of edges of G. We define T_S as the set of all $(k + z)$ trees in G that do not contain any edge in S. Note that we can compute $|T_S|$

by deleting all edges in S and computing the number of $(k + z)$-trees in G by posing an oracle query. Furthermore, it holds that

$$T_{S_1} \cap T_{S_2} = T_{S_1 \cup S_2} \tag{1}$$

for any two subsets of edges S_1 and S_2. Let T be the set of all $(k + z)$ trees in G, i.e., $T := T_\emptyset$. Now we can express the number of $(k + z)$ trees whose edges contain A as

$$|T \setminus \bigcup_{e \in A} T_{\{e\}}|.$$

Using the inclusion-exclusion principle, we get

$$|T \setminus \bigcup_{e \in A} T_{\{e\}}| = |T| - \sum_{\emptyset \neq J \subseteq A} (-1)^{|J|-1} | \bigcap_{e \in J} T_{\{e\}}| = |T| - \sum_{\emptyset \neq J \subseteq A} (-1)^{|J|-1} |T_J|$$

where the second equality follows from (1). Finally, we observe that there are exactly 2^z summands, each of which can be computed by posing an oracle query for the corresponding T_J. □

This allows us to make the final step in the proof of Theorem 2:

Lemma 3. $\#k$-WApexTrees *is fpt Turing reducible to* $\#k$-Trees.

Proof. Let k be a natural number and $G = (V \cup \{a\}, E \cup V \times \{a\})$ be a graph with apex a and edge weights as in Definition 3. The goal is to compute

$$\mathrm{WT}_k(G) = \sum_{t \in T_k(G)} \prod_{e \in t} w_e,$$

where $w_e = z$ if $a \in e$ and $w_e = 1$ otherwise. First, we point out that $T_k(G)$ can be partitioned into trees (with k edges) that do not contain a and trees that do contain a. We denote the set of the latter as $T_k^a(G)$. Then we have that

$$\mathrm{WT}_k(G) = \sum_{t \in T_k(G)} \prod_{e \in t} w_e = \sum_{t \in T_k(G - \{a\})} \prod_{e \in t} w_e + \sum_{t \in T_k^a(G)} \prod_{e \in t} w_e$$

$$= |T_k(G - \{a\})| + \sum_{t \in T_k^a(G)} \prod_{e \in t} w_e.$$

Now $|T_k(G - \{a\})|$ can be computed by querying the oracle. If $z = 0$ then $\sum_{t \in T_k^a(G)} \prod_{e \in t} w_e = 0$ as well, that is, we are done. Otherwise we need to realise the edges with weight z to compute $\sum_{t \in T_k^a(G)} \prod_{e \in t} w_e$. To do so, we construct the graph $G^z = (V^z, E^z)$ from G as follows:

- Delete the apex and the adjacent edges.
- Add apices $a_1, \ldots, a_z$ (including edges to all vertices in G).
- Add a vertex a and edges $\{a, a_i\}$ for all $i \in [z]$.

We first sketch the idea of the proof: Let $\hat{T}$ be the set of all trees with $k + z$ edges in G^z such that all edges $\{a, a_1\}, \ldots, \{a, a_z\}$ are met and let $t \in \hat{T}$. As t is connected, there is at least one vertex v in $G - \{a\}$ such that $\{v, a_i\}$ is contained in t for an $i \in [z]$. Furthermore, for every vertex v in $G - \{a\}$ at most one of the edges $\{v, a_1\}, \ldots, \{v, a_z\}$ can be contained in t, because otherwise t would have a cycle. This means that taking one of the edges $\{v, a_1\}, \ldots, \{v, a_z\}$ in G^z corresponds to taking edge $\{v, a\}$ of weight z in G. We will prove that

$$|\hat{T}| = \sum_{t \in T_k^a(G)} \prod_{e \in t} w_e.$$

Finally we will use Lemma 2, that is, an application of the inclusion-exclusion principle, to compute $|\hat{T}|$.

Now let $t \in \hat{T}$. As stated above we claim that for every vertex v in $G - \{a\}$ at most one of the edges $\{v, a_1\}, \ldots, \{v, a_z\}$ can be contained in t. Assuming not, there is a vertex v in $G - \{a\}$ and indices $i \neq j$ such that $\{v, a_i\}$ and $\{v, a_j\}$ are contained in t, but this induces the cycle (a, a_i, v, a_j, a) which contradicts the fact that t is a tree. Now we define a mapping $f : \hat{T} \to T_k^a(G)$ as follows: For every edge $e = \{u, v\}$ in $G - \{a\}$, e is contained in $f(t)$ if and only if e is contained in t and for every edge $e_a = \{v, a\}$, e is contained in $f(t)$ if and only if there is an $i \in [z]$ such that $\{v, a_i\}$ is contained in t. Note that $f(t)$ contains a as there is at least one vertex v in $G - \{a\}$ such that $\{v, a_i\}$ is contained in t for an $i \in [z]$. For a tree $t_G \in T_k^a(G)$ we define $g(t_G) := \{t \in \hat{T} \mid f(t) = t_G\}$. Now let $T_k^a(G)[\ell]$ be the set of all trees $t_G \in T_k^a(G)$ such that there are exactly ℓ vertices v in $G - \{a\}$ such that $\{v, a\}$ is contained in t_G. For every such vertex, there are z possibilities to construct a tree $t \in \hat{T}$ such that $f(t) = t_G$ (by taking one of the edges $\{v, a_1\}, \ldots, \{v, a_z\}$). Hence $|g(t_G)| = z^\ell$ if $t_G \in T_k^a(G)[\ell]$. Furthermore we claim that

$$\hat{T} = \dot{\bigcup}_{t_G \in T_k^a(G)} g(t_G).$$

This follows from the observation that the sets $g(t_G)$ are the classes of the equivalence relation $t \sim t' \Leftrightarrow f(t) = f(t')$. Putting everything together we have

$$\sum_{t \in T_k^a(G)} \prod_{e \in t}^{k} w_e = \sum_{\ell=1}^{k} \sum_{t \in T_k^a(G)[\ell]} \prod_{e \in t} w_e = \sum_{\ell=1}^{k} \sum_{t \in T_k^a(G)[\ell]} z^\ell$$

$$= \sum_{\ell=1}^{k} \sum_{t \in T_k^a(G)[\ell]} |g(t)| = \sum_{t \in T_k^a(G)} |g(t)| = \left| \dot{\bigcup}_{t_G \in T_k^a(G)} g(t_G) \right| = |\hat{T}|$$

It remains to show how to compute $|\hat{T}|$ with an oracle for #k-TREES. This can be done by applying Lemma 2 with $A = \{\{a, a_1\}, \ldots, \{a, a_z\}\}$.   □

We can thus state:

Theorem 2. #k-TREES *is* #W[1]-*hard when parameterized by* k.

Proof. In Lemma 1 it was shown that $\#k$-WApexTrees is $\#W[1]$-hard. In Lemma 3 we proved that $\#k$-WApexTrees is fpt Turing reducible to $\#k$-Trees. It follows that $\#k$-Trees is $\#W[1]$-hard as well.

4   Counting Forests

In this section, we will prove that counting k-forests is $\#W[1]$-hard. This result will be used to show hardness of counting matroid bases in fields of fixed characteristic.

Definition 4. *Let $G = (V, E)$ be a multigraph with edges labeled with formal variables $\{w_e\}_{e \in E}$. Then the* multivariate forest polynomial *of G is defined as*

$$F(G; \{w_e\}_{e \in E}) = \sum_{A \subseteq E \text{ acyclic}} \prod_{e \in A} w_e.$$

The polynomial obtained by replacing all weights w_e with a fresh variable x, i.e., setting $w_e = x$ for all $e \in E$ and $x \notin \{w_e \mid e \in E\}$, is called the univariate forest polynomial *of the graph and is simply denoted $F(G; x)$.*

For a forest A in G, let $c(A)$ be the family of all sets $T \subseteq V(G)$ such that $T \neq \emptyset$ and T is a maximal connected component in A. Adding an apex, that is, a new vertex that is connected to all other vertices, to a graph $G = (V, E)$ and labeling each of the new edges with a new variable z makes the univariate forest polynomial into a bivariate one, namely $F(G'; x, z)$, where G' is the described graph with an added apex. In the following, G will always be the original graph, and G' will be the graph obtained in this way.

Note that $F(G'; x, z) \in \mathbb{Z}[x, z] \cong (\mathbb{Z}[z])[x]$. In particular, the coefficient of x^k in $F(G'; x, z)$ is an element of $\mathbb{Z}[z]$. To make this very clear in the following, we shall refer to this element of $\mathbb{Z}[z]$ as the *coefficient polynomial* of x^k in $F(G'; x, z)$.

Lemma 4. *There is a polynomial-time Turing reduction from counting matchings of size k in a graph G to computing the coefficient polynomial of x^k of the bivariate forest polynomial $F(G'; x, z)$ of the graph G', parameterized by k in both problems. In particular, this reduction retains the parameter k and is thus even an fpt Turing reduction.*

Proof. The coefficient polynomial $C_k(z) \in \mathbb{Z}[z]$ of x^k in $F(G'; x, z)$ can be expressed in terms of G through

$$C_k(z) = \sum_{A \in \binom{E}{k} \text{ acyclic in } G} \prod_{T \in c(A)} (1 + |T|z),$$

as follows immediately from Lemma 7 in [1] after specializing to x and z. Since a forest in G with k edges can cover at most $2k$ nodes of G, at least $n - 2k$ nodes of G are left uncovered by T, and are thus present in $c(A)$ as components T with

$|T| = 1$. This shows that the product $\prod_{T \in c(A)}(1 + |T|z)$ of each summand (and hence also $C_k(z)$) is a multiple of $(1 + z)^{n-2k}$. Thus, the polynomial quotient

$$Q_k(z) := C_k(z)/(1 + z)^{n-2k}$$

is a well-defined element of $\mathbb{Z}[z]$.

Observe that it is precisely the k-matchings of G that will have $2k$ covered and $n - 2k$ uncovered nodes, and such a k-matching has k components with $|T| = 2$, and $n - 2k$ components with $|T| = 1$. Therefore, the summand corresponding to some A in $C_k(z)$ is of the form $(1 + z)^{n-2k}(1 + 2z)^k$ if and only if A is a k-matching. Likewise, the number of k-matchings is precisely the number of such monomials in $C_k(z)$. In all other monomials, the factor $(1 + z)$ is hence present with degree at least $n - 2k + 1$. Denote with M_k the number of k-matchings in G. After substituting $z \mapsto y - 1$ (hence, $y = z + 1$), this can be stated as follows:

$$Q_k(y) = C_k(y)/y^{n-2k} = M_k \cdot (2y - 1)^k + y \cdot R(y)$$

for some polynomial $R(y)$. We see that for $y = 0$, $y \cdot R(y) = 0$, and hence, keeping in mind that $z = y - 1$, it follows

$$Q_k(y = 0) = Q_k(z = -1) = M_k \cdot (-1)^k.$$

We now argue why this is an fpt Turing reduction. Note that in the coordinates y^i, the polynomial division is just a shift of coefficients. Therefore, an oracle to the k-th coefficient polynomial of $F(G'; x, z)$, as provided in a Turing reduction, yields the polynomial $C_k(z)$. After a change of basis from z to $y - 1$ and a corresponding shift of coefficients to perform the division by y^{n-2k}, we can evaluate the resulting polynomial $Q_k(y)$ at $y = 0$ and obtain $M_k \cdot (-1)^k$ and thus M_k. This can clearly be done in polynomial time in the size of G (and k, for that matter) once $C_k(z)$ was obtained, and the only oracle query involved does not alter the parameter and is hence valid for an fpt Turing reduction. $\square$

This proves that the coefficient polynomial of x^k in the bivariate polynomial $F(G'; x, z)$ is hard to compute. We now want to show that this implies that the k-th coefficient (which is a natural number, not a polynomial) of the univariate polynomial is hard to compute. We do this by reducing the computation the coefficient polynomial of x^k in $F(G'; x, z)$ to computing the k-th coefficient in a suitable univariate forest polynomial.

We first show that, although the degree of the coefficient polynomial $C_k(z)$ (in the bivariate case) is not bounded by $f(k)$, but $\Omega(n)$, it suffices to know $O(k)$ coefficients of the coefficient polynomial $C_k(z)$ in order to reconstruct the whole coefficient polynomial. This is essentially an application of the Chinese Remainder Theorem for polynomials, and is given in detail in the full version [2].

Lemma 5. *There is an fpt Turing reduction from computing the coefficient polynomial $C_k(z)$ of x^k in $F(G'; x, z)$ to computing the first k coefficients of univariate forest polynomials on multigraphs.*

Combining the above proves:

Theorem 3. *Given a graph G and a number k, it is #W[1]-hard to compute the number of acyclic subsets of edges of size k in G, parameterized by k.*

Proof. Combining Theorem 1 with Lemmas 4 and 5 yields that computing the first k coefficients of the univariate forest polynomial of multigraphs is #W[1]-hard. Using Lemma 9 from [1] allows to express the forest polynomial of the multigraph G as $F(G;x) = p(x) \cdot F(G';g(x))$, where $p,g : \mathbb{R} \to \mathbb{R}$ are functions such that g is invertible and G' is a simple graph. Now, observe that computing the mapping $G \mapsto F_k(G;x)$ is #W[1]-hard for each fixed x, where $F_k(G;x)$ is the forest polynomial $F(G;x)$ evaluated at x only over the first k coefficients: It is easy to see that $F_k(G;ax) = F_k(G(a);x)$, where G is the graph obtained from G by replacing each edge with a copies of weight x. By using k different values for a, this would allow polynomial interpolation of the first k coefficients of $F(G;x)$. Employing now the equation $F_k(G;x) = p(x) \cdot F_k(G';g(x))$ and the properties of g, this shows that computing the mapping $G' \mapsto F_k(G';x)$ on *simple graphs* G' is #W[1]-hard for all x, and hence also for $x = 1$, where it coincides with counting forests. $\square$

5   Counting Matroid Bases

Definition 5. *The problem of computing the number of bases of a matroid parameterized by its rank (nullity) is denoted as #RANK-BASES (#NULLITY-BASES).*

Lemma 6. *The problem of counting k-forests in a simple graph is fpt Turing reducible to the problem #RANK-BASES, even when the matroid is restricted to be representable over a field of characteristic 2.*

Proof. Given a graph $G = (V,E)$ with $|V| = n$ and $|E| = m$ and a natural number k, we want to count the k-forests of G. Therefore we first construct the incidence matrix $M[G] \in \mathrm{Mat}(n \times m, \mathbb{F}_2)$ of G. Recall that the linearly independent k-subsets of columns of $M[G]$ correspond one-to-one to k-forests in G. In the next step, we compute the reduced row echelon form of $M[G]$ by applying elementary row operations. As stated in the beginning, these operations do not change the linear dependency of the column vectors. Then, we delete the zero rows which also does not change the linear dependency of the columns. We denote the resulting matrix as $M^{\mathrm{red}}[G]$. Now, let r be the rank of $M^{\mathrm{red}}[G]$, which equals the rank of $M[G]$. Note that $M^{\mathrm{red}}[G] \in \mathrm{Mat}(r \times m, \mathbb{F}_2)$. If $r < k$, then we output 0, as G does not have any k-forests in this case. Otherwise, we k-truncate $M^{\mathrm{red}}[G]$ in polynomial time by the deterministic algorithm of Lokshtanov et al. [10] and end up with the matrix $M^{(k)}[G] \in \mathrm{Mat}(k \times m, \mathbb{F}_{2^{rk}})$. Observe that the linear dependency of the column vectors is preserved, i.e., whenever columns $c_1, \ldots, c_k$ are linearly independent in $M[G]$, they are also linearly independent in $M^{(k)}[G]$ and vice versa. Therefore, the rank of $M^{(k)}[G]$ is at least k, since $M[G]$ has rank greater or equal k. As $M^{(k)}[G]$ has only k rows, it follows that the rank is *exactly* k, i.e., $M^{(k)}[G]$ has full rank. Furthermore, the number of linearly independent k-subsets of columns of $M^{(k)}[G]$ equals the number of k-forests in G. As the rank of $M^{(k)}[G]$ is full, we conclude that the number of bases of the matroid that is represented by $M^{(k)}[G]$ equals the number of k-forests in G. Finally, this matroid is representable over $\mathbb{F}_{2^{rk}}$—a field of characteristic 2—by construction. $\square$

Lemma 7. *The problem of counting k-forests in a simple graph is fpt Turing-reducible to the problem #NULLITY-BASES, even when the matroid is restricted to be representable over a field of characteristic 2.*

Proof. We proceed as in the proof of Lemma 6. Having $M^{(k)}[G]$, we construct its dual matroid $M^*[G]$, which can be done in polynomial time (see e.g. [11]). It holds that the number of bases of $M^*[G]$ equals the number of bases of $M^{(k)}[G]$. Furthermore, the rank of $M^*[G]$ is $n - k$, i.e., its nullity is k, which concludes the proof. $\square$

Theorem 4. *#RANK-BASES and #NULLITY-BASES are #W[1]-hard, even when restricted to matroids representable over a field of characteristic 2.*

Proof. Follows from Lemmas 6, 7 and Theorem 3. $\square$

One might ask whether the same is true for matroids that are representable over a fixed finite field. Due to Vertigan [15], it is known that the classical problem of counting bases in binary matroids is #P-hard. However, it is fixed-parameter tractable for each fixed finite field.

Theorem 5. *For every fixed finite field $\mathbb{F}$, the problems #RANK-BASES and #NULLITY-BASES are fixed parameter tractable for matroids given in a linear representation over $\mathbb{F}$.*

Proof (of Theorem 5). We give an fpt algorithm for #RANK-BASES. For #NULLITY-BASES, an algorithm follows by computing the dual matroid as in the proof of Lemma 7.

Let s be the size of the finite field, M be the representation of the given matroid and let k be its rank. We can assume that M only has k rows. For otherwise, we can compute the reduced row echelon form and delete zero rows, which does not change the linear dependencies of the column vectors. If M has only k rows, then there are at most s^k different column vectors. Therefore, we remember the muliplicity of every column vector and delete multiple occurences afterwards. We end up with a matrix with at most s^k columns. Then, we can check for every k-subset of columns whether they are linearly independent. If this is the case, we just multiply the multiplicities of the columns and in the end, we output the sum of all those terms. The running time of this procedure is bounded by $\binom{s^k}{k} \cdot \mathrm{poly}(n)$, where n is the number of columns of the matrix. $\square$

Acknowledgements. The authors wish to thank Markus Bläser, Radu Curticapean, Holger Dell and Petr Hliněný for helpful comments on this work.

References

1. Brand, C., Dell, H., Roth, M.: Fine-grained dichotomies for the tutte plane and boolean #CSP. In: 11th International Symposium on Parameterized and Exact Computation (IPEC 2016), 24–26 August 2016, Aarhus, Denmark, pp. 9:1–9:14 (2016)
2. Brand, C., Roth, M.: Parametcrized counting of trees, forests and matroid bases. CoRR, abs/1611.01823 (2016)
3. Curticapean, R.: Counting matchings of size k Is ♯W[1]-Hard. In: Fomin, F.V., Freivalds, R., Kwiatkowska, M., Peleg, D. (eds.) ICALP 2013. LNCS, vol. 7965, pp. 352–363. Springer, Heidelberg (2013). doi:10.1007/978-3-642-39206-1_30
4. Curticapean, R., Marx, D.: Complexity of counting subgraphs: only the boundedness of the vertex-cover number counts. In: 55th IEEE Annual Symposium on Foundations of Computer Science (FOCS 2014), Philadelphia, PA, USA, 18–21 October 2014, pp. 130–139 (2014)
5. Cygan, M., Fomin, F.V., Kowalik, L., Lokshtanov, D., Marx, D., Pilipczuk, M., Pilipczuk, M., Saurabh, S.: Parameterized Algorithms. Springer, Heidelberg (2015)
6. Eppstein, D.: Subgraph isomorphism in planar graphs and related problems. In: Graph Algorithms and Applications, p. 283 (2002)
7. Flum, J., Grohe, M.: Parameterized Complexity Theory. Texts in Theoretical Computer Science. An EATCS Series. Springer, Heidelberg (2006)
8. Gebauer, H., Okamoto, Y.: Fast exponential-time algorithms for the forest counting and the Tutte polynomial computation in graph classes. Int. J. Found. Comput. Sci. 20(1), 25–44 (2009)
9. Jerrum, M.: Counting trees in a graph is #P-complete. Inf. Process. Lett. 51(3), 111–116 (1994)
10. Lokshtanov, D., Misra, P., Panolan, F., Saurabh, S.: Deterministic truncation of linear matroids. In: Halldórsson, M.M., Iwama, K., Kobayashi, N., Speckmann, B. (eds.) ICALP 2015. LNCS, vol. 9134, pp. 922–934. Springer, Heidelberg (2015). doi:10.1007/978-3-662-47672-7_75
11. Marx, D.: A parameterized view on matroid optimization problems. Theor. Comput. Sci. 410(44), 4471–4479 (2009)
12. Maurer, S.B.: Matrix generalizations of some theorems on trees, cycles and cocycles in graphs. SIAM J. Appl. Math. 30(1), 143–148 (1976)
13. Oxley, J.G.: Matroid Theory. Oxford University Press, New York (1992)
14. Snook, M.: Counting bases of representable matroids. Electr. J. Comb. 19(4), P41 (2012)
15. Vertigan, D.: Bicycle dimension and special points of the Tutte polynomial. J. Comb. Theory Ser. B 74(2), 378–396 (1998)
16. Vertigan, D., Welsh, D.J.A.: The compunational complexity of the Tutte plane. Comb. Probability Comput. 1, 181–187 (1992)

Generalized Dyck Shifts

Marie-Pierre Béal$^{(\boxtimes)}$ and Pavel Heller

Laboratoire d'informatique Gaspard-Monge, UMR 8049 CNRS,
Université Paris-Est, Marne-la-Vallée Cedex, France
{beal,pavel.heller}@u-pem.fr

Abstract. We introduce a new class of subshifts of sequences, called generalized Dyck shifts, which extends the class of Dyck shifts introduced by Krieger. The finite factors of these shifts are factors of generalized Dyck words. Generalized Dyck words were introduced by Labelle and Yeh who exhibited unambiguous algebraic grammars generating these context-free languages. Other unambiguous algebraic grammars for generalized Dyck languages were found by Duchon. We define a coding of periodic patterns of generalized Dyck shifts which allows to compute their zeta function. We prove that the zeta function of a generalized Dyck shift is the commutative image of the generating function of an unambiguous context-free language and is thus an ℕ-algebraic series.

1 Introduction

The Dyck shift introduced by Krieger in [9] is the set of bi-infinite sequences of symbols whose finite factors are factors of Dyck words, or well-parenthesized words. To be well-parenthesized, a word needs to have exactly as many opening parentheses (represented here by the letter a) as closing parentheses (represented by the letter b) with the added condition that each opening parenthesis is matched with a closing parenthesis. If one gives the height value $+1$ to the letter a and the height value -1 to the letter b, this condition means that the total height of a Dyck word is 0 and the height of each prefix of a Dyck word is nonnegative.

Dyck shifts are symbolic dynamical systems which are not sofic and belong to larger classes of shifts like Markov-Dyck shifts (see [10,13]), or sofic-Dyck shifts (see [1]).

In [11] Labelle and Yeh introduced the notion of generalized Dyck words where potentially a larger set of height values are used. They proved the unambiguous context-free nature of generalized Dyck words and exhibited unambiguous context-free grammars for these languages. In [6], Duchon gave new unambiguous context-free grammars for them. Generalized Dyck words were also studied from the point of view of Lyndon words by Melançon and Jacquet in [7].

In this paper we show how to define a shift from generalized Dyck words. This shift is called a generalized Dyck shift. We assign class values to letters

This work is supported by the French National Agency (ANR) through "Programme d'Investissements d'Avenir" (Project ACRONYME no. ANR-10-LABX-58) and by the region of Île-de-France through the DIM RDM-IdF.

© Springer International Publishing AG 2017
P. Weil (Ed.): CSR 2017, LNCS 10304, pp. 99–111, 2017.
DOI: 10.1007/978-3-319-58747-9_11

(for instance, we assign the class value α to the letters a and b). In order to get a nontrivial shift we need to have at least two class values. Generalized Dyck words and factor-free generalized Dyck words are defined recursively as follows. Factor-free generalized Dyck words are the nonempty sequences w of letters *of a same class* such that $h(w) = 0, h(w_1) > 0$ for any proper prefix w_1 of w, and which have no proper factor with these properties. This includes the letters of height 0. Generalized Dyck words are defined as either the empty word or sequences $a_1 d_1 \cdots a_k d_k$ where d_i are generalized Dyck words and $a_1 \cdots a_k$ is a factor-free generalized Dyck word. The generalized Dyck shift is the set of bi-infinite sequences of symbols whose finite factors are factors of generalized Dyck words. These shifts extend the Dyck shift of Krieger.

We give a computation of the zeta function of generalized Dyck shifts which counts the periodic sequences of the shift. We prove that the multivariate zeta function of a generalized Dyck shift is the commutative image of a product of the generating series of the stars of unambiguous context-free circular codes, the codes being cyclically disjoint. The result is based on an encoding of the periodic patterns of the shift. As a consequence the zeta function of a generalized Dyck shift is an $\mathbb{N}$-algebraic series.

Section 2 provides some background on shifts. In Sect. 3 we define the notions of generalized Dyck words and generalized Dyck shifts. We give unambiguous context-free grammars generating several languages linked to generalized Dyck words. The computation of the multivariate and ordinary zeta functions of a generalized Dyck shift is given in Sect. 4. This section contains the decomposition of the multivariate zeta function of a generalized Dyck shift into the commutative image of a product of the generating series of the stars of two unambiguous context-free circular codes.

2   Background on Shifts

We refer to [12] for basic notions in symbolic dynamics. Let A be a finite alphabet. We denote by A^* the set of words over A and by A^+ the set of nonempty words over A.

A *factor* of a word w is a word u such that $w = vuz$ for some words v, z. A *proper factor* of a word w is a factor distinct from w and the empty word.

A *shift* of sequences X is defined as the set X_F of bi-infinite sequences of symbols of A avoiding some set F of finite words (*i.e.* having no finite factor in F). The set F is called a set of *forbidden factors* of X. We denote by $\mathcal{B}(X)$ the set of finite blocks of X, that is the set of allowed finite factors of X.

When F can be chosen finite (resp. regular, visibly pushdown), X is called a *shift of finite type* (resp. a *sofic shift*, a *sofic-Dyck shift*). The *full shift* over A is the set $A^{\mathbb{Z}}$.

Shifts of sequences may be defined as closed subsets of $A^{\mathbb{Z}}$ invariant by the *shift transformation* σ, where $\sigma((x_i)_{i \in \mathbb{Z}}) = (x_{i+1})_{i \in \mathbb{Z}}$. Sets of bi-infinite sequences which are invariant by the shift transformation without being necessarily closed subsets of $A^{\mathbb{Z}}$ are called σ-*invariant sets*. The *orbit* of a sequence

$x \in A^{\mathbb{Z}}$ is the set of all $\sigma^i(x)$ for $i \in \mathbb{Z}$. A *period* of a sequence $x \in A^{\mathbb{Z}}$ is a positive integer p such that $\sigma^p(x) = x$.

A *(topological) conjugacy* from $X \subseteq A^{\mathbb{Z}}$ to $Y \subseteq B^{\mathbb{Z}}$ is a bijective continuous map from X onto Y which commutes with the shift transformation. Observe that a conjugacy preserves the periods of a sequence.

3 Generalized Dyck Words

In this paper, we consider a finite alphabet $A \subset \mathbb{Z} \times \Sigma$, where Σ is a finite alphabet, equipped with two functions: a *height* function h from A to $\mathbb{Z}$ and a *class* function c from A to Σ. Letters with positive height will be denoted by A_+ and letters with negative height by A_-. The set of letters of class α will be denoted by A_α. The set of letters of class α with positive (resp. negative) height is denoted by $A_{\alpha,+}$ (resp. $A_{\alpha,-}$). We assume that all sets A_α have both letters with a positive and with a negative height. We set $(i_\alpha, \alpha) \in A_{\alpha,+}$ and $(-j_\alpha, \alpha) \in A_{\alpha,-}$. The height of a nonempty word is the sum of the height of its letters. The height of the empty word is 0.

A *factor-free generalized Dyck word* is a nonempty sequence w of letters *of a same class* such that $h(w) = 0, h(w_1) > 0$ for any proper prefix w_1 of w, and which has no proper factor with these properties. This includes the letters of height 0. Note that it is a sequence of letters in a same class. We denote by $\tilde{D}_\alpha$ the set of factor-free generalized Dyck words in A_α^+ and $\tilde{D} = \sqcup_\alpha \tilde{D}_\alpha$. A *generalized Dyck word* is defined recursively as follows. It is either the empty word or a sequence $a_1 d_1 \cdots a_k d_k$ where each d_i is a generalized Dyck word and $a_1 \cdots a_k$ is a factor-free generalized Dyck word, or a concatenation of generalized Dyck words. We denote by D_α the set of generalized Dyck words built from factor-free sequences $a_1 \cdots a_k$ in A_α^+. We denote by D the set of generalized Dyck words. Note that $D = \cup_\alpha D_\alpha \cup \{\varepsilon\}$.

Hence a nonempty generalized Dyck word can be obtained by inserting after each letter of a factor-free generalized Dyck word, other generalized Dyck words. Further (see [6, Theorem 7]) this decomposition is unique.

A word is *factor-free* if no proper factor of this word belongs to D. The set of factor-free words of a language L is denoted by $\tilde{L}$. Generalized Dyck words (resp. factor-free generalized Dyck words) will be simply called Dyck words (resp. factor-free Dyck words).

A *prime Dyck word* over A is a Dyck word which is not empty and not the product of shorter Dyck words. Note that the empty word is a Dyck word which is not prime. We denote by P the set of prime Dyck words.

Observe that P is a prefix and suffix code. A factor-free Dyck word is prime but the converse is not true. If w is a Dyck word over A then $h(w) = 0$ and $h(w_1) \geq 0$ for each prefix w_1 of w. If w is a prime Dyck word over A then $h(w) = 0$ and $h(w_1) > 0$ for each proper prefix w_1 of w.

Example 1. Let $\Sigma = \{\alpha, \beta\}$ and $A = \{a = (+3, \alpha), b = (-2, \alpha), a' = (+3, \beta), b' = (-2, \beta)\}$. The word $abab$ is a factor-free Dyck word over A,

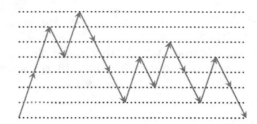

Fig. 1. The prime Dyck word $a(a'b'a'b'b')b(a'b'a'b'b')abb$ of Example 1. Symbols a or a' are represented by up edges while symbols b or b' by down edges according to the height of the symbols. Symbols in A_α (resp. A_β) are represented by red (resp. blue) edges. (Color figure online)

$a(a'b'a'b'b')b$ $(a'b'a'b'b')$ abb (see Fig. 1) is a prime Dyck word over A which is not factor-free.

The *generalized Dyck shift* over A is the set of bi-infinite sequences whose blocks are factors of a Dyck word over A. We denote this shift by X_A. It is thus a coded system as defined by Blanchard and Hansel [4].

Example 2. If Σ is a singleton the generalized Dyck shift is just the full shift, *i.e.* the set of all bi-infinite sequences over A. So the notion of generalized Dyck shift is interesting only for alphabets Σ of size at least two.

Example 3. If $\Sigma = \{\alpha, \beta\}$ and $A = \{$"(" $= (+1, \alpha),")" = (-1, \alpha),$ "[" $= (+1, \beta),"]" = (-1, \beta)\}$, the shift X_A is the Dyck shift with two kinds of parentheses.

Example 4. If $\Sigma = \{\alpha, \beta\}$ and $A = \{a = (+3, \alpha), b = (-2, \alpha), a' = (+3, \beta), b' = (-2, \beta)\}$, for instance the sequences $\cdots bbb.aaaa \cdots$, $^\omega(aba'b'abbaba'b'b'abb)^\omega$ belong to X_A.

Following Duchon [6] we set $m = \max_{a \in A_+} h(a)$, $n = -\min_{a \in A_-} h(a)$. We define for $\alpha \in \Sigma$, $i > 0$, $j > 0$,

- $\tilde{L}_{i,\alpha}$ the set of factor-free words $w \in A_\alpha^+$ with height i such that each proper prefix w_1 of w has a height $h(w_1) > i$.
- $\tilde{R}_{j,\alpha}$ the set of factor-free words $w \in A_\alpha^+$ with height $-j$ such that each proper prefix w_1 of w has a height $h(w_1) > 0$.
- $L_{i,\alpha}{}^1$ the set of nonempty words $w = a_1 d_1 a_2 \cdots d_{k-1} a_k$ with $k \geq 1$, $d_i \in D$, $a_1 a_2 \ldots a_k \in \tilde{L}_{i,\alpha}$.
- $R_{j,\alpha}$ the set of nonempty words $w = a_1 d_1 a_2 \cdots d_{k-1} a_k$ with $k \geq 1$, $d_i \in D$, $a_1 a_2 \ldots a_k \in \tilde{R}_{j,\alpha}$.
- P_α the set of nonempty words $w = a_1 d_1 a_2 \cdots d_{k-1} a_k$ with $k \geq 1$, $d_i \in D$, $a_1 a_2 \ldots a_k \in \tilde{D}_\alpha$.

---

[1] The definition of L_i differs here from the one given in [6].

We set $L_i = \bigcup_\alpha L_{i,\alpha}$, $R_j = \bigcup_\alpha R_{j,\alpha}$, $L = \bigcup_{i=1}^m L_i$, $R = \bigcup_{j=1}^n R_j$. Note that a word in $L_{i,\alpha}$ or $R_{j,\beta}$ does not end nor start with a nonempty Dyck word by definition.

In terms of lattice paths, $L_{i,\alpha}$ is a set of paths that start in $(0,0)$ and end on the line $y = i$ without having a step ending on or under this line before the last step. The set $R_{j,\alpha}$ is a set of paths that start in $(0,0)$ and end on the level $-j$ without having a step ending on or going below the line $y = 0$ before the last step (see Fig. 2).

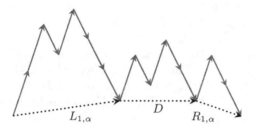

Fig. 2. The prime Dyck word $a(a'b'a'b'b')b(a'b'a'b'b')abb$ of Example 1 in $L_{1,\alpha}DR_{1,\alpha}$.

By definition $L_{i,\alpha}$ and $R_{j,\alpha}$ are codes which are both prefix and suffix. The set $L_{i,\alpha}$ does not overlap strictly with $L_{j,\beta}$. Indeed, if $uv \in L_{i,\alpha}$ and $vw \in L_{j,\alpha}$, $h(v) < 0$ unless v is the empty word or $v = uv$. A word of $L_{i,\alpha}$ is neither prefix nor suffix of a word in $L_{j,\beta}$ with $\alpha \neq \beta$. Observe also that $L_{i,\alpha}$ is empty for $i > m$ and $R_{j,\alpha}$ is empty for $j > n$.

Lemma 1. *We have the following unambiguous grammars for L_i, R_j, P:*

$$\tilde{L}_{i,\alpha} = \sum_{h(a)=i,c(a)=\alpha} a + \sum_{k>i} \tilde{L}_{k,\alpha}\tilde{R}_{k-i,\alpha} \tag{1}$$

$$\tilde{R}_{j,\alpha} = \sum_{h(a)=-j,c(a)=\alpha} a + \sum_{k} \tilde{L}_{k,\alpha}\tilde{R}_{k+j,\alpha} \tag{2}$$

$$\tilde{D}_\alpha = \sum_{h(a)=0,c(a)=\alpha} a + \sum_{k} \tilde{L}_{k,\alpha}\tilde{R}_{k,\alpha} \tag{3}$$

$$L_{i,\alpha} = \sum_{h(a)=i,c(a)=\alpha} a + \sum_{k>i} L_{k,\alpha}DR_{k-i,\alpha} \tag{4}$$

$$R_{j,\alpha} = \sum_{h(a)=-j,c(a)=\alpha} a + \sum_{k} L_{k,\alpha}DR_{k+j,\alpha} \tag{5}$$

$$P_\alpha = \sum_{h(a)=0} a + \sum_{k} L_{k,\alpha}DR_{k,\alpha} \tag{6}$$

$$P = \sum_\alpha P_\alpha, \quad D = P^*, \quad L_i = \sum_\alpha L_{i,\alpha}, \quad R_j = \sum_\alpha R_{j,\alpha}. \tag{7}$$

Proof. We have $\tilde{L}_{i,\alpha}\tilde{R}_{j,\beta}$ with $\alpha \neq \beta$ forbidden in $\tilde{L}_{i,\alpha}$, $\tilde{R}_{j,\alpha}$, $\tilde{D}_\alpha$. We have $\tilde{L}_{k,\alpha}\tilde{R}_{k-i,\alpha} \subseteq \tilde{L}_{i,\alpha}$ for any $k > i$. If $w \in \tilde{L}_{i,\alpha}$, if $|w| > 1$, let u be the unique proper prefix of w such that $h(u)$ is minimal. Let $h(u) = k > i$ and $w = uv$. Then $u \in \tilde{L}_{k,\alpha}$ and $v \in \tilde{R}_{k-i,\alpha}$. Further, if $uv = u'v'$ with $u \in \tilde{L}_{k,\alpha}$, $v \in \tilde{R}_{k-i,\alpha}$, $u' \in \tilde{L}_{k',\alpha}$, $v' \in \tilde{R}_{k'-i,\alpha}$. One has for instance u prefix of u'. Let $u' = uu''$. Then $k \geq k'$. If $k > k'$, then $v \notin \tilde{R}_{k-i,\alpha}$. Thus $k = k'$, $u'' \in D$, implying $u'' = \varepsilon$. This proves Eq. (1). Equations (2) and (3) are obtained similarly.

We have $L_{k,\alpha}DR_{k-i,\alpha} \subseteq L_{i,\alpha}$. If $w \in L_{i,\alpha}$ and if $|w| > 1$ let u be the smallest proper prefix of w such that $h(u)$ is minimal and t be the largest proper prefix of w such that $h(t)$ is minimal (see Fig. 2). We have $t = uv$ with $v \in D$ and $w = uvz$. Then $u \in L_{k,\alpha}$ and $z \in R_{k-i,\alpha}$. Further, if $uvz = u'v'z'$ with $u \in L_{k,\alpha}$, $z \in R_{k-i,\alpha}$, $u' \in L_{k',\alpha}$, $z' \in R_{k'-i,\alpha}$, $v, v' \in D$, then for instance u is a prefix of u'. Assume that u is a strict prefix of u'. Let $u' = uu''$. Then $k \geq k'$. If $k > k'$, then $vz \notin DR_{k-i,\alpha}$. Thus $k = k'$. Then $u'' \in D\backslash\{\varepsilon\}$, a contradiction since $u' \in L_{k',\alpha}$ dos not end with a nonempty word of D. Thus $u = u'$. Similarly, $z = z'$ and thus $v = v'$. This proves Eq. (4). Equations (5) and (6) are obtained similarly.

We consider the free monoid generated by A with a zero quotiented by the following relations

$$a_1 \cdots a_k = \mathbf{1} \text{ if } a_1 \cdots a_k \text{ is a factor-free Dyck word}$$

$$w = 0, \text{ if } w \in \tilde{L}_{i,\alpha}\tilde{R}_{j,\beta} \text{ with } \alpha \neq \beta \text{ and } i, j > 0.$$

where $\mathbf{1}$ is the unity of the monoid.

For a word w over A, we denote by $\overline{w} \in A^* \cup \{0, \mathbf{1}\}$ its *reduced form* which is the unique word obtained by applying the above relations.

For instance ( ] reduces to 0 in the Dyck shift.

Observe that a word z in $\tilde{L}_{i,\alpha}\tilde{R}_{j,\beta}$ with $\alpha \neq \beta$ is not factor of a Dyck word. Indeed, if z is a factor of a Dyck word d, it is a factor uv, where $u \in \tilde{L}_{i,\alpha}$, $v \in \tilde{R}_{j,\beta}$, of $a_1d_1 \cdots a_kd_k$ where $d_i \in D$ and $a_1 \cdots a_k$ is a factor-free Dyck word. If none a_i is a factor of uv, then uv is a factor of some d_i whose length is shorter than d. By recurrence on the size of d we obtain that uv is factor of a factor-free Dyck word. Since $uv \in A_\alpha^+ A_\beta^+$, we get a contradiction. Observe that, since $u \in \tilde{L}_{i,\alpha}$, $v \in \tilde{R}_{j,\beta}$ and $d_i \in D$ for $1 \leq i \leq k$, d_i cannot overlap nor be a factor of uv unless d_i is the empty word. Thus if a_i is factor of u or v for some $1 \leq i \leq k$, then $uv = a_1 \cdots a_k$ and $d_1 = d_2 = \cdots d_{k-1} = \varepsilon$. This gives a contradiction since $uv \in A_\alpha^+ A_\beta^+$ with $\alpha \neq \beta$ and $a_1 \cdots a_k \in A_\gamma^+$ for some γ.

The set of words reducing to $\mathbf{1}$ is the set of Dyck words. If two Dyck words overlap, the overlapping word is a Dyck word: if uv and vw are Dyck words, then u, v, w also. Dyck words. Thus if uv and vw reduce to $\mathbf{1}$, then u, v, w also. Further, a factor-free word has no suffix being a prefix of a word in $\tilde{L}_{i,\alpha}\tilde{R}_{j,\beta}$ with $\alpha \neq \beta$. Hence a word reducing to $\mathbf{1}$ has no non trivial overlap with a word reducing to 0. As a consequence the reduced form is unique.

Proposition 1. *The reduced form of a word w is either 0, **1**, u, v or uv where*

$$u \in \tilde{R}_{j_s,\beta_s} \cdots \tilde{R}_{j_1,\beta_1}$$
$$v \in \tilde{L}_{i_1,\alpha_1} \cdots \tilde{L}_{i_r,\alpha_r}$$

for some $j_1, .., j_s, i_1, .., i_r > 0$ and $\beta_1, .., \beta_s, \alpha_1, .., \alpha_r \in \Sigma$.

Proof. Assume that $\overline{w} \neq 0, \mathbf{1}$. Let u be the unique prefix of $\overline{w}$ of minimal height, the unicity coming from the fact that $\overline{w}$ is reduced. This prefix may be the empty word. We set $\overline{w} = uv$. Then v has a unique decomposition into $\tilde{L}_{i_1,\alpha_1} \cdots \tilde{L}_{i_r,\alpha_r}$ where $i_1 + \cdots + i_s = i$ with $h(v) = i > 0$. Indeed let z be the unique nonempty prefix of v such that $h(z) = i_1$ is minimal. The prefix z being reduced, it is a factor-free word. As z does not contain any factor in $\tilde{L}_{i,\alpha}\tilde{R}_{j,\beta}$ with $\alpha \neq \beta$, $i, j > 0$, $z \in \tilde{L}_{i_1,\alpha_1}$ for some $\alpha_1 \in \Sigma$. The whole decomposition of v thus belongs to $\tilde{L}_{i_1,\alpha_1} \cdots \tilde{L}_{i_r,\alpha_r}$ where $i_1 + \cdots + i_r = i$. A symmetrical property holds for u.

   Let $uv \in \tilde{R}_{j_s,\beta_s} \cdots \tilde{R}_{j_1,\beta_1}\tilde{L}_{i_1,\alpha_1} \cdots \tilde{L}_{i_r,\alpha_r}$. Then $uv \neq 0$. Indeed, if uv contains a factor z in $\tilde{L}_{i,\alpha}\tilde{R}_{j,\beta}$ with $\alpha \neq \beta$ and $i, j > 0$, then $z = tt'$ where $t \in \tilde{L}_{i,\alpha}$ and $t' \in \tilde{R}_{j,\beta}$. Since t cannot be a suffix of some word in $\tilde{R}_{j_k,\beta_k} \cdots \tilde{R}_{j_{k'},\beta_{k'}}$ it is a suffix of some word in $\tilde{R}_{j_s,\beta_s} \cdots \tilde{R}_{j_1,\beta_1}\tilde{L}_{i_1,\alpha_1} \cdots \tilde{L}_{i_{k'},\alpha_{k'}}$ and t' cannot be a prefix of a word in $\tilde{L}_{i_{k'+1},\alpha_{k'+1}} \cdots \tilde{L}_{i_r,\alpha_r}$. Thus $uv \neq 0$. Further a word in $\tilde{L}_{i,\alpha}\tilde{R}_{j,\beta}$ with $\alpha \neq \beta$ and $i, j > 0$ is not factor of a Dyck word, thus $0 \neq \mathbf{1}$.

   Words whose reduced form is either **1** or u (resp. v) as above are called *matched-call* (resp. *matched-return*). The set of matched-call (resp. matched-return) words is denoted by $MC(X)$ (resp. $MR(X)$). Thus matched-call words are sequences of words in $R+D$ and mached-return words are sequences of words in $L + D$.

Example 5. For instance, the reduced form of the word ")(( ) [[ ]] [" in the Dyck shift with two kinds of parentheses is ") ([". We have $u =)$ in $\tilde{R}_{1,\alpha}$ and $v = ([$ in $\tilde{L}_{1,\alpha}\tilde{L}_{1,\beta}$. In the shift of Example 1, the word $ba(a'b'a'b'b')ba$ has the reduced form $b(ab)a \in \tilde{R}_{2,\alpha}\tilde{L}_{1,\alpha}\tilde{L}_{3,\alpha}$.

Proposition 2. *The generalized Dyck shift is the set of sequences avoiding the factors whose reduced form is 0.*

Proof. First a word whose reduced form is 0 is not factor of a Dyck word. Conversely let w be a word over A. Its reduced form is either 0, 1 or u, v, uv with $u \in X = \tilde{R}_{j_s,\beta_s} \cdots \tilde{R}_{j_1,\beta_1}$ and $v \in Y = \tilde{L}_{i_1,\alpha_1} \cdots \tilde{L}_{i_r,\alpha_r}$. Hence it is nonnull if and only if it is a factor of a Dyck word. Indeed any word in X, Y or XY is factor of a Dyck word since $((\tilde{L}_{i_{\beta_1},\beta_1})^{j_1}(\tilde{R}_{j_1,\beta_s})^{i_{\beta_1}-1}) \cdots ((\tilde{L}_{i_{\beta_s},\beta_s})^{j_s}(\tilde{R}_{j_s,\beta_s})^{i_{\beta_s}-1}) X \subseteq D$, $Y((\tilde{L}_{i_r,\alpha_r})^{j_{\alpha_r}-1}(\tilde{R}_{j_{\alpha_r},\alpha_r})^{i_r}) \cdots ((\tilde{L}_{i_1,\alpha_1})^{j_{\alpha_1}-1}(\tilde{R}_{j_{\alpha_1},\alpha_1})^{i_1}) \subseteq D$ and $\tilde{L}_{i_{\beta_1},\beta_1}, \ldots, \tilde{R}_{j_{\alpha_1},\alpha_1}$ are non empty. Thus w itself is factor of a Dyck word.

Example 6. We continue with Example 4. Setting in this example $L_i = L_{i,\alpha}$, $R_i = R_{i,\alpha}$, and $L'_i = L_{i,\beta}$, $R'_i = R_{i,\beta}$, $r_i = R_i D$, $\ell_i = L_i D$, we have

$$
\begin{aligned}
L_3 &= a & L'_3 &= a' \\
L_2 &= L_3 D R_1 = a D R_1 & L'_2 &= a' D R'_1 \\
L_1 &= L_2 D R_1 + L_3 D R_2 & L'_1 &= L'_2 D R'_1 + L'_3 D R'_2 \\
R_2 &= b & R'_2 &= b' \\
R_1 &= L_1 D R_2 = L_1 D b & R'_1 &= L'_1 D b'
\end{aligned}
$$
$$
P = L_1 D R_1 + L_2 D R_2 + L'_1 D R'_1 + L'_2 D R'_2
$$

Thus $R_1 = L_1 D b = (a D R_1 D R_1 + a D b) D b = a D((R_1 D)^2 + b D) b$. We can set $U b D = (R_1 D)^2 + b D$ since $R_1 \in A^* b$. We get $r_1 = R_1 D = \ell_1 b D = (a D) U(b D)(b D)$, $\ell_1 = (a D) U(b D)$.

We have

$$
\begin{aligned}
R_1 D &= (a D) U(b D)(b D) = (a D)((R_1 D)^2 + b D) b D \\
&= (a D)(a D U b D b D a D U b D b D + b D) b D.
\end{aligned}
$$

Thus

$$
U = \varepsilon + (a D) U(b D)(b D)(a D) U(b D).
$$

Similarly,

$$
\begin{aligned}
R'_1 D &= (a' D) V(b' D)(b' D) \\
V &= \varepsilon + (a' D) V(b' D)(b' D)(a' D) V(b' D).
\end{aligned}
$$

We have

$$
\begin{aligned}
PD = \ &(a D) U(b D)(a D) U(b D)(b D) + (a D)(a D) U(b D)(b D)(b D) + \\
&(a' D) V(b' D)(a' D) V(b' D)(b' D) + (a' D)(a' D) V(b' D)(b' D)(b' D).
\end{aligned}
$$

The right symbol D in all the above equations may be removed by right multiplication of both sides by $(1 - P)$ which is the inverse of D.

4   Zeta Function of Generalized Dyck Shifts

4.1   Multivariate Zeta Functions

We recall the notion of multivariate zeta function introduced by Berstel and Reutenauer in [3,14].

For $K = \mathbb{Z}$ or $K = \mathbb{N}$ (containing 0) we denote by $K\langle\langle A \rangle\rangle$ the set of noncommutative formal power series over the alphabet A with coefficients in K. For each language L of finite words over a finite alphabet A we define the *characteristic series* of L as the series $\underline{L} = \sum_{u \in L} u$ in $\mathbb{N}\langle\langle A \rangle\rangle$.

Let $K[\![A]\!]$ be the usual commutative algebra of formal power series in the variables of A and $\pi\colon K\langle\!\langle A\rangle\!\rangle \to K[\![A]\!]$ be the natural homomorphism. Let S be a commutative or noncommutative series. One can write $S = \sum_{n\geq 0}[S]_n$ where each $[S]_n$ is the homogeneous part of S of degree n. The notation extends to matrices H with coefficients in $K\langle\!\langle A\rangle\!\rangle$ or $K[\![A]\!]$ with $([H]_n)_{pq} = [H_{pq}]_n$, where p, q are indices of H.

Call *periodic pattern* of a shift X a word u such that the bi-infinite concatenation of u belongs to X and denote $\mathcal{P}(X)$ the set of periodic patterns of X. These definitions are extended to σ-invariant sets of bi-infinite sequences which may not be shifts.

The *multivariate zeta function* $Z(X)$ of a σ-invariant set X is the commutative series in $\mathbb{Z}[\![A]\!]$

$$Z(X) = \exp \sum_{n\geq 1} \frac{\pi[\mathcal{P}(X)]_n}{n}.$$

The *(ordinary) zeta function* of a σ-invariant set X is

$$\zeta_X(z) = \exp \sum_{n\geq 1} p_n \frac{z^n}{n},$$

where p_n is the number of sequences of X of period n, *i.e.* of sequences x such that $\sigma^n(x) = x$.

Let $\theta\colon \mathbb{Z}[\![A]\!] \to \mathbb{Z}[\![z]\!]$ be the homomorphism such that $\theta(a) = z$ for any letter $a \in A$. If $S \in \mathbb{Z}[\![A]\!]$, $\theta(S)$ will also be denoted by $S(z)$. Note that $\zeta_X(z) = \theta(Z(X))$.

It is known that the multivariate zeta function of a shift has nonnegative integer coefficients [12].

4.2   Encoding of Periodic Sequences of a Generalized Dyck Shift

We say that two finite words x, y are *conjugate* if $x = uv$ and $y = vu$ for some words u, v.

If C is a code, we denote by X_C the σ-invariant set containing all bi-infinite concatenation of words in C. This set is not a shift since it may not be closed.

The following proposition gives an encoding of the periodic patterns of a generalized Dyck shift.

Proposition 3. *Let X be the generalized Dyck shift over A. The set of periodic patterns $\mathcal{P}(X)$ of X is*

$$\mathcal{P}(X) = \mathcal{P}(\mathsf{X}_{DL}) \sqcup \mathcal{P}(\mathsf{X}_{R+P}).$$

Proof. Let z be a periodic pattern. Then $x = \cdots zz.zz \cdots$ is a periodic sequence of X. The reduced form of z is $\bar{z} = \mathbf{1}, u, v$ or $u \cdot v$ where

$$u \in \tilde{R}_{j_s,\beta_s} \cdots \tilde{R}_{j_1,\beta_1}$$
$$v \in \tilde{L}_{i_1,\alpha_1} \cdots \tilde{L}_{i_r,\alpha_r}$$

If z is not already in $\mathrm{MC}(X)$ or in $\mathrm{MR}(X)$, then its reduced form is uv. In this case z has a conjugate z' whose reduced form is the reduced form of vu. We have

$$vu \in \tilde{L}_{i_1,\alpha_1} \cdots \tilde{L}_{i_r,\alpha_r} \tilde{R}_{j_s,\beta_s} \cdots \tilde{R}_{j_1,\beta_1}.$$

Since $\cdots z'z'.z'z' \cdots \in X$, we have $\alpha_r = \beta_s$, and $\tilde{L}_{i_r,\alpha_r}\tilde{R}_{j_s,\beta_s}$ is included in either $L_{i_r-j_s,\alpha_r}$ or $R_{j_s-i_r,\alpha_r}$ or D. In the first case, $\alpha_r = \beta_{s-1}$ and $L_{i_r-j_s,\alpha_r}\tilde{R}_{j_{s-1},\beta_{s-1}}$ is then included in either $L_{i_r-j_s-j_{s-1},\alpha_r}$ or $R_{j_{s-1}-(i_r-j_s),\alpha_r}$ or D. In the second case $\alpha_{r-1} = \beta_s$ and $\tilde{L}_{i_{r-1},\alpha_{r-1}}R_{j_s-i_r,\alpha_r}$ is then included in either $L_{j_s-i_r-i_{r-1},\alpha_r}$ or $R_{i_{r-1}+i_r-j_s,\alpha_r}$ or D. In the third case, $\alpha_{r-1} = \beta_{s-1}$ and $\tilde{L}_{i_{r-1},\alpha_{r-1}}D\tilde{R}_{j_{s-1},\alpha_{r-1}}$ is then included in either $L_{i_{r-1}-j_{s-1},\alpha_{r-1}}$ or $R_{j_{s-1}-i_{r-1},\alpha_{r-1}}$ or D. By iterating the reduction, we get that vu is included in some product equal to either $L_{k_1,\gamma_1} \cdots L_{k_n,\gamma_n}$ or $L_{k_1,\gamma_1} \cdots L_{k_n,\gamma_n}D$ or $R_{k_n,\gamma_n} \cdots R_{k_1,\gamma_1}$ or $DR_{k_n,\gamma_n} \cdots R_{k_1,\gamma_1}$ or D. This product vu is thus either in $\mathrm{MC}(X)$ or in $\mathrm{MR}(X)$.

If z' is matched-call, then it is a product of words in P or in R. In this case z is conjugate to a word in $(P + R)^*$.

If z' is matched-return and not matched-call, i.e. $z' \notin D$, we can assume that it does not end with a Dyck word (if $z' = uw$ with w Dyck, we could consider wu instead). In that case it is a product of words in $P^*L = DL$ and z is conjugate to a word in $(DL)^*$. As a consequence $\mathcal{P}(X) = \mathcal{P}(\mathsf{X}_{DL}) \sqcup \mathcal{P}(\mathsf{X}_{R+P})$.

Let us finally show that $\mathcal{P}(\mathsf{X}_{DL}) \cap \mathcal{P}(\mathsf{X}_{R+P}) = \emptyset$. Assume the contrary. Then there are nonempty conjugate words w, w' such that w is in $(DL)^*$ and w' is in $(R + P)^*$.

This implies that the height of w is positive and the height of w' is nonpositive, contradicting the conjugacy of w and w'.

4.3   Computation of the Zeta Function

We recall below the notion of circular codes (see for instance [2]). We say that a subset S of nonempty words over A is *a circular code* if for all $n, m \geq 1$ and $x_1, x_2, \ldots, x_n \in S$, $y_1, y_2, \ldots, y_m \in S$ and $p \in A^*$ and $s \in A^+$, the equalities $sx_2x_3 \cdots x_n p = y_1y_2 \cdots y_m$ and $x_1 = ps$ imply $n = m$, $p = \varepsilon$ and $x_i = y_i$ for each $1 \leq i \leq n$.

Two codes C_1 and C_2 are *cyclically disjoint* if a word of C_1^* which is conjugate to a word of C_2^*, is empty.

Proposition 4. *The sets DL and $P \sqcup R$ are cyclically disjoint circular codes.*

Proof. We first show that $R \sqcup P$ is circular. Keeping the notation of the definition, let $x_1, x_2, \ldots, x_n \in S$, $y_1, y_2, \ldots, y_m \in S$, $p \in A^*$ and $s \in A^+$. We prove the claim by induction on $n + m$. Suppose that $sx_2x_3 \cdots x_n p = y_1y_2 \cdots y_m$ and $x_1 = ps$ imply $n = m$ and $x_i = y_i$ when $n + m < N$. Assume now that $sx_2x_3 \cdots x_n p = y_1y_2 \cdots y_m$ and $x_1 = ps$ for some n, m with $n + m = N$.

If p was nonempty, then, since $x_1 = ps$ where $s \neq \varepsilon$, we have $h(p) > 0$. This would contradict p being a suffix of $y_1y_2 \cdots y_m$, which is clearly matched-call, hence we get $p = \varepsilon$. It follows that x_1 is a prefix of y_1 or the converse, implying $x_1 = y_1$. By induction hypothesis we obtain that $n = m$ and $x_i = y_i$.

Let us show that P^*L is circular. Let us assume that $s \neq x_1$. Since s is a prefix of $y_1 y_2 \cdots y_m$ and is a suffix of x_1, we have $s \in P^*L$ and $p \in P^*$. As $p \neq \varepsilon$, $p \in P^+$. This contradicts the fact that p is a suffix of $y_1 \cdots y_m$. Hence $s = x_1$ and $p = \varepsilon$. Now $x_1 \cdots x_n = y_1 y_2 \cdots y_m$ implies $x_1 = y_1$ since $x_i, y_i \in P^*L$. By induction hypothesis we get $n = m$ and $x_i = y_i$.

We now show that DL and $P + R$ are cyclically disjoint. Let $u \in (DL)^*$ and $v \in (P+R)^*$ such that u and v are two nonempty conjugate words. This implies that the height of u is positive and the height of v is nonpositive, contradicting the conjugacy of u and v.

Proposition 5. *Let X be a generalized Dyck shift over A. The multivariate zeta function of X has the following expression.*

$$Z(X) = \pi((DL)^*(P + R)^*).$$

Proof. From Proposition 3 we get that the multivariate zeta function of X is $Z(X) = Z(\mathsf{X}_{DL})Z(\mathsf{X}_{P+R})$.

From [15, Proposition 4.7.11] (see also [2, Proposition 3.1],[8]), if C is a circular code $Z(\mathsf{X}_C) = \pi(C^*)$. The result follows from the fact that DL and $P + R$ are circular codes.

Example 7. We consider the Dyck shift X with two kinds of parentheses of Example 3 defined by $\Sigma = \{\alpha, \beta\}$ and $A = \{$ "(" $= (+1, \alpha),$ ")" $= (-1, \alpha),$ "[" $= (+1, \beta),$ "]" $= (-1, \beta)\}$.

Setting $a = $ "(", $b = $ ")", $a' = $ "[", $b' = $ "]", $L_1 = L_{1,\alpha}, L_1' = L_{1,\beta}, R_1 = R_{1,\alpha}, R_1' = R_{1,\beta}$, we have

$$L_1 = a \quad L_1' = a' \tag{8}$$
$$R_1 = b \quad R_1' = b' \tag{9}$$
$$P = L_1 D R_1 + L_1' D R_1' = aDb + a'Db' \tag{10}$$
$$D = 1 + PD = 1 + aDbD + a'Db'D \tag{11}$$

Thus

$$Z(X) = \pi((D(a + a'))^* \ (b + b' + aDb + a'Db')^*),$$

where D is defined by Eq. 11. A computation gives the formula of Keller for $\zeta_X(z)$ [8]:

$$\zeta_X(z) = \frac{2(1 + \sqrt{1 - 8z^2})}{(1 - 4z + \sqrt{1 - 8z^2})^2}.$$

Example 8. We consider the shift X_A defined by $\Sigma = \{\alpha, \beta\}$ and $A = \{a = (+2, \alpha), b = (-1, \alpha), a' = (+2, \beta), b' = (-1, \beta)\}$.

Setting $L_i = L_{i,\alpha}$, $L'_i = L_{i,\beta}$, and $R_i = R_{i,\alpha}$, $R'_i = R_{i,\beta}$, we have

$$L_2 = a \quad L'_2 = a' \tag{12}$$

$$R_1 = b \quad R'_1 = b' \tag{13}$$

$$L_1 = L_2 D R_1 = aDb \quad L'_1 = a'Db' \tag{14}$$

$$P = L_1 D R_1 + L'_1 D R'_1 = aDbDb + a'Db'Db' \tag{15}$$

$$D = 1 + PD = 1 + aDbDbD + a'Db'Db'D \tag{16}$$

Thus

$$Z(X) = \pi((D(aDb + a'Db' + a + a'))^* \ (b + b' + aDbDb + a'Db'Db')^*),$$

where D is defined by Eq. 16.

Let S be a multivariate series in $\mathbb{N}\langle\langle A \rangle\rangle$. We denote by $< S, u >$ the coefficient of a word u in S. We say that S is $\mathbb{N}$-*algebraic* if $S- < S, \varepsilon > \varepsilon$ is the multivariate generating series of some unambiguous context-free language. The multivariate zeta function of a shift is $\mathbb{N}$-*algebraic* if it is the commutative image of some multivariate $\mathbb{N}$-algebraic series. In one variable, a series $S(z)$ is $\mathbb{N}$-algebraic if it is the first component $(S_1(z))$ of a system of equations $S_i(z) = P_i(z, S_1(z), .., S_r(z))$, where $1 \le i \le r$ and P_i are multivariate polynomials with coefficients in $\mathbb{N}$ (see for instance [5]).

Corollary 1. *The multivariate zeta function of a generalized Dyck shift is the commutative image of a product of the generating series of the stars of unambiguous context-free circular codes, the codes being cyclically disjoint. The multivariate and ordinary zeta functions of a generalized Dyck shift are $\mathbb{N}$-algebraic series.*

Proof. The result follows from Proposition 5 and the fact that DL and $P \cup R$ are unambiguous context-free circular codes since the languages $P, L_{i,\alpha}, R_{j,\beta}$ are unambiguous context-free. Further DL and $P \cup R$ are cyclically disjoint.

References

1. Béal, M.-P., Blockelet, M., Dima, C.: Sofic-Dyck shifts. Theor. Comput. Sci. **609**, 226–244 (2016)
2. Berstel, J., Perrin, D., Reutenauer, C.: Codes and Automata. Encyclopedia of Mathematics and its Applications, vol. 129. Cambridge University Press, Cambridge (2010)
3. Berstel, J., Reutenauer, C.: Zeta functions of formal languages. Trans. Amer. Math. Soc. **321**, 533–546 (1990)
4. Blanchard, F., Hansel, G.: Systèmes codés. Theor. Comput. Sci. **44**, 17–49 (1986)
5. Bousquet-Mélou, M.: Rational and algebraic series in combinatorial enumeration. In: International Congress of Mathematicians, vol. III, pp. 789–826 (2006). Eur. Math. Soc., Zürich

6. Duchon, P.: On the enumeration and generation of generalized Dyck words. Discrete Math. **225**(1–3), 121–135 (2000)
7. Jacquet, H., Mélançon, G.: Langages de Dyck généralisés et factorisations du monoide libre. Ann. Sci. Math. Québec **97**, 103–122 (1997)
8. Keller, G.: Circular codes, loop counting, and zeta-functions. J. Combin. Theory Ser. A **56**(1), 75–83 (1991)
9. Krieger, W.: On the uniqueness of the equilibrium state. Math. Systems Theory **8**(2), 97–104 (1974/1975)
10. Krieger, W., Matsumoto, K.: Zeta functions and topological entropy of the Markov-Dyck shifts. Münster J. Math. **4**, 171–183 (2011)
11. Labelle, J., Yeh, Y.: Generalized Dyck paths. Discrete Math. **82**(1), 1–6 (1990)
12. Lind, D., Marcus, B.: An Introduction to Symbolic Dynamics and Coding. Cambridge University Press, Cambridge (1995)
13. Matsumoto, K.: On the simple C^*-algebras arising from Dyck systems. J. Operator Theory **58**(1), 205–226 (2007)
14. Reutenauer, C.: ℕ-rationality of zeta functions. Adv. Appl. Math. **18**(1), 1–17 (1997)
15. Stanley, R.P.: Enumerative Combinatorics. Wadsworth Publ. Co., Belmont (1986)

Green's Relations
in Finite Transformation Semigroups

Lukas Fleischer$^{(\boxtimes)}$ and Manfred Kufleitner

FMI, University of Stuttgart, Universitätsstraße 38, 70569 Stuttgart, Germany
{fleischer,kufleitner}@fmi.uni-stuttgart.de

Abstract. We consider the complexity of Green's relations when the semigroup is given by transformations on a finite set. Green's relations can be defined by reachability in the (right/left/two-sided) Cayley graph. The equivalence classes then correspond to the strongly connected components. It is not difficult to show that, in the worst case, the number of equivalence classes is in the same order of magnitude as the number of elements. Another important parameter is the maximal length of a chain of components. Our main contribution is an exponential lower bound for this parameter. There is a simple construction for an arbitrary set of generators. However, the proof for constant alphabet is rather involved. Our results also apply to automata and their syntactic semigroups.

1 Introduction

Let Q be a finite set with n elements. There are n^n mappings from Q to Q. Such mappings are called *transformations* and the elements of Q are called *states*. The composition of transformations defines an associative operation. If Σ is some arbitrary subset of transformations, we can consider the *transformation semigroup S* generated by Σ; this is the closure of Σ under composition.[1] The set of all transformations on Q is called the *full transformation semigroup* on Q. One can view (Q, Σ) as a description of S. Since every element s of a semigroup S defines a transformation $x \mapsto x \cdot s$ on $S^1 = S \cup \{1\}$, every semigroup S admits such a description (S^1, S); here, 1 either denotes the neutral element of S or, if S does not have a neutral element, we add 1 as a new neutral element. Essentially, the description (S^1, S) is nothing but the multiplication table for S. On the other hand, there are cases where a description as a transformation semigroup is much more succinct than the multiplication table. For instance, the full transformation semigroup on Q can be generated by a set Σ with three elements [7]. In addition to the size of S, it would be interesting to know which other properties could be derived from the number n of states.

Green's relations are an important tool for analyzing the structure of a semigroup S. They are defined as follows:

This work was supported by the DFG grants DI 435/5-2 and KU 2716/1-1.
[1] When introducing transformation semigroups in terms of actions, then this is the framework of *faithful* actions.

P. Weil (Ed.): CSR 2017, LNCS 10304, pp. 112–125, 2017.
DOI: 10.1007/978-3-319-58747-9_12

$$s \leqslant_\mathcal{R} t \text{ if } sS^1 \subseteq tS^1, \quad s \leqslant_\mathcal{L} t \text{ if } S^1s \subseteq S^1t, \quad s \leqslant_\mathcal{J} t \text{ if } S^1sS^1 \subseteq S^1tS^1.$$

We write $s \mathcal{R} t$ if both $s \leqslant_\mathcal{R} t$ and $s \leqslant_\mathcal{R} t$; and we set $s <_\mathcal{R} t$ if $s \leqslant_\mathcal{R} t$ but not $s \mathcal{R} t$. The relations $\mathcal{L}$, $<_\mathcal{L}$, $\mathcal{J}$ and $<_\mathcal{J}$ are defined analogously. The relations $\mathcal{R}$, $\mathcal{L}$, and $\mathcal{J}$ form equivalence relations. The equivalence classes corresponding to these relations are called $\mathcal{R}$-*classes* (resp. $\mathcal{L}$-*classes*, $\mathcal{J}$-*classes*) of S. Instead of ideals, one could alternatively also use reachability in the right (resp. left, two-sided) Cayley graph of S for defining $\leqslant_\mathcal{R}$ (resp. $\leqslant_\mathcal{L}$, $\leqslant_\mathcal{J}$). We note that $s <_\mathcal{R} t$ implies $s <_\mathcal{J} t$ and, symmetrically, $s <_\mathcal{L} t$ implies $s <_\mathcal{J} t$. The complexity of deciding Green's relations for transformation semigroups was recently shown to be PSPACE-complete [1]. When considering a transformation semigroup on n states, one of our first results shows that the maximal number of $\mathcal{J}$-classes is in $n^{\Theta(n)}$. In particular, the number of equivalence classes is in the same order of magnitude as the size of the transformation semigroup. Since every $\mathcal{J}$-class contains at least one $\mathcal{R}$- and one $\mathcal{L}$-class, the same bound holds for $\mathcal{R}$ and $\mathcal{L}$.

Another important parameter is the maximal length ℓ such that there are elements $s_1, \ldots, s_\ell$ with $s_1 >_\mathcal{R} \cdots >_\mathcal{R} s_\ell$, called the $\mathcal{R}$-*height*. Similarly, we are interested in the $\mathcal{L}$- and $\mathcal{J}$-height. Many semigroup constructions such as the Rhodes expansion and variants thereof rely on this parameter; see e.g. [2,3,6]. We show that the maximal $\mathcal{R}$-height is in $2^{\Theta(n)}$; for the maximal $\mathcal{L}$-height and $\mathcal{J}$-height we only have $2^{\Omega(n)}$ as a lower bound. Proving the lower bounds for a fixed number of generators is much more involved than for arbitrarily many generators. The exponential lower bounds are quite unexpected in the following sense: If the transformation semigroup is small, then the number of equivalence classes (and hence, the lengths of chains) cannot be big. On the other hand, the transformation semigroup is maximal if it is full. And an equivalence class in the full transformation semigroup only depends on the number of states in the image; this is because we can apply arbitrary permutations. In particular, the number of equivalence classes in these two extreme cases is small.

There is a tight connection between deterministic automata and transformation semigroups. Roughly speaking, a transformation semigroup is an automaton without initial and finial states. The main difference is that for automata, one usually is interested in the syntactic semigroup rather than the transformation semigroup; the syntactic semigroup is the transformation semigroup of the minimal automaton. We show that the above bounds on the number of equivalence classes and heights also apply to syntactic semigroups.

Theorem 1. *For each $n \in \mathbb{N}$, there exists a minimal automaton $\mathcal{A}_n$ with n states over an alphabet of size 5 such that the number of $\mathcal{J}$-classes (resp. $\mathcal{R}$-classes, $\mathcal{L}$-classes) of the transformation semigroup $T(\mathcal{A}_n)$ is at least $(n-4)^{n-4}$.*

Theorem 2. *There exists a sequence of minimal automata $(\mathcal{A}_n)_{n \in \mathbb{N}}$ over a fixed alphabet such that $\mathcal{A}_n$ has n states and the $\mathcal{R}$-height (resp. $\mathcal{L}$-height, $\mathcal{J}$-height) of the transformation semigroup $T(\mathcal{A}_n)$ is in $\Omega(2^n/n^{9.5})$.*

2 Preliminaries

A *semigroup* is a set S equipped with an associative operation $\cdot\colon S \times S \to S$. A *subsemigroup* of S is a subset T such that $s_1 s_2 \in T$ for all $s_1, s_2 \in T$. It is called *completely isolated* if the converse implication holds, i.e., $s_1 s_2 \in T$ implies $s_1 \in T$ and $s_2 \in T$ for all $s_1, s_2 \in S$. The *opposite* semigroup of S is obtained by replacing the operation with its left-right dual $\circ\colon S \times S \to S$ defined by $x \circ y = y \cdot x$.

In general, Green's relations in a subsemigroup T of S do not coincide with the corresponding relations in S. However, if T is a completely isolated subsemigroup, the following property holds:

Proposition 3. *Let S be a semigroup and let T be a completely isolated subsemigroup of S. Let $\mathcal{K}$ be one of the relations $\leqslant_\mathcal{R}$, $\leqslant_\mathcal{L}$, $\leqslant_\mathcal{J}$, $\mathcal{R}$, $\mathcal{L}$ or $\mathcal{J}$. Then, for all $x, y \in T$, we have $x \mathcal{K} y$ in S if and only if $x \mathcal{K} y$ in T.*

Proof. We will only prove the statement for the preorder $\leqslant_\mathcal{R}$. For the implication from right to left, we have $xS^1 \subseteq xT^1S^1 \subseteq yT^1S^1 \subseteq yS^1$. For the converse implication, suppose that $xS^1 \subseteq yS^1$, i.e., there exists some $z \in S^1$ such that $yz = x$. Since T is completely isolated, we have $z \in T^1$, which yields $zT^1 \subseteq T^1$ and thus, $xT^1 = yzT^1 \subseteq yT^1$.   □

An *$\mathcal{R}$-chain* is a sequence $(s_1, \ldots, s_\ell)$ of elements of S such that $s_{i+1} <_\mathcal{R} s_i$ for all $i \in \{1, \ldots, \ell-1\}$; ℓ is called the *length* of the $\mathcal{R}$-chain. The maximal length of an $\mathcal{R}$-chain of S is called the *$\mathcal{R}$-height* of S. The notions *$\mathcal{L}$-chains*, *$\mathcal{J}$-chains*, *$\mathcal{L}$-height* and *$\mathcal{J}$-height* are defined analogously.

A *partial transformation* on a set Q is a partial function $f\colon Q \to Q$. If the domain of f is all of Q, i.e., if f is a total function, f is called a *transformation*. A partial transformation $f\colon Q \to Q$ is called *injective* if $f(p) \neq f(q)$ whenever $p \neq q$ and both $f(p)$ and $f(q)$ are defined. The elements of Q are often called *states*. In the following, we use the notation $q \cdot f$ instead of $f(q)$ to denote the image of an element $q \in Q$ under f. For $R \subseteq Q$ let $R \cdot f = \{q \cdot f \mid q \in R\}$. Note that for all subsets $R \subseteq Q$ and all partial transformations $f\colon Q \to Q$, the inequality $|R \cdot f| \leqslant |R|$ holds; we will implicitly use this property throughout the paper. The *composition* fg of two transformations $f\colon Q \to Q$ and $g\colon Q \to Q$ is defined by $q \cdot fg = (q \cdot f) \cdot g$. The composition is associative.

The set of all partial transformations (resp. transformations) on a fixed set Q forms a semigroup with composition as the binary operation. It is called the *full partial transformation semigroup* (resp. *full transformation semigroup*) on Q. Subsemigroups of full (partial) transformation semigroups are called *(partial) transformation semigroups* and are often specified in terms of generators. Partial transformation semigroups and transformation semigroups are strongly related. On one side, every transformation semigroup also is a partial transformation semigroup. In the other direction a slightly weaker statement holds:

Proposition 4. *Let P be a partial transformation semigroup on n states. Then there exists a transformation semigroup on $n+1$ states which is isomorphic to P.*

A partial transformation semigroup is called *injective* if it is generated by a set of injective partial transformations. An important property of injective partial transformation semigroups is that they have a left-right dual:

Proposition 5. *The opposite semigroup of an injective partial transformation semigroup is a partial transformation semigroup.*

Transformation semigroups naturally arise when considering deterministic finite automata. Let $\mathcal{A} = (Q, \Sigma, \delta, q_0, F)$ be a deterministic finite automaton. Then, each letter $a \in \Sigma$ can be interpreted as a transformation $a \colon Q \to Q$ where $q \cdot a = \delta(q, a)$. The transformation semigroup on Q generated by all letters in Σ is denoted by $T(\mathcal{A})$ and it is called the *transition semigroup* of $\mathcal{A}$. Conversely, given a transformation semigroup T on a finite set Q and a finite set of generators Σ, for each $q_0 \in Q$ and $F \subseteq Q$, one can define a deterministic finite automaton $\mathcal{A} = (Q, \Sigma, \delta, q_0, F)$ where $\delta \colon Q \times \Sigma \to Q$ is defined as $\delta(q, a) = q \cdot a$.

A well-known approach for translating bounds on the size of a transformation semigroup to syntactic monoids is to *make* an automaton minimal. This can be done by introducing a new generator c with $q_i \cdot c = q_{i+1}$ for $Q = \{q_1, \ldots, q_n\}$ and $q_{n+1} = q_1$; moreover, one chooses some arbitrary state to be both initial and final. We adapt this construction to also work with Green's relations.

Proposition 6. *Let T be a transformation semigroup on n states, generated by a finite set Σ. Then there exists a minimal $(n + 1)$-state deterministic finite automaton $\mathcal{A}$ over an alphabet of size $|\Sigma| + 1$ such that T is a completely isolated subsemigroup of $T(\mathcal{A})$.*

Proof. Let T be a transformation semigroup on a set of states $Q = \{q_1, \ldots, q_n\}$, generated by Σ. Let $\mathcal{A} = (Q \cup \{q_0\}, \Sigma \cup \{c\}, \delta, q_0, \{q_n\})$ be the automaton defined by $\delta(q_0, a) = q_0$ and $\delta(q_i, a) = q_i \cdot a$ for $i \geqslant 1$ and all $a \in \Sigma$. The transitions for the letter c are defined by $\delta(q_i, c) = q_{i+1}$ for $i < n$ and $\delta(q_n, c) = q_1$. This automaton is minimal: for two different states $q_i, q_j \in Q \cup \{q_0\}$ with $i > j$, we have $\delta(q_i, c^{n-i}) = q_n$ but $\delta(q_j, c^{n-i}) \neq q_n$.

By construction, T is a subsemigroup of $T(\mathcal{A})$. To see that T is completely isolated within $T(\mathcal{A})$, note that we have $\delta(q_0, u) = q_0$ if and only if $u \in \Sigma^*$. $\quad\square$

3   Bounds for the Number of Classes

Let $\mathcal{K}$ be any of the relations $\mathcal{R}$, $\mathcal{L}$ or $\mathcal{J}$. The naïve upper bound for the number of $\mathcal{K}$-classes of a transformation semigroup T on n states is given by the size of T itself. Since there are n^n different functions from Q to Q, the semigroup T contains at most n^n elements. It is well known that this bound is tight even for a constant number of generators, since for each $n \geqslant 1$ there exists a transformation semigroup of size n^n generated by a set Σ with three elements; see e.g. [7].

As each $\mathcal{R}$-class (resp. $\mathcal{L}$-class, $\mathcal{J}$-class) consists of at least one element, the number of such classes is also bounded by n^n. We now show that this upper bound is tight up to a constant factor.

Proposition 7. *Let T be a transformation semigroup on n states, generated by a finite set Σ. Then there exists a transformation semigroup on $n + 3$ states which is generated by $|\Sigma| + 1$ elements and has at least $|T|$ different $\mathcal{J}$-classes.*

Proof. Let T be a semigroup of transformations on a set of states Q, generated by a finite set Σ, and let q_0 be an arbitrary element from Q. Let q_1, q_2, q_3 be new states not in Q and let c be a new generator not in Σ. Let U be the transformation semigroup on $Q \cup \{q_1, q_2, q_3\}$ obtained by extending the transformations of T as follows: for each $a \in \Sigma$ and $q \in Q$, let $q \cdot c = q$, $q_1 \cdot a = q_3 \cdot a = q_3 \cdot c = q_0$, $q_1 \cdot c = q_2 \cdot a = q_2$, and $q_2 \cdot c = q_3$.

Let $u, v \in \Sigma^*$ be different elements of T. Then cuc and cvc are different in U. We claim that $cuc \not\leqslant_{\mathcal{J}} cvc$ in U. For the sake of contradiction, suppose that there exist $x, y \in (\Sigma \cup \{c\})^*$ such that $cuc = xcvcy$ in U. Clearly, $q_1 \cdot cuc = q_3 \notin Q$. Moreover, at least one of the words x or y must be non-empty and therefore $q_1 \cdot xcucy \in Q$. This shows that $cuc \neq xcvcy$, as desired. $\qquad\square$

Combining the result with statements from the previous section, we obtain a lower bound for the number of $\mathcal{J}$-classes of the transition semigroup of an automaton.

Proof (Theorem 1). As we mentioned before, it is well known that there exists a 3-generator transformation semigroup on n states of size n^n. If we first apply Proposition 7 and then Proposition 6 to T, we obtain the claim by Proposition 3. The statement extends to $\mathcal{R}$-classes (resp. $\mathcal{L}$-classes) because each $\mathcal{J}$-class contains at least one $\mathcal{R}$-class (resp. $\mathcal{L}$-class). $\qquad\square$

4   Bounds for the Length of Chains

Let $\mathcal{K}$ be any of the relations $\mathcal{R}$, $\mathcal{L}$ or $\mathcal{J}$. As with the number of $\mathcal{K}$-classes, the naïve upper bound for the length of $\mathcal{K}$-chains is given by the maximal size n^n of the transformation semigroup on n states. In this section, we improve this upper bound for $\mathcal{R}$-chains and later give a lower bound that matches up to a polynomial gap.

Lemma 8. *Let P be a partial transformation semigroup on a finite set Q of cardinality n. Let $x, y \in P$ such that $Q \cdot x = Q \cdot xy$. Then $x \mathrel{\mathcal{R}} xy$.*

Proof. Let $\omega = n!$ and let $z = y^{\omega-1}$. It suffices to show that $xyz = x$ in P, i.e., for all $q \in Q$, we have $q \cdot x = q \cdot xyz$. By assumption, the restriction of y to the set $Q \cdot x$ is bijective. Thus, the mapping y^ω acts as identity on $Q \cdot x$. This yields $q \cdot xyz = q \cdot xy^\omega = (q \cdot x) \cdot y^\omega = q \cdot x$. $\qquad\square$

Proposition 9. *Let P be a partial transformation semigroup on n states. Then the $\mathcal{R}$-height of P is at most 2^n.*

Proof. Let P be a partial transformation semigroup on a set of states Q with $|Q| = n$. Let $(u_1, u_2, \ldots, u_\ell)$ be an $\mathcal{R}$-chain of P. We show that all sets $Q \cdot u_i$ must be pairwise distinct which yields the desired bound. Suppose that $Q \cdot u_i = Q \cdot u_j$ for $1 \leqslant i < j \leqslant \ell$. Since $u_j <_{\mathcal{R}} u_i$, there exists $v \in P$ with $u_i v = u_j$. Lemma 8 yields $u_j \mathrel{\mathcal{R}} u_i$ which implies $u_{i+1} \mathrel{\mathcal{R}} u_i$, a contradiction. $\qquad\square$

4.1 Token Computations in Transformation Semigroups

In this subsection, we introduce the building blocks for the lower bound on the height. A *token machine* is a pair (C, I) where C is a finite set and I is a set of partial transformations on C. The elements of the set C are called *cells*, subsets of C are called *configurations* and the generators I are called *instructions*.

A *program* is a finite word over the alphabet I and a *computation* is a sequence

$$R_0 \xrightarrow{\iota_1} R_1 \xrightarrow{\iota_2} R_2 \cdots \xrightarrow{\iota_\ell} R_\ell$$

where all $R_i \subseteq C$ have the same cardinality and $R_{i-1} \cdot \iota_i = R_i$. The configuration R_0 is called *initial configuration* and R_ℓ is called the *final configuration* of the computation. The program $\iota_1 \iota_2 \cdots \iota_\ell$ is the *label* of the computation and ℓ is its *length*. It is *progressing* if all configurations appearing in the computation are pairwise distinct and for each $i \in \{1, \ldots, \ell\}$ and each $\iota \in I \setminus \{\iota_i\}$, we have $|R_{i-1} \cdot \iota| < |R_i|$. It is *maximal* if $|R_\ell \cdot \iota| < |R_\ell|$ for all $\iota \in I$.

A language over programs $L \subseteq I^*$ is called *deterministic* on a configuration $R \subseteq C$ if $|R \cdot u_1| = |R| = |R \cdot u_2|$ implies $u_1 = u_2$ for all $u_1, u_2 \in L$.

The focal idea of token machines is captured in the following proposition which states that computations in token machines naturally yield lower bounds for the length of $\mathcal{R}$-chains.

Proposition 10. *Let (C, I) be a token machine and let P be the partial transformation semigroup on C generated by I. If there exists a maximal progressing computation of length ℓ, then the $\mathcal{R}$-height of P is at least ℓ.*

Proof. Let $R_0 \xrightarrow{\iota_1} R_1 \xrightarrow{\iota_2} R_2 \cdots \xrightarrow{\iota_\ell} R_\ell$ be a maximal progressing computation. For each $i \in \{1, \ldots, \ell\}$, we let $u_i = \iota_1 \iota_2 \cdots \iota_i$. It remains to show that $(u_1, \ldots, u_\ell)$ is an $\mathcal{R}$-chain. By definition, we immediately obtain $u_{i+1} \leqslant_\mathcal{R} u_i$. Assume, for the sake of contradiction, that $u_i \leqslant_\mathcal{R} u_{i+1}$ for some $i \in \{1, \ldots, \ell - 1\}$, i.e., there exists $v \in I^*$ with $u_i = u_{i+1} v$. Without loss of generality, we may assume that i is maximal with this property. If $|v| = 0$, then $R_i = R_0 \cdot u_i = R_0 \cdot u_{i+1} = R_{i+1}$, contradicting the premise of progression. Thus, $|v| \geqslant 1$ and since the computation is progressing and maximal, we have $i < \ell - 1$ and $v = \iota_{i+2} w$ for some $w \in I^*$. This yields $u_{i+2} w \iota_{i+1} = u_{i+1} \iota_{i+2} w \iota_{i+1} = u_{i+1} v \iota_{i+1} = u_i \iota_{i+1} = u_{i+1}$, contradicting the maximality of i.   □

4.2 Lower Bounds over a Growing Instruction Set

Before describing the technical ingredients required in our main result, we prove a slightly weaker statement. In contrast to the result presented later, it relies on an alphabet that grows exponentially with the number of elements.

Theorem 11. *For all even $n \in \mathbb{N}$, there exists a token machine with n cells which admits a maximal progressing computation of length at least $\binom{n}{n/2} - 1$.*

Proof. Let $C = \{1, 2, \ldots, n\}$. Let $\ell = \binom{n}{n/2} - 1$ and let $\{R_0, R_1, \ldots, R_\ell\}$ be the set of $n/2$-element subsets of C. For each $i \in \{1, \ldots, \ell\}$, let $\iota_i \colon R_{i-1} \to R_i$ be a bijection. Note that in the context of the present proof, it does not matter which of the $(n/2)!$ bijections is chosen; for example, one can always choose the unique bijection ι_i such that $\iota_i(j) < \iota_i(k)$ if and only if $j < k$. Each ι_i can be viewed as a partial transformation on C which is undefined for all $c \in C \backslash R_{i-1}$. We now show that in the token machine (C, I) with $I = \{\iota_i \mid 1 \leqslant i \leqslant \ell\}$, the sequence

$$R_0 \xrightarrow{\iota_1} R_1 \xrightarrow{\iota_2} R_2 \cdots \xrightarrow{\iota_\ell} R_\ell$$

is a maximal progressing computation. It is a valid computation by the definition of the instructions ι_i. Consider $i \in \{0, \ldots, \ell\}$ and $j \in \{1, 2, \ldots, \ell\} \backslash \{i + 1\}$. Since $R_{j-1} \neq R_i$, the instruction ι_j is undefined on at least one element of R_i and thus, $|R_i \cdot \iota_j| < |R_i|$. This shows that the computation is both progressing and maximal. $\qquad\square$

The theorem has a series of interesting consequences which will be outlined in Sect. 4.4, after proving an improved variant of the theorem with fixed alphabet.

4.3   Tapes and Binary Counters

A *sub-machine* of a token machine (C, I) is a subset $S \subseteq C$ such that for each configuration R and for each instruction $\iota \in I$ with $|R \cdot \iota| = |R|$, we also have $|(R \cap S) \cdot \iota| = |R \cap S|$. In other words, each computation stays a computation when restricted to S. The *union* of two token machines (C, I) and (C', I') with $C \cap C' = \emptyset$ is the token machine $(C \cup C', I \cup I')$ where the instructions in $I \backslash I'$ are extended to act as identity on C' and the instructions in $I' \backslash I$ are extended to act as identity on C. The cells C and C' of the original machines are sub-machines of the union.

An *n-bit tape* T is a token machine (C, I) with n cells and an arbitrary (but fixed) order $(c_0, c_1, \ldots, c_{n-1})$. One can interpret configurations $R \subseteq C$ as bit strings $b_{n-1} b_{n-2} \cdots b_0$ where $b_i = 1$ if and only if $c_i \in R$ and $b_i = 0$ otherwise, and think of T as a ring buffer with a read/write head at position 0. An instruction ι_{rotl}^T can be used to move the tape head to the right (or, actually, retain the head position but left-rotate the buffer). For each $i \in \{0, \ldots, n-2\}$, we let $c_i \cdot \iota_{\mathsf{rotl}}^T = c_{i+1}$ and $c_{n-1} \cdot \iota_{\mathsf{rotl}}^T = c_0$. The instruction ι_{rotr}^T is defined analogously and moves in the opposite direction. An instruction $\iota_{=0}^T$ can be used to check whether the head is scanning a zero and halt the program otherwise. It is undefined on c_0 and defined as the identity on $\{c_1, \ldots, c_{n-1}\}$. Conversely, the ι_{sync}^T instruction is defined as the identity on c_0 and undefined on every other cell. An instruction ι_{mvl}^T maps c_0 to c_1, acts as the identity on $\{c_2, c_3, \ldots, c_{n-1}\}$ and is undefined on c_1. Analogously, ι_{mvr}^T maps c_0 to c_{n-1}, acts as the identity on $\{c_1, c_2, \ldots, c_{n-2}\}$ and is undefined on c_{n-1}. The *value* of T under a configuration R is $\sum_{c_i \in R} 2^i$.

An *n-bit binary counter* N is constructed as follows. Three new *n*-bit tapes S, T and $\overline{T}$ are introduced. Their cells are $(d_0, d_1, \ldots, d_{n-1})$, $(c_0, c_1, \ldots, c_{n-1})$ and $(\overline{c}_0, \overline{c}_1, \ldots, \overline{c}_{n-1})$, respectively. Then, the union of S, T and $\overline{T}$ is constructed and the following instructions are added:

- $\iota_{\text{rotl}}^N = \iota_{\text{rotl}}^T \iota_{\text{rotl}}^{\overline{T}} \iota_{\text{rotl}}^S$, $-\ \iota_{=0}^N = \iota_{=0}^T$, $-\ \iota_{\text{sync}}^N = \iota_{\text{sync}}^S$,
- $\iota_{\text{rotr}}^N = \iota_{\text{rotr}}^T \iota_{\text{rotr}}^{\overline{T}} \iota_{\text{rotr}}^S$, $-\ \iota_{=1}^N = \iota_{=0}^{\overline{T}}$, $-\ \iota_{\text{off}}^N = \iota_{=0}^S$,
- ι_{inc}^N with $\overline{c}_0 \cdot \iota_{\text{inc}}^N = c_0$ and $c_0 \cdot \iota_{\text{inc}}^N$ undefined and $c \cdot \iota_{\text{inc}}^N = c$ for all $c \notin \{c_0, \overline{c}_0\}$,
- ι_{dec}^N with $c_0 \cdot \iota_{\text{dec}}^N = \overline{c}_0$ and $\overline{c}_0 \cdot \iota_{\text{dec}}^N$ undefined and $c \cdot \iota_{\text{dec}}^N = c$ for all $c \notin \{c_0, \overline{c}_0\}$.

Following this, the original instructions of S, T and $\overline{T}$ are removed from I. Thus, a binary counter provides exactly eight instructions. A configuration R of N is *valid* if $|R \cap S| = 1$ and for each $i \in \{0, \dots, n-1\}$, we have $c_i \in R$ if and only if $\overline{c}_i \notin R$.

Lemma 12. *Let R be a valid configuration of a binary counter N and let $u \in I^*$ such that $|R \cdot u| = |R|$. Then $R \cdot u$ is a valid configuration of N.*

Proof. By induction on the length of u, it suffices to prove that the action of instructions on R preserves validity.

The instructions ι_{rotl}^N and ι_{rotr}^N cyclically rotate the tapes T, $\overline{T}$ and S. Thus, if R is valid, then $R \cdot \iota_{\text{rotl}}^N$ and $R \cdot \iota_{\text{rotr}}^N$ are valid as well.

For each $\iota \in \{\iota_{=0}^N, \iota_{=1}^N, \iota_{\text{sync}}^N, \iota_{\text{off}}^N\}$, we have either $|R \cdot \iota| < |R|$ or $R \cdot \iota = R$.

If $\left| R \cdot \iota_{\text{inc}}^N \right| = |R|$, then R does not contain c_0. If, moreover, R is a valid configuration, it contains $\overline{c}_0$. But then, $R \cdot \iota_{\text{inc}}^N$ contains c_0 and does not contain $\overline{c}_0$. It coincides with R on all other cells. Thus, $R \cdot \iota_{\text{inc}}^N$ is valid as well. By a symmetric argument, the instruction ι_{dec}^N preserves validity. $\qquad\square$

We now define three regular languages

$$L_{\text{reset}}^N = \iota_{\text{sync}}^N ((\iota_{=0}^N \mid \iota_{\text{dec}}^N) \iota_{\text{rotr}}^N \iota_{\text{off}}^N)^* (\iota_{=0}^N \mid \iota_{\text{dec}}^N) \iota_{\text{rotr}}^N \iota_{\text{sync}}^N,$$

$$L_{\text{inc}}^N = \iota_{\text{sync}}^N (\iota_{\text{dec}}^N \iota_{\text{rotr}}^N \iota_{\text{off}}^N)^* \iota_{\text{inc}}^N (\iota_{\text{off}}^N \iota_{\text{rotr}}^N)^* \iota_{\text{sync}}^N \text{ and}$$

$$L_{\text{dec}}^N = \iota_{\text{sync}}^N (\iota_{\text{inc}}^N \iota_{\text{rotr}}^N \iota_{\text{off}}^N)^* \iota_{\text{dec}}^N (\iota_{\text{off}}^N \iota_{\text{rotr}}^N)^* \iota_{\text{sync}}^N.$$

Lemma 13. *The languages L_{reset}^N, L_{inc}^N and L_{dec}^N are deterministic on all valid configurations.*

Proof. Suppose there are two different words $u_1, u_2 \in L_{\text{reset}}^N$ and a valid configuration R such that $|R \cdot u_1| = |R|$. Since L_{reset}^N is prefix-free, there exist a unique word $p \in I^*$ and different instructions $\iota_1, \iota_2 \in I$ such that $u_1 \in p\iota_1 I^*$ and $u_2 \in p\iota_2 I^*$. A careful analysis of the structure of the regular expression for L_{reset}^N shows that either $\{\iota_1, \iota_2\} = \{\iota_{=0}^N, \iota_{\text{dec}}^N\}$ or $\{\iota_1, \iota_2\} = \{\iota_{\text{off}}^N, \iota_{\text{sync}}^N\}$.

In the first case, we may assume without loss of generality that $\iota_1 = \iota_{=0}^N$ and $\iota_2 = \iota_{\text{dec}}^N$. From $|R \cdot p\iota_1| = |R \cdot p|$, we deduce $c_0 \notin R \cdot p$ because $\iota_{=0}^N$ is undefined on c_0. This implies $\overline{c}_0 \in R \cdot p$ since $R \cdot p$ is a valid configuration by Lemma 12. Since ι_{dec}^N is undefined on $\overline{c}_0$, it follows that $|R \cdot u_2| \leqslant |R \cdot p\iota_2| < |R \cdot p| \leqslant |R|$.

In the second case, we may assume that $\iota_1 = \iota_{\text{off}}^N$ and $\iota_2 = \iota_{\text{sync}}^N$. Since $|R \cdot p\iota_1| = |R \cdot p|$ and since ι_{off}^N is undefined on d_0, we have $d_0 \notin R \cdot p$. This implies $d_i \in R \cdot p$ for some $i \in \{1, \dots, n-1\}$ because $R \cdot p$ is valid by Lemma 12. The instruction ι_{sync}^N is undefined on $\{d_1, d_2, \dots, d_{n-1}\}$ which yields $|R \cdot u_2| < |R|$, as above.

The proofs for L_{inc}^N and L_{dec}^N follow by a similar reasoning. $\qquad\square$

Let R be a configuration of N. We say that the counter is *synchronized* under R if $d_0 \in R$. The *value* of N under R is the value of T under $R \cap \{c_0, c_1, \ldots, c_{n-1}\}$.

In addition to the eight counter instructions defined above, for any fixed constant $k \in \{0, \ldots, 2^n - 1\}$ one can define an instruction $\iota^N_{\mathsf{val}=k}$ which asserts that the value of the counter equals k as follows. For each $i \in \{0, \ldots, n-1\}$ with $k \bmod 2^{i+1} \geqslant 2^i$, we let $c_i \cdot \iota^N_{\mathsf{val}=k} = c_i$ and let $\overline{c}_i \cdot \iota^N_{\mathsf{val}=k}$ be undefined. Symmetrically, we let $\overline{c}_i \cdot \iota^N_{\mathsf{val}=k} = \overline{c}_i$ and $c_i \cdot \iota^N_{\mathsf{val}=k}$ undefined if $k \bmod 2^{i+1} < 2^i$.

Lemma 14. *Let R be a valid configuration and let $u \in L^N_{\mathsf{reset}}$ such that $|R \cdot u| = |R|$. Then, under $R \cdot u$, the counter is synchronized and its value is zero.*

Proof. It is easy to see that each word $u \in L^N_{\mathsf{reset}}$ with $|R \cdot u| = |R|$ cyclically rotates the three tapes of N exactly n times and after each cyclic rotation, either $\iota^N_{=0}$ or ι^N_{dec} is applied. The codomains of both $\iota^N_{=0}$ and ι^N_{dec} do not contain c_0 and thus, we have $R \cdot u \cap \{c_0, c_1, \ldots, c_{n-1}\} = \emptyset$ which is equivalent to saying that the value under $R \cdot u$ is zero. To see that the counter is synchronized, note that applying ι^N_{sync} to a valid configuration preserves the number of elements if and only if the configuration is synchronized. $\square$

Lemma 15. *Let R be a valid configuration and let $u \in L^N_{\mathsf{inc}}$ such that $|R \cdot u| = |R|$. If v is the value of the counter under R and v' is its value under $R \cdot u$, we have $v' = v + 1 \leqslant 2^n - 1$.*

Proof. Let us first assume that $v < 2^n - 1$. Let $i \in \{0, \ldots, n-1\}$ be minimal such that $c_i \notin R$ and let

$$w = \iota^N_{\mathsf{sync}} (\iota^N_{\mathsf{dec}} \iota^N_{\mathsf{rotr}} \iota^N_{\mathsf{off}})^i \iota^N_{\mathsf{inc}} (\iota^N_{\mathsf{off}} \iota^N_{\mathsf{rotr}})^{n-i} \iota^N_{\mathsf{sync}}.$$

We claim that $u = w$. By Lemma 13, it suffices to show that $|R \cdot w| = |R|$. Let us first investigate the instructions operating on S. The word starts with an ι^N_{sync} instruction, each ι^N_{off} instruction is applied after R has been rotated cyclically 1 to $n-1$ times and the second ι^N_{sync} instruction is applied after exactly n cyclic rotations. We deduce $|R \cdot \iota^N_{\mathsf{sync}}| = |R|$ from $|R \cdot u| = |R|$, and thus, the counter is synchronized on both R and on the configuration reached before the last ι^N_{sync} instruction. Moreover, whenever a ι^N_{off} instruction is applied to a configuration R', we have $d_i \in R'$ for some $i \in \{1, \ldots, n-1\}$. Note that the case $v = 2^n - 1$ can be excluded since in order for the ι^N_{inc} instruction to preserve the number of elements in the configuration, it would have to be preceded by at least n $\iota^N_{\mathsf{rotr}} \iota^N_{\mathsf{off}}$-factors and one of those factors would reduce the number of elements.

The instruction ι^N_{dec} is applied exactly once before each of the first i cyclic rotations. Since $\{c_0, c_1, \ldots, c_{i-1}\} \subseteq R$, we have $c_0 \in R \cdot \iota^N_{\mathsf{sync}} (\iota^N_{\mathsf{dec}} \iota^N_{\mathsf{rotr}} \iota^N_{\mathsf{off}})^j$ for all $j \in \{0, \ldots, i-1\}$. Moreover, since $c_i \notin R$, we have $c_0 \notin R \cdot \iota^N_{\mathsf{sync}} (\iota^N_{\mathsf{dec}} \iota^N_{\mathsf{rotr}} \iota^N_{\mathsf{off}})^i$ which implies $\overline{c}_0 \in R \cdot \iota^N_{\mathsf{sync}} (\iota^N_{\mathsf{dec}} \iota^N_{\mathsf{rotr}} \iota^N_{\mathsf{off}})^i$ by Lemma 12. Consequently, the occurrences of ι^N_{dec} and ι^N_{inc} in w do not reduce the number of elements in the configuration. The above observations also show that

$$R \cdot u = R \cdot w = \{c_i\} \cup (R \cap \{c_{i+1}, c_{i+2}, \ldots, c_{n-1}\})$$

which is equivalent to the claim $v' = v + 1$. $\square$

For the ι_{dec}^N instruction, a symmetric version of the lemma holds.

Lemma 16. *Let R be a valid configuration and let $u \in L_{\mathrm{dec}}^N$ such that $|R \cdot u| = |R|$. If v is the value of the counter under R and v' is its value under $R \cdot u$, we have $v' = v - 1 \geqslant 0$.*

4.4   Main Result

Let $n \in \mathbb{N}$ be an even number. Let T be an n-bit tape with cells $(t_0, t_1, \ldots, t_{n-1})$. The union of T with three $\lceil \log_2 n \rceil$-bit counters P, Q and Z forms a token machine, henceforth referred to as U. A configuration of U is *valid* if it is valid when restricted to each of the three counters.

Informally, the idea of our construction is the following: as in the proof of Theorem 11, we enumerate all $n/2$-element subsets of an n-element set on the tape T. In order to do so with a constant number of generators, this enumeration needs to be done in a very specific way. We say that a word $Y \in \{0, 1\}^*$ is a *successor* of $X \in \{0, 1\}^*$ if there exist $p \in \{0, 1\}^*$, $i \geqslant 1$ and $j \geqslant 0$ such that $X = p01^i0^j$ and $Y = p10^{j+1}1^{i-1}$. For each $m \in \{0, 1, \ldots, n\}$ one can define a sequence of bit strings $(X_0, X_1, \ldots, X_\ell)$ as stated in the following lemma:

Lemma 17. *For all $n \in \mathbb{N}$ and $m \in \{0, 1, \ldots, n\}$, there exists a unique sequence $(X_0, X_1, \ldots, X_\ell)$ such that*

- *$X_0 = 0^{n-m}1^m$,*
- *for each $k \in \{1, \ldots, \ell\}$, X_k is a successor of X_{k-1} and*
- *X_ℓ does not have a successor.*

The terms of this sequence are pairwise distinct, each term contains exactly m occurrences of the letter 1, and we have $\ell = \binom{n}{m}$ as well as $X_\ell = 1^m0^{n-m}$.

Proof. First observe that if a word $X \subset \{0, 1\}^*$ can be factorized as $X = p01^i0^j$ with $p \in \{0, 1\}^*$ and $i \geqslant 1$ and $j \geqslant 0$, then this factorization is unique. As a consequence, the sequence defined above is unique and its terms are pairwise distinct. It is also easy to see that if Y is a successor of X, then X and Y contain the same number of 1's. The remaining two properties $\ell = \binom{n}{m}$ and $X_\ell = 1^m0^{m-n}$ clearly hold if $n = 0$ or $m \in \{0, n\}$.

We now assume $n \geqslant 1$, as well as $m \in \{1, \ldots, n-1\}$, and proceed by induction on n. Let $s \in \{0, \ldots, n\}$ such that $X_0, X_1, \ldots, X_s \in 0\{0, 1\}^{n-1}$ and $X_{s+1}, X_{s+2}, \ldots, X_\ell \in 1\{0, 1\}^{n-1}$. Applying the induction hypothesis to the suffixes of length $n - 1$ of $X_0, X_1, \ldots, X_s$, we know that $s = \binom{n-1}{m}$ and $X_s = 01^m0^{(n-1)-m}$. This yields $X_{s+1} = 10^{n-m}1^{m-1}$ and by applying induction again to the suffixes of $X_{s+1}, X_{s+2}, \ldots, X_\ell$, we obtain $\ell - s = \binom{n-1}{m-1}$ as well as $X_\ell = 11^{m-1}0^{(n-1)-(m-1)} = 1^m0^{n-m}$. Note that by Pascal's rule, $\ell = \ell - s + s = \binom{n-1}{m-1} + \binom{n-1}{m} = \binom{n}{m}$ which concludes the proof. $\square$

Note that the sequence corresponds to binary counting and deleting all counter values not having m bits equal 1. Since we are interested in enumerating $n/2$-element subsets, we only consider the case $m = n/2$. Interpreting the bit strings X_k as $n/2$-element subsets of an n-element set, the sequence $(X_0, X_1, \ldots, X_\ell)$ describes our enumeration order. Thus, all configurations appearing in the computation always contain $n/2$ elements when restricted to T. The counter P keeps track of the position of the head on T. It is needed for moving a block of 1-bits as far to the right as possible when transitioning from X_{k-1} to X_k. The volatile counters Q and Z are only used by the following macro that checks whether the bit below the tape head of T is 1.

$$L_{=1} = \iota_{\text{rotr}}^T((\varepsilon \mid \iota_{=0}^T L_{\text{inc}}^Z)\iota_{\text{rotr}}^T L_{\text{inc}}^Q)^* \iota_{\text{val}=n-1}^Q \iota_{\text{val}=n/2}^Z L_{\text{reset}}^Q L_{\text{reset}}^Z.$$

Roughly speaking, a word from $L_{=1}$, which preserves the cardinality of the configuration, rotates the tape T cyclically n times. The counter Q is used to ensure that neither more nor less rotations are performed. After each rotation, except for the last one, the counter Z is increased non-deterministically if the bit under the tape head is 0. Then, the value of Z is checked to be exactly $n/2$. Since we know that the number of 0-bits on T is $n/2$ and since the bit under the tape head cannot contribute to the value of Z, this is only possible if the bit under the tape head is set. More precisely, the following lemma holds.

Lemma 18. *Let R be a valid configuration such that $|R \cap T| = n/2$, the counters P and Q are synchronized and the values of P and Q are zero. Then there exists a word $u \in L_{=1}$ with $|R \cdot u| = |R|$ if and only if $t_0 \in R$. Moreover, if such a word u exists, it is unique and we have $R \cdot u = R$.*

Proof. For $i \in \{0, 1, \ldots, n-1\}$, let $m_i = 1$ if $t_i \notin R$ and let $m_i = 0$ otherwise.

By Lemma 15, the $\iota_{\text{val}=n-1}^Q$ instruction in a word $w \in L_{=1}$ preserves the number of elements in a valid configuration if and only if w contains exactly $n - 1$ occurrences of L_{inc}^Q. Therefore, each word that preserves the number of elements when applied to R contains the instruction ι_{rotr}^N exactly n times. Since each occurrence of L_{inc}^Z is paired with a $\iota_{=0}^T$ instruction, L_{inc}^Z is applied at most m_i times after the i-th rotation, i.e., every program that does not reduce the number of elements when applied to R has the form

$$\iota_{\text{rotr}}^T \prod_{i=1}^{n-1}((\iota_{=0}^T L_{\text{inc}}^Z)^{k_i} \iota_{\text{rotr}}^T L_{\text{inc}}^Q)\iota_{\text{val}=n-1}^Q \iota_{\text{val}=n/2}^Z L_{\text{reset}}^Q L_{\text{reset}}^Z$$

for some $k_i \in \{0, 1\}$ with $k_i \leqslant m_i$. Moreover, the $\iota_{\text{val}=n/2}^Z$ instruction preserves the cardinality of the configuration if and only if the sum of all k_i with $1 \leqslant i \leqslant n-1$ equals $n/2$. Therefore, any choice of values k_i must also satisfy

$$n/2 = \sum_{i=1}^{n-1} k_i \leqslant \sum_{i=1}^{n-1} m_i = n/2 - m_0$$

where the last equality follows from the assumption that $|R \cap T| = n/2$. This is only possible if $m_0 = 0$, i.e., $t_0 \in R$, and $k_i = m_i$ for all $i \in \{1, 2, \dots, n-1\}$. By letting $k_i = m_i$ in the program above, we obtain the unique word u such that $|R \cdot u| = |R|$. To see that $R \cdot u = R$, note that after n cyclic rotations, the tape T returns to its original state. Moreover, by Lemma 14, both Q and Z are synchronized and have value zero.   □

We also let $L_{\mathsf{rotl}} = L_{\mathsf{dec}}^P \iota_{\mathsf{rotl}}^T$ and $L_{\mathsf{rotr}} = L_{\mathsf{inc}}^P \iota_{\mathsf{rotr}}^T$. The language L is now defined as $L = L_{\mathsf{reset}}^P L_{\mathsf{reset}}^Q L_{\mathsf{reset}}^Z (L_{=1} L_{\mathsf{rotr}})^* \iota_{=0}^T L_{\mathsf{rotl}} (\iota_{\mathsf{mvl}}^T (L_1 \mid L_2 \mid L_3))^* \iota_{\mathsf{val}=n-1}^P$ with

$$L_1 = (\iota_{\mathsf{val}=0}^P \mid L_{\mathsf{rotl}} \iota_{=0}^T L_{\mathsf{rotr}}) L_{\mathsf{rotr}} (L_{=1} L_{\mathsf{rotr}})^* \iota_{=0}^T L_{\mathsf{rotl}},$$

$$L_2 = (L_{\mathsf{rotl}} L_{=1})^+ \iota_{\mathsf{val}=0}^P (L_{=1} L_{\mathsf{rotr}})^+ \iota_{=0}^T L_{\mathsf{rotl}},$$

$$L_3 = (L_{\mathsf{rotl}} L_{=1})^+ L_{\mathsf{rotl}} \iota_{=0}^T L_{\mathsf{rotr}} L_{\mathsf{rotr}} (K_1 \mid K_2 K_3^* K_4),$$

$$K_1 = \iota_{=0}^T L_{\mathsf{rotl}} (\iota_{\mathsf{mvr}}^T L_{\mathsf{rotl}})^* \iota_{\mathsf{val}=0}^P,$$

$$K_2 = L_{=1} L_{\mathsf{rotl}} (\iota_{\mathsf{mvr}}^T L_{\mathsf{rotl}})^* \iota_{\mathsf{val}=0}^P L_{\mathsf{rotr}} (\iota_{=0}^T L_{\mathsf{rotr}})^* L_{=1} L_{\mathsf{rotr}},$$

$$K_3 = L_{=1} L_{\mathsf{rotl}} (\iota_{\mathsf{mvr}}^T L_{\mathsf{rotl}})^* L_{\mathsf{rotl}} L_{=1} L_{\mathsf{rotr}} L_{\mathsf{rotr}} (\iota_{=0}^T L_{\mathsf{rotr}})^* L_{=1} L_{\mathsf{rotr}},$$

$$K_4 = \iota_{=0}^T L_{\mathsf{rotl}} (\iota_{\mathsf{mvr}}^T L_{\mathsf{rotl}})^* L_{\mathsf{rotl}} L_{=1} L_{\mathsf{rotr}}.$$

The following lemma is the technical main ingredient for Theorem 20.

Lemma 19. *There exists a valid initial configuration R such that L is deterministic on R. Moreover, there exists a word $u \in L$ of length at least $\binom{n}{n/2}$ such that $|R \cdot u| = |R|$.*

A proof of the lemma can be found in the full version of this paper [5]. Here, we only give a sketch of the arguments. To show that L is deterministic, one can use case distinctions similar to those in the proof of Lemma 13. It then suffices to prove the existence of a word which enumerates the subsets corresponding to the sequence $(X_0, X_1, \dots, X_\ell)$ defined above. An important invariant is that after each application of a factor from $\iota_{\mathsf{mvr}}^T (L_1 \mid L_2 \mid L_3)$, the tape head points at the leftmost bit of the rightmost 1-block of T. Each such factor replaces the subset corresponding to X_{k-1} by the subset corresponding to X_k on T.

The last missing piece is a component that imposes the language L on the labels of valid computations. To this end, let $\mathcal{A} = (Q, I, \delta, q_0, F)$ be the minimal deterministic automaton of L. We remove the sink state from Q and let all transitions leading to that state be undefined instead. Then, as long as there exists a state which has two ingoing transitions labeled by the same letter, we create a copy of the state and redirect one of the transitions to the copy. When interpreting the letters of I as actions on Q, the tuple (Q, I) then forms a token machine which we call *control unit*. By construction, all instructions are injective.

Theorem 20. *For all $n \in \mathbb{N}$, there exists a token machine with $n + 9 \lceil \log n \rceil + \mathcal{O}(1)$ cells and 32 instructions which admits a maximal progressing computation of length at least $\binom{n}{\lfloor n/2 \rfloor}$.*

Proof. It suffices to prove the theorem for even numbers n. Let V be the union of U and the control unit. Any word, which is not a prefix of a word in L, empties the configuration when applied to the initial configuration $\{q_0\}$ in the control unit. Thus, by taking the union of the initial configuration from Lemma 19 and $\{q_0\}$, we obtain a maximal progressing computation of the desired length in V.

The only instructions required in the construction are ι_{rotl}^T, ι_{rotr}^T, ι_{mvl}^T, ι_{mvr}^T, $\iota_{=0}^T$, $\iota_{\mathrm{val}=0}^P$, $\iota_{\mathrm{val}=n-1}^Q$, $\iota_{\mathrm{val}=n/2}^Z$ and eight additional instructions for each of the three binary counters. Since L is a fixed language, the control unit has c cells for a constant $c \in \mathbb{N}$ (independent of n), and U has $n + 9 \lceil \log n \rceil$ cells: n cells for the tape T and $\lceil \log n \rceil$ cells for each of the three tapes of the three binary counters. Therefore, the number of cells of V is $n + 9 \lceil \log n \rceil + c$. $\qquad\square$

Corollary 21. *There exists a sequence of transformation semigroups $(T_n)_{n \in \mathbb{N}}$ with a fixed number of generators such that T_n has n states and the $\mathcal{R}$-height (resp. $\mathcal{L}$-height, $\mathcal{J}$-height) of T_n is in $\Omega(2^n/n^{9.5})$.*

Proof. For the $\mathcal{R}$-height, the result is an immediate consequence of Theorem 20, Propositions 10 and 4. The statement also holds for $\mathcal{J}$-height because every $\mathcal{R}$-chain also is a $\mathcal{J}$-chain; see e.g. [8, Proposition 1.4]. An equivalent statement for the $\mathcal{L}$-height follows from Proposition 5 and the fact that all instructions used in the construction are injective. By Stirling's formula, we have $\binom{n}{n/2} \in \Omega(2^n/n^{0.5})$; see [4,9]. Thus, we obtain the desired bound. Note that the bound in Theorem 20 is for $n + 9 \lceil \log n \rceil + \mathcal{O}(1)$ cells and not just n cells. This yields the factor n^9 in the denominator. $\qquad\square$

We can now prove our second main result.

Proof (Theorem 2). In view of Proposition 6 and Proposition 3, the theorem immediately follows from Corollary 21. $\qquad\square$

Acknowledgments. We thank the anonymous referees for several useful suggestions which helped to improve the presentation of this paper.

References

1. Brandl, C., Simon, H.U.: Complexity analysis: transformation monoids of finite automata. In: Potapov, I. (ed.) DLT 2015. LNCS, vol. 9168, pp. 143–154. Springer, Cham (2015). doi:10.1007/978-3-319-21500-6_11
2. Carton, O., Michel, M.: Unambiguous Büchi automata. Theoret. Comput. Sci. **297**(1), 37–81 (2003)
3. Eilenberg, S.: Automata, Languages, and Machines, vol. B. Academic Press, New York (1976)
4. Feller, W.: An Introduction to Probability Theory and its Applications, vol. 1. New York, Wiley (1957)
5. Fleischer, L., Kufleitner, M.: Green's Relations in Finite Transformation Semigroups. CoRR, abs/1703.04941 (2017)

6. Ganardi, M., Hucke, D., Lohrey, M.: Querying regular languages over sliding windows. In: FSTTCS 2016, Proceedings, vol. 65. LIPIcs, pp. 18:1–18:14. Dagstuhl Publishing (2016)
7. Holzer, M., König, B.: On deterministic finite automata and syntactic monoid size. Theoret. Comput. Sci. **327**(3), 319–347 (2004)
8. Pin, J.É.: Varieties of Formal Languages. North Oxford Academic, London (1986)
9. Robbins, H.: A remark on Stirling's formula. Am. Math. Monthly **62**, 26–28 (1955)

Nondeterministic Unitary OBDDs

Aida Gainutdinova[1](✉) and Abuzer Yakaryılmaz[2]

[1] Kazan Federal University, Kazan, Russia
aida.ksu@gmail.com
[2] Faculty of Computing, University of Latvia, Rīga, Latvia
abuzer@lu.lv

Abstract. We investigate the width complexity of nondeterministic unitary OBDDs (NUOBDDs). Firstly, we present a generic lower bound on their widths based on the size of strong 1-fooling sets. Then, we present classically "cheap" functions that are "expensive" for NUOBDDs and vice versa by improving the previous gap. We also present a function for which neither classical nor unitary nondeterminism does help. Moreover, based on our results, we present a width hierarchy for NUOBDDs. Lastly, we provide the bounds on the widths of NUOBDDs for the basic Boolean operations negation, union, and intersection.

1 Introduction

Branching Programs (BPs) are one of the well known computational models, which are important not only theoretically but also practically, such as hardware verification, model checking and others [21]. The main complexity measures for BPs are the size of BPs – its number of nodes and length (time complexity). It is well–known that BPs of polynomial size are equivalent to non-uniform log-space Turing machines.

The important restricted variant of BPs is Ordered Binary Decision Diagrams (OBDDs), which are oblivious read-once branching programs [21]. Time complexity for OBDD is at most n (the length of the input), and so the natural complexity measure for OBDD is its width. Different variants of OBDDs such as deterministic, probabilistic, nondeterministic, and quantum have been considered (e.g. [3,5,6,15,18]) and they have been compared in term of their widths. For example, it was shown that randomized OBDDs can be exponentially more efficient than deterministic and nondeterministic OBDDs [6], and, quantum OBDDs can be exponentially more efficient than deterministic and stable probabilistic OBDD and this bound is tight [3]. In [18] some simple functions were presented such that unitary OBDDs (the known most restricted quantum

The arXiv number is 1612.07015.

A. Gainutdinova—Some parts of this work was done during Gainutdinova's visit to National Laboratory for Scientific Computing (Brazil) in June 2015 supported by CAPES with grant 88881.030338/2013-01.

A. Yakaryılmaz—Partially supported by CAPES with grant 88881.030338/2013-01 and ERC Advanced Grant MQC.

P. Weil (Ed.): CSR 2017, LNCS 10304, pp. 126–140, 2017.
DOI: 10.1007/978-3-319-58747-9_13

OBDD) need exponential size for computing these functions with bounded error, while deterministic OBDDs can represent these functions in linear size. Quantum and classical nondeterminism for OBDD models were considered in [5], where the superiority of quantum OBDDs over classical counterparts was shown. In particular, an explicit function was presented, which is computed by a quantum nondeterministic OBDD of constant width, but any classical nondeterministic OBDD for this function needs non-constant width.

The OBDDs of constant width can also be considered as a nonuniform analog of one-way finite automata [1]. It is well known that classical nondeterministic automata recognize precisely regular languages. There are different variants of nondeterministic quantum finite automata (NQFA) in literature [9,16,22]. Nakanishi et al. [16] considered quantum finite automata of Kondacs-Watrous type [12], which use measurement at each step of a computation. They showed that (unlike the case of classical finite automata) the class of languages recognizable by NQFAs properly contains the class of all regular languages. A full characterization of the class of languages recognized by all NQFA variants that are at least as general as the Kondacs-Watrous type was presented in [22]: they define the class of exclusive stochastic languages.

Bertoni and Carpentieri [9] considered a weaker model – nondeterministic quantum automata of Moore-Crutchfield type [14] with a single measurement at the end of a computation. They showed that the class of languages recognized by this model does not contain any finite nonempty language but contains a nonregular language.

In this paper we investigate nondeterministic quantum OBDDs where the model can evolve unitarily, followed by a projective measurement at the end. We call the model as nondeterministic unitary OBDD (NUOBDD). It can be seen as OBDD counterparts of unitary space bounded circuits [10] or Moore-Crutchfield (measure-once) quantum finite automata [8,14].

Section 2 presents the necessary background. We present our results in Sect. 3. We start by presenting a generic lower bound on the widths of NUOBDD based on the size of strong 1-fooling sets (Sect. 3.1). Then, we present (i) quantumly "cheap" but classical "expensive" functions by improving the previous gap (Sect. 3.2) and (ii) classically "cheap" functions that are "expensive" for NUOBDDs (Sect. 3.3). We also present a function for which neither classical nor unitary nondeterminism does help (Sect. 3.4). Moreover, based on our results, we present a width hierarchy for NUOBDDs (Sect. 3.5). Lastly, we provide the bounds on the widths of NUOBDDs for the basic Boolean operations negation, union, and intersection (Sect. 3.6). We close the paper by Sect. 4. Due to limited space, some proofs are omitted, which can be found in [11].

2   Preliminaries

We start with the definitions of the models. Then, we present some basic facts from linear algebra which will be used in the proofs.

2.1   Definitions

We use superscripts for enumerating vectors and strings, and, subscripts for enumerating the elements of vectors and strings. A d-state quantum system (QS) can be described by a d-dimensional Hilbert space $(\mathcal{H}^d)$ over the field of complex numbers with the norm $|| \cdot ||_2$. A pure (quantum) state of the QS is described by a column vector $|\psi\rangle \in \mathcal{H}^d$, whose length is one (unitary ket-vector), i.e. $\sqrt{\langle\psi|\psi\rangle} = 1$. As long as it is a closed system, the evolution of the QS is described by some unitary matrices U. In order to retrieve information from the system, we can apply a projective measurement (then the system is no longer closed). We refer the reader to [19] for more details on the finite dimensional QSs (see [17] for a complete reference on quantum computing).

Definition 1. *A branching program (BP) on the variable set $X = \{x_1, \ldots, x_n\}$ is a finite directed acyclic graph with one source node and sink nodes partitioned into two sets – Accept and Reject. Each non-sink node is labelled by a variable x_i and has two outgoing edges labelled 0 and 1, respectively.*

An input σ is *accepted* if and only if it induces a chain of transitions leading to a node in *Accept*, otherwise σ is rejected. A BP P computes a Boolean function $f : \{0,1\}^n \to \{0,1\}$ iff P accepts each $\sigma \in f^{-1}(1)$ and P rejects each $\sigma \in f^{-1}(0)$.

Definition 2. *A BP is* oblivious *if its nodes can be partitioned into levels $V_0, \ldots, V_\ell$ such that nodes in V_ℓ are sink nodes, nodes in each level V_j with $0 \le j < \ell$ have outgoing edges only to nodes in the next level V_{j+1}, and all nodes in the level V_j query the same bit $\sigma_{i_{j+1}}$ of the input. If on each computational path from the source node to a sink node each variable from X is tested at most once, then such BP is called read-once BP.*

In this paper, we investigate read-once oblivious BPs that are commonly called as *Ordered Binary Decision Diagrams* (OBDDs). Since the lengths of OBDDs are fixed, the main complexity measure for them is their widths, i.e. for OBDD P, $width(P) = \max_j |V_j|$. The width of OBDDs can be seen as the number of states of finite automata and so we can refer the widths also as the sizes of OBDDs.

A nondeterministic OBDD (NOBDD) can have the ability of making more than one outgoing transition for each tested input bit from each node and so the program can follow more than one computational path and if one of the paths ends with an accepting node, then the input is accepted. Otherwise (all computation paths end with some rejecting nodes), the input is rejected.

Quantum OBDDs (QOBDDs) are non-trivial generalizations of classical OBDDs [5] when using general quantum operators like superoperators [20]. Here we focus on a restricted version of QOBDDs that evolves only unitarily followed by a projective measurement at the end [2]: unitary OBDDs (UOBDDs).

Definition 3. *A UOBDD M_n, defined on the variable set $X = \{x_1, \ldots, x_n\}$, with width d (operating on $\mathcal{H}^d$) is a quadruple $M_n = (Q, |\psi^0\rangle, T, Q_{acc})$, where $Q = \{q_1, \ldots, q_d\}$ is the set of states such that the set $\{|q_1\rangle, \ldots, |q_d\rangle\}$ forms a* *basis for* $\mathcal{H}^d$, $|\psi^0\rangle \in \mathcal{H}^d$ *is the initial quantum state,* $Q_{acc} \subseteq Q$ *is the set of accepting states, and* $T = \{(i_j, U_j(0), U_j(1))\}_{j=1}^n$ *is a sequence of instructions such that* i_j *determines a variable* x_{i_j} *tested at the step* j, $U_j(0)$ *and* $U_j(1)$ *are unitary transformations defined over* $\mathcal{H}^d$.

For any given input $\sigma \in \{0,1\}^n$, the computation of M_n can be traced by a unitary vector, which is initially $|\psi^0\rangle$. At the j-th step $(j = 1, \ldots, n)$ the input bit x_{i_j} is tested and then the corresponding unitary operator is applied:

$$|\psi^j\rangle = U_j(\sigma_{i_j})|\psi^{j-1}\rangle,$$

where $|\psi^{j-1}\rangle$ and $|\psi^j\rangle$ represent the quantum states after the $(j-1)^{th}$ and j^{th} steps, respectively.

After all input bits are read, the following projective measurement is applied: $P = \{P_{acc}, P_{rej}\}$, where both P_{acc} and P_{rej} are diagonal 0–1 matrices such that $P_{acc}[j,j] = 1$ iff $q_j \in Q_{acc}$ and $P_{rej} = I - P_{acc}$. Here P_{acc} (P_{rej}) projects any quantum state into the subspace spanned by accepting (non-accepting/rejecting) basis states. Then, the accepting probability of M_n on σ is calculated from the final state vector $|\psi^n\rangle$ as follows: $Pr_{accept}^{M_n}(\sigma) = ||P_{acc}|\psi^n\rangle||^2$.

It is clear that M_n defines a probability distribution over the inputs from $\{0,1\}^n$. By picking some threshold between 0 and 1, we can classify the inputs as the ones accepted with probability greater than the threshold and the others. Picking threshold as 0 is a special case and also known as nondeterministic acceptance mode for probabilistic and quantum models [7,22].

Definition 4. *Nondeterministic UOBDD (NUOBDD) is a UOBDD, say N_n, that is restricted to compute the Boolean function f with threshold 0: each member of $f^{-1}(1)$ is accepted with non-zero probability by N_n and each member of $f^{-1}(0)$ is accepted with zero probability by N_n. Then it is said that f is computed by NUOBDD N_n.*

Definition 5. *A probabilistic OBDD (POBDD) P_n can be defined in the same way as UOBDD M_n with the following modifications: the initial state is a stochastic vector (v^0), each transformation is a stochastic matrix (the ones at the j-th levels are $A_j(0)$ and $A_j(1)$).*

The computation of P_n is traced by a stochastic vector: at the j-th step $(j = 1, \ldots, n)$ the input bit x_{i_j} is tested and then the corresponding stochastic operator is applied: $v^j = A_j(\sigma_{i_j})v^{j-1}$, where v^{j-1} and v^j represent the probabilistic states after the $(j-1)^{th}$ and j^{th} steps, respectively. Lastly, the accepting probability is calculated from the final vector as follows: $Pr_{accept}^{P_n}(\sigma) = \sum_{q_i \in Q_{acc}} v_i^n$. If the initial probabilistic state and each stochastic matrix in P_n is restricted to have only 0s and 1s, then all the computations become deterministic and so P_n is called a deterministic OBDD. If we do the same restriction to M_n, then we obtain again a deterministic OBDD but its computation must be reversible (0–1 unitary matrices are also known as permutation matrices) and so it is called a (classical) reversible OBDD (ROBDD). Similar to quantum nondeterminism, P_n

with threshold 0 forms an NOBDD. Besides a POBDD or UOBDD is called exact if it accepts any input with probability either 1 or 0. Then, the corresponding function is called to be computed exactly.

The classes $OBDD_n^d$, $NOBDD_n^d$, and $NUOBDD_n^d$ are formed by the Boolean functions defined on $\{0,1\}^n$ that can be respectively computed by OBDDs, NOBDDs, and NUOBDDs with width at most d.

2.2   Some Facts from Linear Algebra

Let V be a vector space over the field $\mathbb{C}$ of complex numbers with the norm $|| \cdot ||_2$. We denote zero element of V by $\mathbf{0}$. Here are the properties of norm:

1. $||\psi|| = 0 \Leftrightarrow \psi = \mathbf{0}$;
2. $\forall \psi, \phi \in V, ||\psi + \phi|| \leq ||\psi|| + ||\phi||$ (triangle inequality); and,
3. $\forall \alpha \in \mathbb{C}, \forall \psi \in V, ||\alpha \psi|| = |\alpha| \cdot ||\psi||$.

If Ψ is a set of linearly independent vectors and $\psi \notin \Psi$ can not be expressed as a linear combination of the vectors from Ψ, then the set $\Psi \cup \{\psi\}$ obtained by adding ψ to the set Ψ is linearly independent.

Lemma 6. *Let $\{\psi_1, \psi_2, \ldots, \psi_d\} \in V$ be a linearly independent set of vectors and U be a unitary transformation of the space V. Then, the set of vectors $\{U\psi_1, U\psi_2, \ldots, U\psi_d\}$ is linearly independent.*

Lemma 7. *Let $\psi_1, \ldots, \psi_m, \psi \in V$ be such that $\psi_1, \ldots, \psi_m$ are linearly independent and U be a linear map in V such that $||U|\psi_i\rangle|| = 0$ ($i = 1, \ldots, m$) and $||U|\psi\rangle|| > 0$. Then the set $\{\psi_1, \ldots, \psi_m, \psi\}$ is linearly independent.*

3   Our Results

We present our results under six subsections.

3.1   A Lower Bound for NUOBDDs

Let $f : \{0,1\}^n \to \{0,1\}$ be an arbitrary function and $\pi = (i_1, \ldots, i_n)$ be a permutation of $\{1, \ldots, n\}$. For a given $X = \{x_1, \ldots, x_n\}$, an integer k ($0 < k < n$) and a permutation π, X_k^π denotes $\{x_{i_1}, \ldots, x_{i_k}\}$. Any possible assignment on X_k^π, say $\sigma \in \{0,1\}^k$, is denoted by $\rho_{\pi,k}^\sigma : X_k^\pi \to \sigma$. Then $f|_{\rho_{\pi,k}^\sigma}$ is called a subfunction obtained from f by applying $\rho_{\pi,k}^\sigma$.

A set $S_k^\pi = \{(\sigma, \gamma) : \sigma \in \{0,1\}^k, \gamma \in \{0,1\}^{n-k}\}$ is called a *strong 1-fooling set* for f if

– $f|_{\rho_{\pi,k}^\sigma}(\gamma) = 1$ for each $(\sigma, \gamma) \in S_k^\pi$, and,
– $f|_{\rho_{\pi,k}^\sigma}(\gamma') = 0$ and $f|_{\rho_{\pi,k}^{\sigma'}}(\gamma) = 0$ for each $(\sigma, \gamma), (\sigma', \gamma') \in S_k^\pi$.

Let $\sigma, \sigma' \in \{0,1\}^k$. We say that the string $\gamma \in \{0,1\}^{n-k}$ distinguishes the string σ from the string σ', if $f|_{\rho^{\sigma}_{\pi,k}}(\gamma) > 0$ and $f|_{\rho^{\sigma'}_{\pi,k}}(\gamma) = 0$. Note that this definition is not symmetric.

Theorem 8. *Let NUOBDD (NOBDD) N_n computes a function $f : \{0,1\}^n \to \{0,1\}$ reading variables in an order $\pi = (i_1, \ldots, i_n)$. Then*

$$Width(N_n) \geq \max_k |S^{\pi}_k|.$$

Proof. Let $d = \max_k |S^{\pi}_k|$ and l be an index satisfying $|S^{\pi}_l| = d$, and $S^{\pi}_l = \{(\sigma^1, \gamma^1), \ldots, (\sigma^d, \gamma^d)\}$. Let N_n be an NUOBDD computing f. Consider the l-th level of N_n. Let $\Psi = \{|\psi(\sigma^j)\rangle \mid j = 1, \ldots, d\}$ be a set of state vectors of program N_n after processing inputs $\sigma^1, \ldots, \sigma^d$, i.e. $|\psi(\sigma^j)\rangle = U(\sigma^j)|\psi^0\rangle$.

Claim. The set Ψ is linearly independent.

Proof. Assume that Ψ is not linearly independent. Then there is a quantum state $|\psi\rangle = |\psi(\sigma^i)\rangle \in \Psi$ expressed as a linear combination of the others in Ψ:

$$|\psi(\sigma^i)\rangle = \sum_{\substack{j-1 \\ j \neq i}}^d \alpha_j |\psi(\sigma^j)\rangle,$$

and $\alpha_j \neq 0$ for some j.

Let γ^i be a string such that $(\sigma^i, \gamma^i) \in S^{\pi}_l$. Then, by definition, for every input σ^j $(j \neq i)$, we have $f|_{\rho^{\sigma^j}_{\pi,k}}(\gamma^i) = 0$, and program N_n accepts the inputs $\sigma^j \gamma^i$ with zero probability:

$$Pr^{N_n}_{accept}(\sigma^j \gamma^i) = ||P_{acc} U(\gamma^i)|\psi(\sigma^j)\rangle||^2 = 0.$$

That means $||P_{acc} U(\gamma^i)|\psi(\sigma^j)\rangle|| = 0$.

The final quantum state for the input $\sigma^i \gamma^i$ is

$$|\psi(\sigma^i \gamma^i)\rangle = U(\gamma^i)|\psi(\sigma^i)\rangle = U(\gamma^i) \sum_{\substack{j=1 \\ j \neq i}}^d \alpha_j |\psi(\sigma^j)\rangle$$

and by linearity we can follow that

$$|\psi(\sigma^i \gamma^i)\rangle = \sum_{\substack{j=1 \\ j \neq i}}^d \alpha_j U(\gamma^i)|\psi(\sigma^j)\rangle = \sum_{\substack{j=1 \\ j \neq i}}^d \alpha_j |\psi(\sigma^j \gamma^i)\rangle.$$

Then, the accepting probability of the input $\sigma^i \gamma^i$ can be calculated as

$$Pr^{N_n}_{accept}(\sigma^i \gamma^i) = ||P_{acc}\psi(\sigma^i \gamma^i)\rangle||^2 =$$

$$||\sum_{\substack{j=1 \\ j \neq i}}^k \alpha_j P_{acc}|\psi(\sigma^j \gamma^i)\rangle||^2 \leq (\sum_{\substack{j=1 \\ j \neq i}}^k |\alpha_j| \, ||P_{acc}\psi(\sigma^j \gamma^i)||)^2 = 0.$$

However, $f|_{\rho^{\sigma^i}_{\pi,k}}(\gamma^i) > 0$ and N_n must accept this input with nonzero probability. Since this is a contradiction, the set Ψ is linearly independent. ◄

Since the set Ψ of the state vectors of N_n at the l-th level is linearly independent and its size is d ($|\Psi| = d$), then the dimension of the space of states of N_n cannot be less than d: $Width(N_n) \geq d$.

The proof for the case when N_n is an NOBDD is the similar. The difference is that in this case we need to consider the norm $||\cdot||_1$ instead of $||\cdot||_2$. If N_n is an NOBDD computing f, then the acceptance probability is defined from the final vector v^n as $Pr^{N_n}_{accept}(v^n) = ||P_{acc}v^n||_1$. Recall that for a vector $v = (v_1, \ldots, v_d)$ $||v||_1 = \sum_{i=1}^{d} |v_i|$. □

3.2   Function notPerm

In [4] some functions were presented that are computed by NUOBDDs with constant width but NOBDDs need at least logarithmic width ($\Omega(\log n)$). Here, we present a Boolean function based on which we obtain a better bound.

Let $n = m^2$ for some $m > 0$. We define function $\texttt{notPERM}_n : \{0,1\}^n \to \{0,1\}$ as

$$\texttt{notPERM}_n(\sigma) = \begin{cases} 0, & \text{if } A(\sigma) \text{ is a permutation matrix,} \\ 1, & \text{otherwise,} \end{cases}$$

where the input bits are indexed as

$$x_{1,1}, \ldots, x_{1,m}, x_{2,1}, \ldots, x_{2,m}, \ldots, \ldots, x_{m,1}, \ldots, x_{m,m}$$

and $x_{i,j}$ is $\sigma_{i,j}$, the (i,j)-th entry of A. Note that A is a permutation matrix if and only if it contains exactly one 1 in every row and in every column.

The column and row summations of A can be represented by a $2m$ digit integer in base $(m+1)$: $T(A) = (c_m c_{m-1} \cdots c_1 r_m r_{m-1} \cdots r_1)$, where c_i and r_i are the summations of the entries in i-th column and j-th row, respectively, for $1 \leq i, j \leq m$. Then $T(A)$ can be a value between 0 and $T_{max} = (m+1)^{2m} - 1$, i.e. between $(0 \cdots 0)$ and $(m \cdots m)$. It can be easily verified that A is a permutation matrix if and only if $T(A) = (1 \cdots 1) = \sum_{i=0}^{2m-1}(m+1)^i = T_{perm}$.

Theorem 9. *Function* $\texttt{notPERM}_n$ *is computed by a width-2 NUOBDD* N_n.

Proof. The *NUOBDD* N_n has two states $\{q_1, q_2\}$, q_2 is the only accepting state, and N_n operates on $\mathbb{R}^2$. Let α be the angle of $\frac{\pi}{T_{max}}$. The initial state is

$$\cos(-T_{perm}\alpha)|q_1\rangle + \sin(-T_{perm}\alpha)|q_2\rangle,$$

the point on the unit circle away from $|q_1\rangle$ by angle $T_{perm}(A)\alpha$ in clockwise direction. After reading the input, N_n makes a counter clockwise rotation with angle $T(A)\alpha$, i.e., it rotates with angle $\alpha((m+1)^i + (m+1)^{m+j})$ if $x_{i,j} = 1$ and it applies the identity operator if $x_{i,j} = 0$.

If A is a permutation matrix, it makes a total rotation with angle $T_{perm}\alpha$ and so the final quantum state becomes $|q_1\rangle$. Thus, the input is accepted with zero probability.

If A is not a permutation matrix, then the amplitude of $|q_2\rangle$ in the final quantum state always takes a nonzero value and so the input is always accepted with nonzero probability. Note that N_n can make at most π degree rotation. □

It is known that Function PERM_n ($\neg\text{notPERM}_n$) cannot be computed efficiently by deterministic read-once BPs, where $\text{PERM}_n(\sigma) = 1$ iff $A(\sigma)$ is a permutation matrix. By using a known lower bound given for BPs, we can obtain the following.

Fact 1 [13]. *The size of any nondeterministic read-once BP, computing* PERM_n, *cannot be less than* $2^m/(2\sqrt{m})$, *where* $m = \sqrt{n}$.

Theorem 10. *The width of any NOBDD computing* notPERM_n *cannot be less than* $\sqrt{n} - \frac{5}{4}\log n - 1$.

Remark that any NOBDD can be simulated by a nondeterministic QOBDD with the same width if quantum model can use superoperators [4]. However, as shown here, NOBDDs and NUOBDDs with the same widths are incomparable under certain bounds.

3.3 Function EXACT

We present a classically "cheap" but *unitarily* "expensive" function: $\text{EXACT}_n^k : \{0,1\}^n \to \{0,1\}$:

$$\text{EXACT}_n^k(\sigma) = \begin{cases} 1, & \text{if } \#_1(\sigma) = k, \\ 0, & \text{otherwise,} \end{cases}$$

where $\#_1(\sigma)$ is a number of 1s in σ. If $k = n$, then we have the function $\text{AND}_n : \{0,1\}^n \to \{0,1\}$ that equals 1 iff the input does not contain any 0.

Theorem 11. *There exists a UOBDD* M_n *with width* $d = \max\{k+1, n-k+1\}$ *that computes* EXACT_n^k *exactly (and so nondeterministically).*

Theorem 12. *The width of any NUOBDD computing* EXACT_n^k *cannot be less than* $\max\{k+1, n-k+1\}$.

Proof. Let $N_n = (Q, |\psi^0\rangle, T, Q_{acc})$ be an NUOBDD that computes EXACT_n^k, $\pi = (i_1, \ldots, i_n)$ be an order of reading variables used by N_n, and $d = \max\{k, n-k\}$.

The computation begins from the initial configuration $|\psi^0\rangle$. The input is of the form $\sigma = \sigma_1 \cdots \sigma_n$. After the l-th step of the computation ($1 \leq l \leq n-1$), the variables $x_{i_1}, \ldots, x_{i_l}$ are read by N_n and the configuration is $|\psi^l(\sigma_{i_1} \cdots \sigma_{i_l})\rangle$. At the $(l+1)$-th step, N_n reads the next variable $x_{i_{l+1}} = \sigma_{i_{l+1}}$ and the new configuration becomes $|\psi^{l+1}(\sigma_{i_1} \cdots \sigma_{i_l}\sigma_{i_{l+1}})\rangle = U_{l+1}(\sigma_{i_{l+1}})|\psi^l(\sigma_{i_1} \cdots \sigma_{i_l})\rangle$. At the end of the computation, the projective measurement is applied to the resulting configuration $|\psi^n(\sigma_{i_1} \cdots \sigma_{i_n})\rangle$, and then, the probability of accepting the input is calculated as $Pr_{acc}^{N_n}(\sigma) = ||P_{acc}|\psi^n(\sigma_{i_1} \cdots \sigma_{i_n})\rangle||^2$.

The idea behind our proof is as follows. For each level l ($l = 0, \ldots, d$) of N_n, we consider the set of all possible quantum states and then focus on a maximal subset that is linearly independent. Then we can give a lower bound on the size of this subset.

Let $\Psi_l = \{|\psi^l(\sigma)\rangle : \sigma \in \{0,1\}^l\}$ be the set of all possible quantum states after the l-th step, i.e. $|\psi^l(\sigma)\rangle = U_l(\sigma_l) \cdots U_1(\sigma_1)|\psi^0\rangle$.

Lemma 13. *Let* $|\psi^1\rangle, \ldots, |\psi^m\rangle, |\psi\rangle \in \Psi_l$ *and* $|\psi^1\rangle, \ldots, |\psi^m\rangle$ *be linearly independent for some* $m \geq 1$, *where* $|\psi^i\rangle = |\psi^l(\sigma^i)\rangle$ *for* $i = 1, \ldots, m$ *and* $|\psi\rangle = |\psi^l(\sigma)\rangle$. *If there exists a string* $\gamma \in \{0,1\}^{n-l}$ *that distinguishes the string* σ *from each of the strings* $\sigma^1, \ldots, \sigma^m$, *then the set* $\{|\psi^1\rangle, \ldots, |\psi^m\rangle, |\psi\rangle\}$ *is linearly independent.*

Proof. Let $U = U_n(\gamma_{n-l}) \cdots U_{l+1}(\gamma_1)$. It is given that $||P_{acc}U|\psi^i\rangle|| = 0$ for each $i = 1, \ldots, m$, and $||P_{acc}U|\psi\rangle|| > 0$. Due to Lemma 7, we can follow that the set $\{|\psi^1\rangle, \ldots, |\psi^m\rangle, |\psi\rangle\}$ is linearly independent. ◄

Let Φ_l ($\Phi_l \subseteq \Psi_l$) be the maximal set of linearly independent vectors. We will estimate the cardinality of Φ_l by induction on l ($l = 0, \ldots, d$). We will consider two cases: when $k \geq n/2$ and when $k < n/2$.

Case 1. First we assume $k \geq n/2$ that is $d = k$.

Initial step: At the level $l = 0$, the set Ψ_0 consists of a single vector $|\psi^0\rangle$. So we have $|\Phi_0| = 1$. At the level $l = 1$, the set Ψ_1 contains two vectors $|\psi^1(0)\rangle, |\psi^1(1)\rangle$. It is clear that these vectors are linearly independent since the string $\gamma = 1^{k-1}0^{n-k}$ distinguishes the string 1 from the string 0.

Induction step (for $l = 2, \ldots, d$): At the $(l-1)$-th level, we assume that $\Phi_{l-1} \subseteq \Psi_{l-1}$ has at least l elements, say $|\psi^{j_0}\rangle, \ldots, |\psi^{j_{l-1}}\rangle$, where the corresponding inputs are $\sigma^{j_0}, \ldots, \sigma^{j_{l-1}} \in \{0,1\}^{l-1}$, respectively.

At the l-th step, N_n reads the value $x_{i_l} = \sigma_{i_l}$. Due to Lemma 6 (Sect. 2.2), we know that the set $\Phi_l^0 = \{U_l(0)|\psi^{j_0}\rangle, \ldots, U_l(0)|\psi^{j_{l-1}}\rangle\}$ is linearly independent. It is clear that $|\psi^l(1^l)\rangle = U_l(1)U_{l-1}(1) \cdots U_1(1)|\psi^0\rangle$ is not a member of Φ_l^0. Moreover, the string $1^{k-l}0^{n-k}$ distinguishes 1^l from each of $\sigma^{j_1}0, \ldots, \sigma^{j_l}0$. Therefore, due to Lemma 13, we can follow that the set $\Phi_l^0 \cup \{|\psi^l(1^l)\rangle\}$ is linearly independent. Thus, Φ_l contains at least $(l+1)$ elements, i.e. $|\psi^{j_0}\rangle, \ldots, |\psi^{j_l}\rangle$, and $|\psi^l(1^l)\rangle$.

Therefore, Φ_d has at least $d+1$ elements and so the dimension of the space of quantum states must be at least $d+1$.

Case 2. Now assume that $k < n/2$ and therefore $d = n - k$. It is clear that $\text{EXACT}_n^k(\sigma) = 1$ iff $\#_0(\sigma) = n - k$, where $\#_0(\sigma)$ denotes the number of 0s in σ and we have $n - k \geq n/2$. We can apply the same reasoning as in the previous case by interchanging 0 and 1.

Therefore, in both cases Φ_d has at least $d+1$ elements and so the dimension of quantum system must be at least $d+1$, where $d = \max\{k, n-k\}$. Since there is an NUOBDD with width $(d+1)$ to solve $\texttt{EXACT}_n^k$, we can also conclude that $|\Phi_d| = d+1$. $\qquad\square$

Theorem 14. *The function $\texttt{EXACT}_n^k$ is computed by an OBDD D_n with width $\min(k+1, n-k+1)+1$.*

Theorem 15. *The width of any NOBDD computing $\texttt{EXACT}_n^k$ cannot be less than $\min(k+1,\ n-k+1)+1$.*

Proof. Let $d = \min(k, n-k)$. Assume $k \leq n/2$ that is $d = k$. For any order π of reading variables we can construct the following strong 1-fooling set for the function $\texttt{EXACT}_n^k$:

$$S_k^\pi = \{(\sigma^i, \gamma^i) : i = 0, \ldots, k, \sigma^i = \underbrace{0\cdots0}_{k-i}\underbrace{1\cdots1}_{i}, \gamma^i = \underbrace{0\cdots0}_{n-2k+i}\underbrace{1\cdots1}_{k-i}\}.$$

Due to Theorem 8, we follow the result.

If $k > n/2$, then $n - k \leq n/2$. In this case

$$S_k^\pi = \{(\sigma^i, \gamma^i) : i = 0, \ldots, k, \sigma^i = \underbrace{0\cdots0}_{k-i}\underbrace{1\cdots1}_{i}, \gamma^i = \underbrace{0\cdots0}_{n-2k+i}\underbrace{1\cdots1}_{k-i}\}.$$

Let $d = \min(k, n-k)$. Assume $k \leq n/2$ that is $d = k$. Let P_n be an NOBDD that computes $\texttt{EXACT}_n^k$ and has width $< d+2$. Consider the k-th level V_k of P_n and a set of partial inputs $\Sigma = \{\sigma^j \in \{0,1\}^k : \sigma^i = \underbrace{0\cdots0}_{k-i}\underbrace{1\cdots1}_{i}, j = 0, \ldots, k\}$. Let $path(\sigma^j)$ be one of the paths after reading σ^j that can also lead the computation to an accepting node after reading $(k-j)$ more 1s. Due to the Pigeonhole principle, each $path(\sigma^j)$ must be in a different node of the k-th level and so V_k contains at least $k+1$ different nodes, say $v_0, \ldots, v_k$.

The level V_{k+1} contains $k+1$ different nodes, say $v_0', \ldots, v_k'$, that can be accessed from $v_0, \ldots, v_k$ by reading a single 0, because from these nodes the computation can still go to some accepting nodes. If a single 1 is read, then v_k must switched to a node other than $v_0', \ldots, v_k'$. If it switches to v_j', then the non-member input $1^k 1 1^{k-j} 0^*$ with length n is accepted since the computation from v_j' can go to an accepting node (the input $0^{k-j} 1 1^j 0 1^{k-j} 0^*$ with length n is a member). Therefore, there must be at least $(k+2)$ nodes.

If $k > n/2$, then $n - k \leq n/2$ and so we can use the same proof by interchanging 0s and 1s. $\qquad\square$

Corollary 16. *The function $\texttt{AND}_n$ is computed by an NOBDD P_n with width 2. The function $\texttt{AND}_n$ is computed by an NUOBDD N_n with width $n+1$ and there is no NUOBDD computing $\texttt{AND}_n$ with width less than $n+1$.*

Now we show that negation of the function $\texttt{EXACT}_n^k$ is *cheap* for NUOBDD:

$$\texttt{notEXACT}_n^k(\sigma) = \begin{cases} 0, & \text{if } \#_1(\sigma) = k, \\ 1, & \text{otherwise.} \end{cases}$$

Theorem 17. *For any positive integer k $(k \leq n)$ the function $\texttt{notEXACT}_n^k$ can be computed by an NUOBDD N_n with width 2.*

3.4   Function MOD

Here we present a series of results for Boolean function $\text{MOD}_n^p : \{0,1\}^n \to \{0,1\}$, which is defined as:

$$\text{MOD}_n^p(\sigma) = \begin{cases} 1, & \text{if } \#_1(\sigma) \equiv 0 \pmod{p}, \\ 0, & \text{otherwise.} \end{cases}$$

It is clear that MOD_n^p can be solved by reversible OBDDs and so by exact UOBDDs with width p.

Theorem 18. *There is a width-p ROBDD R_n computing the function MOD_n^p.*

Now, we show that nondeterminism does not help neither classically nor quantumly in order to solve MOD_n^p.

Theorem 19. *If $p \leq n/2$, then the width of any NOBDD computing MOD_n^p cannot be less than p. For any p ($p \leq n$) the width of any NUOBDD computing MOD_n^p cannot be less than p.*

Proof. For the case $p \leq n/2$ the proof is the following. For any order π of reading variables we can construct the following strong 1-fooling set for the function MOD_n^p:

$$S_{n-p+1}^\pi = \{(\sigma^i, \gamma^i) : i = 0, \ldots, p-1, \sigma^i = \underbrace{0 \cdots 0}_{n-p+1-i} \underbrace{1 \cdots 1}_{i}, \gamma^i = \underbrace{0 \cdots 0}_{i-1} \underbrace{1 \cdots 1}_{p-i}\}.$$

Due to Theorem 8, we follow the result.

Let consider the case $p > n/2$ and let N_n be any NUOBDD computing MOD_n^p. Using the same arguments as in the proof of Theorem 12 we can show that on the $(p-1)$-th level of N_n the set of linear independent vectors, which are achievable quantum states, contains at least p elements. They are $|\psi(\sigma^0)\rangle, \ldots, |\psi(\sigma^{p-1})\rangle$, where $\sigma^j = 1^j 0^{p-j-1}$ and $j = 0, \ldots, p-1$. $\square$

Currently we do not know whether using more general QOBDD models can narrow the width for MOD_n^p.

3.5   Hierarchy for NUOBDDs

In [4,5], the following width hierarchy for OBDDs and NOBDDs was presented. For any integer $n > 3$ and $1 < d \leq \frac{n}{2}$, we have

$$\text{OBDD}_n^{d-1} \subsetneq \text{OBDD}_n^d \quad \text{and} \quad \text{NOBDD}_n^{d-1} \subsetneq \text{NOBDD}_n^d.$$

For any integer n, $d = d(n)$, $16 \leq d \leq 2^{n/4}$, we have

$$\text{OBDD}^{\lfloor d/8 \rfloor - 1} \subsetneq \text{OBDD}^d \quad \text{and} \quad \text{NOBDD}^{\lfloor d/8 \rfloor - 1} \subsetneq \text{NOBDD}^d.$$

Here we obtain a complete hierarchy result for NUOBDDs with width up to n.

Theorem 20. *For any integer $n > 1$ and $1 < d \leq n$, we have*

$$\text{NUOBDD}_n^{d-1} \subsetneq \text{NUOBDD}_n^d.$$

Proof. It is obvious that $\text{NUOBDD}_n^{d-1} \subseteq \text{NUOBDD}_n^d$. If $d \leq n/2$, we know that $\text{MOD}_n^d \in \text{NUOBDD}_n^d$ and $\text{MOD}_n^d \notin \text{NUOBDD}_n^{d-1}$ due to Theorems 18 and 19. If $d > n/2$, we know that $\text{EXACT}_n^{d-1} \in \text{NUOBDD}_n^d$ and $\text{EXACT}_n^{d-1} \notin \text{NUOBDD}_n^{d-1}$ due to Theorems 11 and 12. $\qquad\square$

Theorem 21. *(1) For any (d_1, d_2) satisfying $1 < d_1, d_2 \leq n$, $\text{NOBDD}_n^{d_2} \not\subseteq \text{NUOBDD}_n^{d_1}$. (2) For any (d_1, d_2) satisfying $1 < d_1, d_2 < \sqrt{n} - \frac{5}{4}\log n - 1$, $\text{NUOBDD}_n^{d_2} \not\subseteq \text{NOBDD}_n^{d_1}$.*

Proof. Let d_1, d_2 be arbitrary integers satisfying $1 < d_1, d_2 \leq n$. By Corollary 16, we know that $\text{AND}_n \in \text{NOBDD}_n^2 \subseteq \text{NOBDD}_n^{d_2}$ and $\text{AND}_n \notin \text{NUOBDD}_n^n$ and so $\text{AND}_n \notin \text{NUOBDD}_n^{d_1}$. Therefore, $\text{NOBDD}_n^{d_2} \not\subseteq \text{NUOBDD}_n^{d_1}$.

Let d_1, d_2 be arbitrary integers satisfying $1 < d_1, d_2 < \sqrt{n} - \frac{5}{4}\log n - 1$. By Theorem 9 and Corollary 10, we know that $\text{notPERM}_n \in \text{NUOBDD}_n^2 \subseteq \text{NUOBDD}_n^{d_1}$ and $\text{notPERM}_n \notin \text{NOBDD}_n^{d_2}$. Therefore, $\text{NUOBDD}_n^{d_2} \not\subseteq \text{NOBDD}_n^{d_1}$. $\qquad\square$

3.6   Union, Intersection, and Complementation

Let $f, g : \{0, 1\}^n \to \{0, 1\}$. We call a function $h = f \cup g$ the union of the functions f and g iff $h(\sigma) = f(\sigma) \bigvee g(\sigma)$ for all $\sigma \in \{0, 1\}^n$. We call a function $h = f \cap g$ the intersection of the functions f and g iff $h(\sigma) = f(\sigma) \bigwedge g(\sigma)$ for all $\sigma \in \{0, 1\}^n$. We call h the negation of the function f iff $h(\sigma) = \neg f(\sigma)$ for all $\sigma \in \{0, 1\}^n$.

Theorem 22. *Let f and g be Boolean functions defined on $\{0, 1\}^n$ computed by an NUOBDD N_n with width c and an NUOBDD N_n' with width d respectively such that N_n and N_n' use the same order π of reading variables. Then, the Boolean function $f \cup g$ can be computed by an NUOBDD, say N_n'', with width $c + d$.*

Proof. Let $N_n = (Q = \{q_1, \ldots, q_c\}, |\psi^0\rangle, T, Q_{acc})$, $N_n' = (Q' = \{q_1', \ldots, q_d'\}, |\psi'^0\rangle, T', Q_{acc}')$, where $T = \{(i_j, U_j(0), U_j(1))\}_{j=1}^n$, $T' = \{(i_j, U_j'(0), U_j'(1))\}_{j=1}^n$. The NUOBDD N_n'' can be constructed based on N_n and N_n' as follows.

$N_n'' = (Q'' = Q \cup Q' = \{q_1, \ldots, q_c, q_1', \ldots, q_d'\}, |\psi''^0\rangle, T'', Q_{acc}'' = Q_{acc} \cup Q_{acc}')$, where the initial quantum state is $|\psi''^0\rangle = \frac{1}{\sqrt{2}}(|\psi^0\rangle \oplus |\psi'^0\rangle)$. The sequence of instructions $T'' = \{i_j, U_j''(0), U_j''(1)\}_{j=1}^n$, where $U_j''(\sigma) = \begin{pmatrix} U_j(\sigma) & 0 \\ 0 & U_j'(\sigma) \end{pmatrix}$. Here 0 denotes zero matrix.

By construction, N_n'' executes both N_n and N_n' in parallel with equal amplitude, and so it accepts a given input with zero probability iff both N_n and N_n' accept it with zero probability. In other words, it accepts an input with non-zero probability iff N_n or N_n' accepts it with non-zero probability. Thus, N_n'' computes the function $f \cup h$. $\qquad\square$

Theorem 23. *Let f and g be Boolean functions defined on $\{0,1\}^n$ computed by an NUOBDD N_n with width c and an NUOBDD N'_n with width d, respectively, such that N_n and N'_n use the same order π of reading variables. Then, the Boolean function $f \cap g$ can be computed by an NUOBDD, say N''_n, with width $c \cdot d$.*

Proof. Let $N_n = (Q = \{q_1, \ldots, q_c\}, |\psi^0\rangle, T, Q_{acc})$, $N'_n = (Q' = \{q'_1, \ldots, q'_d\}, |\psi'^0\rangle, T', Q'_{acc})$, where $T = \{(i_j, U_j(0), U_j(1))\}_{j=1}^n$, $T' = \{(i_j, U'_j(0), U'_j(1))\}_{j=1}^n$. The NUOBDD N''_n can be constructed by tensoring N_n and N'_n as follows.

$N''_n = (Q'' = Q \times Q' = \{q_{1,1}, \ldots, q_{c,d}\}, |\psi^0\rangle \otimes |\psi'^0\rangle, T'', Q''_{acc})$, where the sequence of instructions $T'' = \{i_j, U_j(0) \otimes U'_j(0), U_j(1) \otimes U'_j(1)\}_{j=1}^n$ and the set of accepting states contains all the states $q_{i,j}$ satisfying $q_i \in Q_{acc}$ and $q_j \in Q'_{acc}$.

From this construction it follows that $Pr^{N''_n}_{accept}(\sigma) = Pr^{N_n}_{accept}(\sigma) \cdot Pr^{N'_n}_{accept}(\sigma)$. If the input σ is such that $f(\sigma) = 1$ and $g(\sigma) = 1$, then both N_n and N'_n accept it with non-zero probability and therefore N''_n also accepts this input with non-zero probability. If the input σ is such that $f(\sigma) = 0$ or $g(\sigma) = 0$ then $Pr^{N''_n}_{accept}(\sigma) = 0$ for this input. □

The bound for intersection can be shown to be tight in certain cases.

Theorem 24. *There exist functions f and g computed by NUOBDDs $N_{f,n}$ with width c and $N_{g,n}$ with width d, respectively, such that the width of any NUOBDD computing the function $h = f \cap g$ cannot be less than $lcm(c \cdot d)$, where $lcm(c \cdot d) \leq n$.*

The bounds given in Theorems 22 and 23 are also valid for NOBDDs. Deterministic OBDDs, on the other hand, requires $c \cdot d$ for union operation.

Classically, if a function, say f, solved by an NOBDD with width d, then the negation of f can be solved by another NOBDD with width at most 2^d. By using Corollary 16 and the result below we conclude that in case of NUOBDD, we cannot provide such a bound.

Corollary 25. *(from Theorem 17) The function $\neg AND_n$ is computable by NUOBDD with width 2.*

4   Concluding Remarks

In this paper we investigate the width complexity of nondeteministic unitary OBDDs and compare them with their classical counterpart. Our results are mainly for linear and sublinear widths. As a future work, we plan to investigate the superlinear widths. Here we present a width hierarchy and a similar result is not known for nondeterministic quantum OBDDs using general quantum operators. We also find interesting possible applications of our results to some other models like quantum finite automata.

Acknowledgments. We thank the anonymous reviewers for their very helpful comments and corrections.

References

1. Ablayev, F., Gainutdinova, A.: Complexity of quantum uniform and nonuniform automata. In: Felice, C., Restivo, A. (eds.) DLT 2005. LNCS, vol. 3572, pp. 78–87. Springer, Heidelberg (2005). doi:10.1007/11505877_7
2. Ablayev, F., Gainutdinova, A., Karpinski, M.: On computational power of quantum branching programs. In: Freivalds, R. (ed.) FCT 2001. LNCS, vol. 2138, pp. 59–70. Springer, Heidelberg (2001). doi:10.1007/3-540-44669-9_8
3. Ablayev, F.M., Gainutdinova, A., Karpinski, M., Moore, C., Pollett, C.: On the computational power of probabilistic and quantum branching program. Inf. Comput. **203**(2), 145–162 (2005)
4. Ablayev, F., Gainutdinova, A., Khadiev, K., Yakaryılmaz, A.: Very narrow quantum OBDDs and width hierarchies for classical OBDDs. In: Jürgensen, H., Karhumäki, J., Okhotin, A. (eds.) DCFS 2014. LNCS, vol. 8614, pp. 53–64. Springer, Cham (2014). doi:10.1007/978-3-319-09704-6_6
5. Ablayev, F.M., Gainutdinova, A., Khadiev, K., Yakaryılmaz, A.: Very narrow quantum OBDDs and width hierarchies for classical OBDDs. Lobachevskii J. Math. **37**(6), 670–682 (2016)
6. Ablayev, F., Karpinski, M.: On the power of randomized branching programs. In: Meyer, F., Monien, B. (eds.) ICALP 1996. LNCS, vol. 1099, pp. 348–356. Springer, Heidelberg (1996). doi:10.1007/3-540-61440-0_141
7. Adleman, L.M., DeMarrais, J., Huang, M.D.A.: Quantum computability. SIAM J. Comput. **26**(5), 1524–1540 (1997)
8. Ambainis, A., Yakaryılmaz, A.: Automata and quantum computing. Technical report. arXiv 1507.01988 (2015)
9. Bertoni, A., Carpentieri, M.: Analogies and differences between quantum and stochastic automata. Theoret. Comput. Sci. **262**(1–2), 69–81 (2001)
10. Fefferman, B., Lin, C.Y.Y.: A complete characterization of unitary quantum space. Technical report, arXiv 1604.01384 (2016)
11. Gainutdinova, A., Yakaryılmaz, A.: Nondeterministic unitary OBDDs. Technical report, arXiv 1612.07015 (2016)
12. Kondacs, A., Watrous, J.: On the power of quantum finite state automata. In: FOCS, pp. 66–75. IEEE Computer Society (1997)
13. Krause, M., Meinel, C., Waack, S.: Separating the eraser turing machine classes L_e, NL_e, co-NL_e and P_e. Theor. Comput. Sci. **86**(2), 267–275 (1991)
14. Moore, C., Crutchfield, J.P.: Quantum automata and quantum grammars. Theoret. Comput. Sci. **237**(1–2), 275–306 (2000)
15. Nakanishi, M., Hamaguchi, K., Kashiwabara, T.: Ordered quantum branching programs are more powerful than ordered probabilistic branching programs under a bounded-width restriction. In: Du, D.-Z.-Z., Eades, P., Estivill-Castro, V., Lin, X., Sharma, A. (eds.) COCOON 2000. LNCS, vol. 1858, pp. 467–476. Springer, Heidelberg (2000). doi:10.1007/3-540-44968-X_46
16. Nakanishi, M., Indoh, T., Hamaguchi, K., Kashiwabara, T.: On the power of nondeterministic quantum finite automata. IEICE Trans. Inf. Syst. **E85–D**(2), 327–332 (2002)
17. Nielsen, M.A., Chuang, I.L.: Quantum Computation and Quantum Information. Cambridge University Press, Cambridge (2000)
18. Sauerhoff, M., Sieling, D.: Quantum branching programs and space-bounded nonuniform quantum complexity. Theoret. Comput. Sci. **334**(1–3), 177–225 (2005)

19. Say, A.C.C., Yakaryılmaz, A.: Quantum finite automata: a modern introduction. In: Calude, C.S., Freivalds, R., Kazuo, I. (eds.) Computing with New Resources. LNCS, vol. 8808, pp. 208–222. Springer, Cham (2014). doi:10.1007/978-3-319-13350-8_16
20. Watrous, J.: Quantum computational complexity. In: Meyers, R.A. (ed.) Encyclopedia of Complexity and System Science. Springer, New York (2009)
21. Wegener, I.: Branching Programs and Binary Decision Diagrams. SIAM (2000)
22. Yakaryılmaz, A., Say, A.C.C.: Languages recognized by nondeterministic quantum finite automata. Quantum Inf. Comput. **10**(9–10), 747–770 (2010)

Unary Coded PSPACE-Complete Languages in ASPACE(loglog n)

Viliam Geffert[✉]

Department of Computer Science, P.J. Šafárik University,
Jesenná 5, 04154 Košice, Slovakia
viliam.geffert@upjs.sk

Abstract. We show that there exists a binary PSPACE-complete language $\mathcal{L}$ such that its unary coded version $\mathcal{L}'$ is in $\mathrm{ASPACE}^{\mathrm{dm}}(\log \log n)$, i.e., $\mathcal{L}'$ is accepted by an alternating Turing machine using an initially delimited worktape of size $\log \log n$. As a consequence, the standard translation between unary languages accepted with $\log \log n$ space and binary languages accepted with $\log n$ space works for alternating machines *if and only if* $\Gamma = \mathrm{PSPACE}$.

In general, if a binary language $\mathcal{L}$ is in $\mathrm{DTIMESPACE}(2^n \cdot n^{O(1)}, n^{O(1)})$, i.e., if $\mathcal{L}$ is accepted by a deterministic Turing machine in $2^n \cdot n^{O(1)}$ time and, simultaneously, in $n^{O(1)}$ space, then its unary coded version $\mathcal{L}'$ is in $\mathrm{ASPACE}^{\mathrm{dm}}(\log \log n)$. In turn, if a unary $\mathcal{L}'$ is in $\mathrm{ASPACE}^{\mathrm{dm}}(\log \log n)$, then its binary coded version $\mathcal{L}$ is in $\mathrm{DTIME}(2^n \cdot n^{O(1)}) \cap \mathrm{DSPACE}(n^{O(1)})$, and also in $\mathrm{NTIMESPACE}(2^n \cdot n^{O(1)}, n^{O(1)})$.

This unexpected power of sublogarithmic space follows from the fact that, with a worktape of size $\log \log n$ on a unary input 1^n, an alternating machine can simulate a *stack* with $\log n$ bits, representing the contents of the stack by its input head position. The standard push/pop operations are implemented by moving the head along the input.

Keywords: Computational complexity · Alternation · Sublogarithmic space

1 Introduction and Preliminaries

In computational complexity theory, an important role plays the relation between binary inputs and their unary coded counterparts. As an example, we know that $\mathrm{NTIME}(2^{O(n)})$ separates from $\mathrm{DTIME}(2^{O(n)})$ if and only if NP separates from P on unary languages [13]. Consequently, the first separation is much more difficult than the separation of NP from P, since we must provide a witness language $\mathcal{L}'$ in $\mathrm{NP} - \mathrm{P}$ which is, in addition, *unary*. The same holds for the

---

V. Geffert—Supported by the Slovak grant contracts VEGA 1/0142/15 and APVV-15-0091.

P. Weil (Ed.): CSR 2017, LNCS 10304, pp. 141–153, 2017.
DOI: 10.1007/978-3-319-58747-9_14

relation $\mathrm{DSPACE}(n) \overset{?}{=} \mathrm{NSPACE}(n)$ versus $\mathrm{DSPACE}(\log n) \overset{?}{=} \mathrm{NSPACE}(\log n)$ [18]. Such results can be obtained by the use of the following translation [18]:[1]

Theorem 1. *If a unary language $\mathcal{L}_1 \in \mathrm{TIMESPACE}(t(N), s(N))$, then its binary coded version $\mathcal{L}_{1\to2} \in \mathrm{TIMESPACE}(t(2^n) \cdot n, s(2^n))$. Vice versa, if a binary language $\mathcal{L}_2 \in \mathrm{TIMESPACE}(t(2^n), s(2^n))$, then its unary coded version $\mathcal{L}_{2\to1} \in \mathrm{TIMESPACE}(t(N), s(N))$. This holds for deterministic, nondeterministic, and alternating complexity classes with simultaneous bounds on time and space, for any monotone functions $t(N) \geq N$ and $s(N) \geq \Omega(\log N)$ satisfying $t(2N) \leq O(t(N))$ and $s(2N) \leq O(s(N))$.*

Thus, by taking $s(N) = \log N$, we get that a unary language $\mathcal{L}_1$ can be accepted with $O(\log N)$ space if and only if its binary coded version $\mathcal{L}_{1\to2}$ with $O(n)$ space, which holds for the deterministic, nondeterministic, as well as for the alternating space complexity classes. That is,

$$\mathrm{XSPACE}(\log N)_{1\to2} = \mathrm{XSPACE}(n)_2, \text{ for } \mathrm{X} \in \{\mathrm{D}, \mathrm{N}, \mathrm{A}\}.$$

Here $\mathrm{XSPACE}(s(N))_{1\to2}$ denotes the class of binary coded versions of all unary languages in $\mathrm{XSPACE}(s(N))$ and $\mathrm{XSPACE}(s(n))_2$ the class of binary languages in $\mathrm{XSPACE}(s(n))$.

By combination of more sophisticated techniques [1,6,10] (for details and a survey, see Theorem 1 and its proof in [12]), the assumption $s(N) \geq \Omega(\log N)$ in Theorem 1 can be replaced by $s(N) \geq \Omega(\log \log N)$, in the case of deterministic and nondeterministic classes $\mathrm{XSPACE}^{\mathrm{dm}}(s(n))$ that represent a more powerful model, where the worktapes contain initially $\lfloor s(n) \rfloor$ blank cells delimited by end-markers, as opposed to the classical classes $\mathrm{XSPACE}(s(n))$, where the worktapes are initially empty.[2] Thus, taking $s(N) = \log \log N$, we get

$$\mathrm{XSPACE}^{\mathrm{dm}}(\log \log N)_{1\to2} = \mathrm{XSPACE}(\log n)_2, \text{ for } \mathrm{X} \in \{\mathrm{D}, \mathrm{N}\}. \tag{1}$$

On the other hand, for alternating space, it was shown in [12] that

$$\begin{aligned}\mathrm{ASPACE}^{\mathrm{dm}}(\log \log N)_{1\to2} &\supseteq \mathrm{ASPACE}(\log n)_2, \text{ but} \\ \mathrm{ASPACE}^{\mathrm{dm}}(\log \log N)_{1\to2} &\subseteq \mathrm{ASPACE}(\log n)_2 \text{ implies } \mathrm{P} = \mathrm{NP}.\end{aligned} \tag{2}$$

Since $\mathrm{P} = \mathrm{NP}$ is unlikely, we conjecture that (1) does not hold for the alternating classes. Clearly, this does not exclude the possibility that one could disprove the "$\subseteq$" inclusion in (2) without gaining anything about the $\mathrm{P} \overset{?}{=} \mathrm{NP}$ problem.

This brings our attention to the properties of $\mathrm{ASPACE}^{\mathrm{dm}}(\log \log N)_{1\to2}$, the class of binary coded versions of all unary languages that can be accepted by alternating machines with worktapes of size $\lfloor \log \log N \rfloor$, delimited initially.

---

[1] Throughout the paper, we denote the length of a binary input by n, while the length of a unary input by N. This reflects the fact that $n < N$.

[2] $\mathrm{XSPACE}(s(n)) = \mathrm{XSPACE}^{\mathrm{dm}}(s(n))$, if $s(n)$ is fully space constructible. The notation "dm" derives from "Demon" Turing Machines [4].

One of the key observations in this paper is that, with a worktape of size $O(\log \log N)$ on a unary input 1^N, an alternating machine can simulate a *stack* with $\lfloor \log(N+1) \rfloor$ bits, representing the contents of the stack by a "proper" input head position H. The contents of the stack can be changed by moving along the input to a new calculated position H'. Besides the push/pop operations, we have also a read-only access to all bits stored below the top of the stack.

Based on this, we then introduce two machine models that characterize the class $\mathrm{ASPACE}^{\mathrm{dm}}(\log \log N)_{1 \to 2}$ *exactly*. It turns out that this class corresponds to an alternating machine equipped, besides the standard binary two-way read-only input tape, with a worktape using $O(\log n)$ cells and an auxiliary pushdown store capable of containing n bits. The second—but equivalent—model is the same except that, instead of a pushdown store, it uses a binary stack with an additional read-only access to the interior contents, below the top. That is,

$$\mathrm{ASPACE}^{\mathrm{dm}}(\log \log N)_{1 \to 2} = \mathrm{ASPACEBINPUSHDOWN}(\log n, n)_2$$
$$= \mathrm{ASPACEBINSTACK}(\log n, n)_2.$$

The above results allow us to specify the position of $\mathrm{ASPACE}^{\mathrm{dm}}(\log \log N)_{1 \to 2}$ among the standard complexity classes as follows:

$$\mathrm{ASPACE}^{\mathrm{dm}}(\log \log N)_{1 \to 2} \supseteq \mathrm{DTIMESPACE}(2^n \cdot n^{O(1)}, n^{O(1)})_2,$$
$$\mathrm{ASPACE}^{\mathrm{dm}}(\log \log N)_{1 \to 2} \subseteq \mathrm{DTIME}(2^n \cdot n^{O(1)})_2 \cap \mathrm{DSPACE}(n^{O(1)})_2,$$
$$\mathrm{ASPACE}^{\mathrm{dm}}(\log \log N)_{1 \to 2} \subseteq \mathrm{NTIMESPACE}(2^n \cdot n^{O(1)}, n^{O(1)})_2.$$

Thus, translation from unary languages accepted by alternating machines with delimited $\lfloor \log \log N \rfloor$ space to their binary counterparts brings us to a quite strong class. After that, we show that

- there exists a binary PSPACE-complete language such that its unary coded version is in $\mathrm{ASPACE}^{\mathrm{dm}}(\log \log N)$.

This allows us to derive a much stronger version of (2):

$$\mathrm{ASPACE}^{\mathrm{dm}}(\log \log N)_{1 \to 2} = \mathrm{ASPACE}(\log n)_2 \text{ if and only if } \mathrm{P} = \mathrm{PSPACE}. \quad (3)$$

Solved in either way, the left "$=$" in (3) makes a great impact. We do not expect a positive answer—this requires a deterministic polynomial-time simulation of an alternating machine with $O(\log \log N)$ space, but on an *expanded unary counterpart* of the original binary input, of length $N \geq 2^n - 1$. Every less efficient simulation/separation counts, e.g.: $\mathrm{ASPACE}^{\mathrm{dm}}(\log \log N)_{1 \to 2} \subseteq \mathrm{NP}$ if and only if $\mathrm{NP} = \mathrm{PSPACE}$. Thus, the alternating machines with $O(\log \log n)$ space may bring answers to questions related to higher complexity classes.

We assume the reader is familiar with the basics of the standard deterministic, nondeterministic, and alternating Turing machines. (See, e.g., [3, 14, 19]).

We only recall here that the length of an integer $K > 0$ written in binary is $k = 1 + \lfloor \log K \rfloor$. However, a binary input $\vdash b_{n-1} \ldots b_0 \dashv$ enclosed in between two endmarkers will be interpreted as $K = \sum_{i=0}^{n} b_i \cdot 2^i$, with $b_n \stackrel{\mathrm{df.}}{=} 1$. Thus, the left endmarker plays the role of the hidden most significant bit.[3] Similarly, a unary input $\vdash 1^N \dashv$ will be interpreted as $K = N+1$, so also here the left endmarker plays the role of the hidden digit. Thus, for $K = 1, 2, 3, 4, 5, \ldots$, we obtain the respective binary inputs $\varepsilon, 0, 1, 00, 01, \ldots$ and the corresponding unary inputs $\varepsilon, 1, 11, 111, 1111, \ldots$. Taking into account the role of the left endmarker, the lengths of the corresponding binary and unary coded inputs are related as follows:

$$n = \lfloor \log(N+1) \rfloor. \tag{4}$$

2   Input Head as an Additional Memory

In this section, we first show how an alternating machine, on a unary input tape containing $\vdash 1^N \dashv$, can compute $\mathrm{bit}(H, \ell)$, the ℓ-th bit in the binary representation of the current input head position H, using only $O(\log\log N)$ worktape space. The position H itself should be preserved for future. Once this is possible, we can remember a *stack* capable of containing $\lfloor \log(N+1) \rfloor$ bits by the current input head position H. The standard push/pop operations will be implemented by moving the head along the input to a new calculated position H', using an auxiliary worktape of size $O(\log\log N)$. Besides the standard push/pop operations, we shall also have a read-only access to all bits stored below the top of the stack.

Such task is beyond the power of nondeterministic machines with space bounded by $o(\log N)$: if the head moves too far along the unary input, the original tape position H is lost [9,10]. It should be pointed out that the bits in the binary representation of $N+1$ (the value $N+1$ is given in unary—the special case, corresponding to H at the end of the input) can be computed with $O(\log\log N)$ space even by a *deterministic* machine, if $\lfloor \log\log N \rfloor$ worktape cells are delimited initially.[4]

Recall that, on the unary input tape with $\vdash 1^N \dashv$, the input head position is in the range $H \in \{0, \ldots, N+1\}$. Thus, the meaningful bit positions in the value H written in binary are in the range $\{0, \ldots, \lfloor \log(N+1) \rfloor\}$, and hence the binary representation of any meaningful bit position $\ell \leq \lfloor \log(N+1) \rfloor$ is of length $1 + \lfloor \log \ell \rfloor \leq O(\log\log N)$.

---

[3] This ensures that the same number cannot be represented by two different binary strings, using a different number of leading zeros.

[4] This is based on the following facts. First, it is quite trivial to see that the machine can compute $m_i = (N+1) \bmod p_i$ for any given prime $p_i \leq O(\log N)$, by counting modulo p_i while traversing across the unary input tape with $\vdash 1^N \dashv$. Thus, the machine has a read-only access to $(m_1, m_2, m_3, \ldots)$, the first $O(\log N / \log\log N)$ remainders in the *Chinese Residual Representation* of $N+1$. With access to these remainders, the ℓ-th bit in the binary representation of $N+1$ can be computed by using $O(\log\log N)$ worktape space. This was shown in [6], building on ideas presented in [5,7]. (See also [1, Theorem 4.5]. Some related topics and other applications can be found in [2,16]).

Theorem 2. *For any given* $v \in \{0,1\}$, $\ell \geq 0$, *and any starting input head position* $H \in \{0, \ldots, N+1\}$ *on the unary input* 1^N, *the question of whether* $v = \mathrm{bit}(H, \ell)$ *can be decided by an alternating machine with* $O(\log \ell)$ *space.*

Let us present just the main idea. Our alternating Turing machine implements an alternating procedure $\mathsf{test\_bit}(v, \ell)$ that starts in a special finite control state q_{test} with v and ℓ written in binary in two separate worktape tracks and the input head placed at the starting position H. Depending on whether $v = \mathrm{bit}(H, \ell)$, the subtree of all computation paths in rooted in q_{test} will be accepting or rejecting. If $H = 0$, i.e., if the input head scans the left endmarker, then $\mathrm{bit}(H, \ell) = 0$ for each ℓ, so we just compare v with zero. Now, for $H > 0$, we use the fact that $\mathrm{bit}(H, \ell) = \mathrm{bit}(H-1, \ell)$ if and only if $\mathrm{bit}(H-1, k) = 0$, for some $k < \ell$. Thus, the question of whether $v = \mathrm{bit}(H, \ell)$ can be decided by deciding the same kind of questions, namely, whether $b = \mathrm{bit}(H-1, k)$, for "properly" chosen values $b \in \{0, 1\}$ and $k \leq \ell$. Therefore, the computation proceeds by assignment $H := H-1$, that is, by moving the input head one position to the left. After that, the question is decided by activating new instances of the alternating procedure $\mathsf{test\_bit}(b, k)$ running in parallel, from the very beginning. Each of these instances is activated by switching to the state q_{test}, but with its own parameters b and k on the worktape and with the new starting position of the input head, namely, with $H' = H-1$.

Thus, we have constructed the alternating procedure $\mathsf{test\_bit}(v, \ell)$ that decides whether $v = \mathrm{bit}(H, \ell)$ but, as a side effect, it destroys the original input tape position H. Now we are going to design a function that returns the value $\mathrm{bit}(H, \ell)$ in a straightforward way, without side effects. We shall also implement a function that returns $\mathrm{bit}(N+1, \ell)$, preserving the current input head position H. These two functions are implemented in an alternating fashion, so the statement of the next theorem should be a little bit more careful.

Theorem 3. *For any given* $\ell \geq 0$ *and any input head position* $H \in \{0, \ldots, N+1\}$ *on the unary input* 1^N, *the value* $\mathrm{bit}(H, \ell)$ *can be computed by an alternating machine with* $O(\log \ell)$ *space. The same holds for computing the value* $\mathrm{bit}(N+1, \ell)$. *Moreover, in both these cases,*

(I) *The machine has a unique computation path returning the correct value. Along this path, all universal decisions stepping aside lead to alternating subtrees that are accepting and all existential decisions stepping aside to subtrees that are rejecting. In addition, this path preserves the current input head position* H.

(II) *The remaining paths may return wrong results, halt without returning any results at all, or destroy the input head position. The outcome of such paths is overridden, by item (I) above.*

The alternating machine implementing a function $\mathsf{head\_bit}(\ell)$ is quite simple. Starting in a special finite control state, with ℓ on the worktape and the input head at the position H, the machine first guesses $b = \mathrm{bit}(H, \ell)$, that is, branching existentially, it chooses between $b = 0$ and $b = 1$. Next, the machine branches universally. The first branch returns the value b as a result and exits the function call, *without changing* the original position H. The second branch verifies the guess, by calling the procedure test_bit(b, ℓ) presented in Theorem 2. This destroys the original position H, but after stepping aside from the path returning the result b. The machine for the function input_bit(ℓ) proceeds in the same way as for head_bit(ℓ) but, before verifying the guessed value b, we move the input head to the right endmarker. Thus, the procedure test_bit(b, ℓ) verifies $b = \mathrm{bit}(H', \ell)$ for $H' = N+1$. The original input head position is lost, but after stepping aside from the path returning b.

We are now ready to implement a *binary stack of size* $n = \lfloor \log(N+1) \rfloor$. Note that, by (4), the size of the binary stack is exactly equal to the length of the binary coded version of $\vdash 1^N \dashv$. To be more precise, let *current contents of the stack* be a binary string $v_{k-1} \ldots v_0$, with the bit v_0 on *top*. For this data structure, we shall provide the following package of operations:

- stack_bit(ℓ), for $\ell \in \{0, \ldots, k-1\}$: a function that returns v_ℓ, the ℓ-th bit from top, not changing the contents of the stack $v_{k-1} \ldots v_0$.
- stack_pop: removes the topmost bit, changing the contents of the stack from $v_{k-1} \ldots v_0$ to $v_{k-1} \ldots v_1$.
- stack_push(v), for $v \in \{0, 1\}$: adds the bit v on top, changing the stack contents from $v_{k-1} \ldots v_0$ to $v_{k-1} \ldots v_0 v$.

Theorem 4. *The package of operations handling the binary stack of size* $n = \lfloor \log(N+1) \rfloor$, *namely,* stack_bit$(\ell)$, stack_pop, *and* stack_push(v), *can be implemented as a package of alternating procedures keeping the current contents of the stack by the current position of the input head on the unary input* 1^N *and using* $O(\log n) \leq O(\log \log N)$ *space on the worktape.*

Moreover, each of these alternating procedures works with the contents of the stack correctly along the unique computation path, such that all universal decisions stepping aside from this path lead to alternating subtrees that are accepting while all existential decisions stepping aside to subtrees that are rejecting. Thus, even though some computation paths do not work with the contents of the stack correctly, the outcome of such paths is overridden.

Proof. By (4), we have that $2^n - 1 = 2^{\lfloor \log(N+1) \rfloor} - 1 \leq N$. Thus, for each binary string $b_{n-1} \ldots b_0$, there does exist a position $H \leq 2^n - 1$ along the unary input 1^N such that H is binary written as $b_{n-1} \ldots b_0$. (For $b_{n-1} \ldots b_0 = 0^n$, the corresponding position is $H = 0$ at the left endmarker.) This allows us to represent the current contents of the binary stack of size n by two quantities:

- the current position of the input head $H \in \{0, \ldots, 2^n - 1\}$ and
- a global variable $g \in \{0, \ldots, n\}$, stored on a separate worktape track.

These two values are interpreted as follows: if the current input head position H is binary written as $b_{n-1} \ldots b_g \ldots b_0$, the current contents of the stack is $b_{n-1} \ldots b_g$, with the bit b_g on top. The remaining bits $b_{g-1} \ldots b_0$ are ignored. That is, we can represent the given stack contents $v_{k-1} \ldots v_0$ by *any* input head

```
structure stack                    • — global values:
   n   = ⌊log(N+1)⌋                 • — stack size limit
   H   ∈ {0,...,2^n−1}              • — input head position: b_{n−1}...b_0 in binary
   g   ∈ {0,...,n}                  • — stack contents: b_{n−1}...b_g, with b_g on top

function stack_bit(ℓ)
      return head_bit(g+ℓ)          • — return the ℓ-th bit from top of the stack
end

procedure stack_pop
      g := g + 1                    • — remove one bit from top
end

procedure stack_push(v)
   1: g := g − 1                    • — add one bit on top
   2: while head_bit(g) < v ∧ H < N+1 do H := H + 1
                                    • — ensure that bit(H, g) ≥ v
   3: while head_bit(g) > v ∧ H > 0 do H := H − 1
                                    • — ensure that bit(H, g) ≤ v
end
```

Fig. 1. A package of procedures implementing the binary stack of size $\lfloor \log(N+1) \rfloor$.

position H the binary representation of which ranges between $v_{k-1} \ldots v_0 0^{n-k}$ and $v_{k-1} \ldots v_0 1^{n-k}$. Let us now implement this stack.

First, it is obvious that the value stack_bit(ℓ) can be obtained by the use of head_bit($g+\ell$). (Displayed as the function stack_bit in Fig. 1.) An important detail is that head_bit has a unique computation path returning the result that is correct, not changing the current input head position, and hence preserving the structure of the stack. All remaining paths are overridden. (See also items (I) and (II) in Theorem 3).

Second, removing the topmost bit from the stack is trivial, by assignment $g := g+1$. Clearly, if the current input head position H is binary written as $b_{n-1} \ldots b_g \ldots b_0$, the stack contents changes from $b_{n-1} \ldots b_{g+1} b_g$ to $b_{n-1} \ldots b_{g+1}$. (The procedure for this is stack_pop, displayed in Fig. 1).

Finally, let us consider pushing the bit $v \in \{0,1\}$ on top the stack. Let $b_{n-1} \ldots b_g b_{g'} \ldots b_0$ be the binary representation of the current input head position H. First, we replace g on the worktape by the new value $g' = g-1$. This increases the stack height by one (Fig. 1, procedure stack_push, line 1). Now, if the value $b_{g'} = \text{bit}(H, g')$ is equal to v, we are done, the bit v has already been pushed on top of the stack. For $b_{g'} \neq v$, we have two cases to consider.

First, if $b_{g'} = 0 < 1 = v$, the binary representation of H is in the form $b_{n-1} \ldots b_g 0 b_{g'-1} \ldots b_0$. In this case, we run a loop (line 2) and move the input head to the right, searching for the nearest input position H' with $\text{bit}(H', g') = 1$. Clearly, the binary representation of H' is $b_{n-1} \ldots b_g 1 0^{g'}$. Thus, we have updated the stack contents from $b_{n-1} \ldots b_g$ to $b_{n-1} \ldots b_g 1$, with the bit $v = 1$ on top. It should be pointed out that we move the input head to the right in the loop that recomputes the g'-th bit for the current input position again and again, by calling head_bit. This function is implemented in an alternating way which, by a wrong existential guess, may return a wrong result. However, by items (I) and (II) in Theorem 3, the outcome of such paths is overridden. For this reason, the outcome of the loop executed in line 2 depends only on the unique path working with the correct bits in all iterations, which overrides all paths that stop too early or too late—before or behind the correct position H'. Nevertheless, by a wrong sequence of existential choices—guessing always that the g'-th bit is equal to zero—the machine may try to traverse the entire input tape, to the right of the right endmarker. For this reason, the iteration is stopped when we reach this endmarker (condition $H < N+1$ in line 2).

Second, if $b_{g'} = 1 > 0 = v$, the binary representation of H is in the form $b_{n-1} \ldots b_g 1 b_{g'-1} \ldots b_0$. Here we run a loop that moves the input head to the left (line 3), searching for the nearest input position H' with $\mathrm{bit}(H', g') = 0$. Clearly, H' is binary represented by $b_{n-1} \ldots b_g 0 1^{g'}$. This updates the stack contents from $b_{n-1} \ldots b_g$ to $b_{n-1} \ldots b_g 0$, which pushes the bit $v = 0$ onto the stack. The reasoning about correctness is the same as in the case of $b_{g'} < v$. Here we pay a special attention to the path that, making wrong guesses, tries to traverse the entire input tape, to the left of the left endmarker. Thus, we stop the iteration when we reach the left endmarker (condition $H > 0$ in line 3).

Since both g and ℓ are bounded by $n = \lfloor \log(N+1) \rfloor$, all stack operations are implemented with $O(\log n) \le O(\log \log N)$ worktape space. $\square$

3   Machine Models for Binary Inputs

Here we shall establish, in terms of machine models, an exact characterization of $\mathrm{ASPACE}^{\mathrm{dm}}(\log \log N)_{1 \to 2}$, the class of binary coded versions of all unary languages that can be accepted by alternating Turing machines starting with $\lfloor \log \log N \rfloor$ worktape cells, delimited initially. It turns out that this class is equivalent to the class of binary languages accepted by the following two types of machines:

Definition 5. *An* alternating auxiliary binary pushdown automaton *is an alternating Turing machine equipped with a finite state control, a two-way read-only input tape, a separate two-way read-write worktape that is initially empty, and a pushdown store containing initially a bottom-of-the-pushdown endmarker* $\vdash$. *During the computation, only zeros and ones can be pushed on top.*

An alternating auxiliary binary stack automaton *is similar except that, instead of a pushdown store, it is equipped with a binary stack which permits also a read-only access to the interior contents below the top. The stack is implemented as a special worktape with a two-way read-only head, but the standard push/pop operations are permitted whenever this head is scanning the topmost symbol. Initially, the stack contains only the bottom-of-the-stack endmarker* $\vdash$.

The class of languages accepted by alternating auxiliary pushdown automata keeping at most $O(\log n)$ symbols on the worktape and at most n bits in the pushdown store will be denoted by ASPACEBINPUSHDOWN($\log n, n$). The corresponding language class for machines equipped with the binary stack of size n will be denoted by ASPACEBINSTACK($\log n, n$).

Theorem 6. *If a binary language $\mathcal{L}_2$ is in ASPACEBINSTACK($\log n, n$), then its unary coded version $\mathcal{L}_{2\to 1}$ is in* ASPACE^{dm}($\log \log N$).

Proof. Let $\mathcal{L}_2$ be accepted by an alternating auxiliary stack machine A keeping at most $O(\log n)$ symbols on the worktape and at most n bits in the stack. We need a machine A' that decides, for a given input $\vdash 1^N \dashv$, whether the string $b_{n-1} \ldots b_0$, representing the number $N+1$ in binary, is in $\mathcal{L}_2$. The machine A' uses a worktape of length $s = \lfloor \log \log N \rfloor$, delimited by two endmarkers.

We begin with computing n, the length of the virtual binary input, which is also the position of the most significant bit in the binary represented $N+1$. Using (4), this value is bounded by $n = \lfloor \log(N+1) \rfloor \leq \log(2N) = 1 + 2^{\log \log N} < 1 + 2^{1+\lfloor \log \log N \rfloor} = 1 + 2^{1+s}$. Therefore, we can compute n by running a loop iterated for $\ell := 2^{1+s}, 2^{1+s}-1, 2^{1+s}-2, \ldots, 0$ in which we search for the first ℓ satisfying bit$(N+1, \ell) = 1$. When this happens, take $n := \ell$ and save this value in a separate worktape track. On the unary input $\vdash 1^N \dashv$, the machine A' computes bit$(N+1, \ell)$ by calling the function input_bit(ℓ), described in Theorem 3. The used space is bounded by $O(\log \ell) \leq O(\log 2^{1+s}) \leq O(s) \leq O(\log \log N)$.

During the simulation, A' maintains the following data about A:

First, $n \leq 2^{1+s}$ is saved in a separate worktape track. This value does not change in the course of the simulation.

Second, the current state of A is kept in the finite state control and the current contents of the worktape of A is kept in a separate worktape track. During the simulation, these data are manipulated in a straightforward way.

Third, $h_\mathsf{i} \in \{n+1, \ldots, 0\}$, representing the current head position along the virtual input tape $\vdash b_{n-1} \ldots b_0 \dashv$, is kept in a separate worktape track. A' does not keep the virtual input itself — this would require $\Omega(\log N)$ space. The corresponding input bit is computed "on demand": each time the simulation of a single step of A needs it, A' calls the alternating function input_bit$(h_\mathsf{i}-1)$. For $h_\mathsf{i} = 0$, the input head of A scans the *right* endmarker and no function call is needed. Similarly, for $h_\mathsf{i} = n+1$, the input head of A scans the *left* endmarker. Thus, if the simulation requires to move the input head of A one position to the right (left), A' decreases (increases, respectively) h_i by one. For this reason, this value is initialized to $h_\mathsf{i} = n + 1$.

Fourth, $g \in \{n, \ldots, 0\}$ and $h_\mathsf{s} \in \{n-g, \ldots, 0\}$ are kept in two separate worktape tracks, and $H \in \{0, \ldots, N+1\}$ is represented by the current head position along the unary input $\vdash 1^N \dashv$, initialized to $g = n, h_\mathsf{s} = 0$, and $H = 0$. These values implement the stack of A, capable of containing n bits, by the use of the alternating procedures described in Theorem 4. Namely, if H is written in binary as $v_{n-1} \ldots v_g \ldots v_0$, the current contents in the stack is $\vdash v_{n-1} \ldots v_g$, with v_g on top, and the stack head is scanning $v_{h_\mathsf{s}+g}$. The corresponding bit in the stack is computed on demand: each time the simulation of A needs it, A' calls the alternating function stack_bit(h_s). For $h_\mathsf{s}+g = n$, the stack head scans the bottom-of-the-stack endmarker $\vdash$ and no function call is needed. If the simulation requires to move the stack head to the right (left), h_s is decreased (increased, respectively). Moreover, if $h_s = 0$, i.e., if the stack head is at the topmost symbol, the simulation may require the standard pop/push operations, by the use of the alternating procedures stack_pop or stack_push, respectively. Finally, A' aborts the simulation, if A tries a push operation when the stack is full.

During the simulation, A' mirrors existential/universal branching of A, the same holds for all accepting/rejecting decisions. Clearly, h_i, g, h_s are all bounded by $O(n)$, and hence, by (4), they are kept with $O(\log n) \leq O(\log \log N)$ bits. The same holds for the current contents of the worktape and the position of the worktape head, using $O(\log n) \leq O(\log \log N)$ bits. $\square$

Theorem 7. *If a unary language $\mathcal{L}_1$ is in $\mathrm{ASPACE}^{\mathrm{dm}}(\log \log N)$, then its binary coded version $\mathcal{L}_{1 \to 2}$ is in $\mathrm{ASPACEBINPUSHDOWN}(\log n, n)$.*

To give an idea, we need, for the given alternating machine A accepting $\mathcal{L}_1$, an alternating auxiliary pushdown machine A' such that, on an input $\vdash b_{n-1} \ldots b_0 \dashv$ representing a number $N+1$ in binary, it decides whether $\vdash 1^N \dashv$ is accepted by A. We first allocate a worktape space of size $s = \lfloor \log \log N \rfloor \leq O(\log n)$. The simulation is straightforward; the only problem is that A' has to keep H, the position of the input head of A on $\vdash 1^N \dashv$. Whenever the input head is not placed at the left endmarker ($H = 0$ is handled separately), H is written in binary in the pushdown store. Similarly as on the binary input tape, the most significant bit is hidden, represented by $\vdash$, the bottom-of-the-pushdown endmarker. Thus, the pushdown store containing only $\vdash$ corresponds to $H = 1$ (the leftmost "1" in $\vdash 1^N \dashv$) while $\vdash b_{n-1} \ldots b_0$ (a perfect copy of the binary input) to $H = N+1$ (the right endmarker). During the simulation, on demand, we need to test whether $H = N+1$, i.e., to compare the contents in the pushdown $\vdash v_{k-1} \ldots v_0$ with the binary input tape $\vdash b_{n-1} \ldots b_0 \dashv$ and to increase or decrease H by one, i.e., to replace, for some $i \geq 0$, a string 01^i on top of the pushdown by 10^i, or vice versa. The length of 01^i can be counted by moving along the input $\vdash b_{n-1} \ldots b_0 \dashv$.

By combining Theorems 6 and 7, we thus get:

Theorem 8. *A unary language $\mathcal{L}_1$ is in $\mathrm{ASPACE}^{\mathrm{dm}}(\log \log N)$ if and only if its binary coded version $\mathcal{L}_{1 \to 2}$ is in $\mathrm{ASPACEBINPUSHDOWN}(\log n, n)$ or, equivalently, if and only if $\mathcal{L}_{1 \to 2}$ is in $\mathrm{ASPACEBINSTACK}(\log n, n)$.*

We conclude this section by exhibiting the power of such machines:

Theorem 9. $\mathrm{DTIMESPACE}(2^n n^{O(1)}, n^{O(1)}) \subseteq \mathrm{ASPACEBINPUSHDOWN}(\log n, n)$.

The argument is an updated version of the proof showing that $\mathrm{DTIME}(n^{O(1)}) \subseteq \mathrm{ASPACE}(\log n)$ [3]. As a starting point, we use the simplified presentation from [8]. Let $\mathcal{L}$ be accepted by a single-tape deterministic Turing machine A working in time $t(n) \leq 2^n \cdot n^{O(1)}$ and space $s(n) \leq n^{O(1)}$. For the given input $w \in \mathcal{L}$, consider the rectangular table of size $t(n) \times s(n)$ describing the accepting computation. The top row describes the initial configuration and the bottom row the final configuration. In general, the i-th row describes the configuration after i steps, in the form $\vdash x_{i,1} \ldots x_{i,j_i-1} x_{i,j_i}^{(q_i)} x_{i,j_i+1} \ldots x_{i,s(n)} \bar{\#}$, which displays the used part of the worktape and the current state q_i placed at the position j_i. Each intermediate square is determined by the three squares in the row directly above. The alternating device A' starts from the bottom left corner of the table, since A accepts in the unique state q_F with the head at the left endmarker. Now, for the given square in the i-th row and j-th column, A' branches existentially and guesses the contents in the three squares above. Then, branching universally, these squares are verified in the same way, all the way up to the top row, where guesses are verified by comparison with the input. Since A is deterministic, all guesses along different branches running in parallel must be globally consistent.

During this simulation, the current square is kept in the finite state control. The horizontal coordinate j is stored on the worktape with $O(\log n)$ bits. The vertical coordinate i needs $k \cdot \log n + n$ bits, for some $k \geq 1$. This value is represented by $i' = \lfloor i/2^n \rfloor$, kept on the worktape with $k \cdot \log n$ bits, and by $i'' = i \bmod 2^n$, kept in the pushdown with n bits. The value i'' is manipulated (decreased, tested for zero) in a similar way as in Theorem 7: using the input head of A' as an auxiliary counter, we can replace a suffix 10^i on top of the pushdown by 01^i.

4   Consequences

By combining the results derived so far with the Savitch's theorem [18] and the fact that $\mathrm{ASPACE}^{\mathrm{dm}}(\log \log N)$ is a subset of $\mathrm{DTIME}(N \cdot (\log N)^{O(1)})$ and also of $\mathrm{NTIMESPACE}(N \cdot (\log N)^{O(1)}, (\log N)^{O(1)})$, shown in [11], we can now specify the position of $\mathrm{ASPACE}^{\mathrm{dm}}(\log \log N)_{1 \to 2}$ among the standard complexity classes.

Theorem 10. *If a binary language $\mathcal{L}_2$ is in $\mathrm{DTIMESPACE}(2^n \cdot n^{O(1)}, n^{O(1)})$, then its unary coded version $\mathcal{L}_{2 \to 1}$ is in $\mathrm{ASPACE}^{\mathrm{dm}}(\log \log N)$. In turn, if a unary language $\mathcal{L}_1$ is in $\mathrm{ASPACE}^{\mathrm{dm}}(\log \log N)$, then its binary coded version $\mathcal{L}_{1 \to 2}$ is in $\mathrm{DTIME}(2^n \cdot n^{O(1)}) \cap \mathrm{DSPACE}(n^{O(1)})$ and in $\mathrm{NTIMESPACE}(2^n \cdot n^{O(1)}, n^{O(1)})$.*

By (2), we already had the "downward" translation $\mathrm{P}_2 = \mathrm{ASPACE}(\log n)_2 \subseteq \mathrm{ASPACE}^{\mathrm{dm}}(\log \log N)_{1 \to 2}$. To shed some light on the reverse inclusion, consider the problem of *quantified boolean formulas*, i.e., QBF. This language consists of expressions of the form $(Q_1 x_{i_1} \ldots Q_m x_{i_m}) f(x_1, \ldots, x_m)$, where each Q_j is a universal or existential quantifier and $f(x_1, \ldots, x_m)$ is a boolean function. Moreover, the boolean formula should have no free variables and should be true. Each variable x_i is represented by its index i, written in binary. For example, $(\forall 10 \; \exists 1 \; \forall 11) \, (\neg 1 \vee \neg 10 \wedge 11)$ represents $(\forall x_2 \exists x_1 \forall x_3) \, (\neg x_1 \vee \neg x_2 \wedge x_3)$. Now, let QBF_2 denote a binary version of QBF in which, in a straightforward way, each letter of the alphabet $\Delta = \{\forall, \exists, (,), \vee, \wedge, \neg, 0, 1\}$ is coded by four bits.

Theorem 11. *There exists a binary PSPACE-complete language, namely, QBF_2, such that its unary coded version $\mathrm{QBF}_{2 \to 1}$ is in $\mathrm{ASPACE}^{\mathrm{dm}}(\log \log N)$.*

Proof. It is well known that QBF is PSPACE-complete [17], which can also be found in many textbooks, e.g., in [8,14]. Moreover, the deterministic algorithm for testing membership in QBF, presented in [14, Theorem 11.10], runs not only in $n^{O(1)}$ space, but also in $2^n \cdot n^{O(1)}$ time, even though this is not stated in [14] explicitly. Basically, given a formula of length n with $m \le n$ quantified variables, this algorithm evaluates the formula by a recursive procedure which examines the binary tree with $O(2^m)$ nodes, in which 2^m leaves correspond to the 2^m possible combinations of the truth values assigned to the m variables. Processing of a single node in this tree takes time $n^{O(1)}$, which gives the total computation time $O(2^m) \times n^{O(1)} \le 2^n \cdot n^{O(1)}$, whereas the space depends on the recursion depth, which gives the total space $m \times n^{O(1)} \le n^{O(1)}$. Thus, QBF is in $\text{DTimeSpace}(2^n \cdot n^{O(1)}, n^{O(1)})$.

All this clearly holds for the binary version QBF_2 as well. But then, by Theorem 10, the unary coded version $\text{QBF}_{2 \to 1}$ is in $\text{ASpace}^{\text{dm}}(\log \log N)$. □

Actually, almost all "natural" PSPACE-complete (also NP-complete) problems in the literature belong to $\text{DTimeSpace}(2^n \cdot n^{O(1)}, n^{O(1)})$, and hence their unary versions to $\text{ASpace}^{\text{dm}}(\log \log N)$. Now, using Theorem 11, we get:

Theorem 12. *The class consisting of the binary coded versions of all unary languages in $\text{ASpace}^{\text{dm}}(\log \log N)$ coincides with the class of binary languages in $\text{ASpace}(\log n)$ if and only if* P = PSPACE.

Proof. Since $\text{ASpace}^{\text{dm}}(\log \log N)_{1 \to 2} \supseteq \text{ASpace}(\log n)_2$, by (2), we have that $\text{ASpace}^{\text{dm}}(\log \log N)_{1 \to 2} = \text{ASpace}(\log n)_2$ if and only if $\text{ASpace}^{\text{dm}}(\log \log N)_{1 \to 2} \subseteq \text{ASpace}(\log n)_2$.

Now, suppose that P = PSPACE. Let $\mathcal{L}_1$ be an arbitrary unary language in $\text{ASpace}^{\text{dm}}(\log \log N)$. But then, by Theorem 10, its binary coded version $\mathcal{L}_{1 \to 2}$ is in $\text{DSpace}(n^{O(1)}) = $ PSPACE and hence, by assumption, it is also in P = $\text{ASpace}(\log n)$. Thus, $\text{ASpace}^{\text{dm}}(\log \log N)_{1 \to 2} \subseteq \text{ASpace}(\log n)_2$.

Conversely, let $\text{ASpace}^{\text{dm}}(\log \log N)_{1 \to 2} \subseteq \text{ASpace}(\log n)_2$. By Theorem 11, we have that QBF_2, the binary version of QBF, is a PSPACE-complete language such that its unary coded version $\text{QBF}_{2 \to 1}$ is in $\text{ASpace}^{\text{dm}}(\log \log N)$. But then, by assumption, the binary language QBF_2 is in $\text{ASpace}(\log n) = $ P. Thus, we have a PSPACE-complete language in P, which gives that P = PSPACE.   □

Corollary 13. $\text{ASpace}^{\text{dm}}(\log \log N)_{1 \to 2} \subseteq$ NP *if and only if* NP=PSPACE.

The argument is an easy modification of the proof of Theorem 12. The same holds if "NP" is replaced by any other complexity class $\mathcal{C} \subseteq$ PSPACE such that $\mathcal{C}$ is closed under polynomial time reductions.

In this context, a promising line of research is studying the computational power of the alternating auxiliary pushdown/stack automata with simultaneous bounds on the worktape space and the number of bits in the pushdown/stack. By Theorem 8, we have that the class $\text{ASpace}^{\text{dm}}(\log \log n)_{1 \to 2}$ is equal to both $\text{ASpaceBinPushdown}(\log n, n)_2$ and $\text{ASpaceBinStack}(\log n, n)_2$. Hence, these two classes are equal. On the other hand, if the size of the pushdown/stack is not restricted, these two computational models are substantially different [15]: using our notation, we have $\text{ASPACEBINPUSHDOWN}(\log n, \infty) = \text{DTIME}(2^{n^{O(1)}})$, while $\text{ASPACEBINSTACK}(\log n, \infty) = \text{DTIME}(2^{2^{n^{O(1)}}})$. However, nothing is known about the general case.

Acknowledgment. The author thanks the reviewers for their suggestions, especially for sending a summary of PC discussions which has stimulated future work in this area.

References

1. Allender, E.: The division breakthroughs. Bull. Eur. Assoc. Theoret. Comput. Sci. **74**, 61–77 (2001)
2. Allender, E., Mix Barrington, D., Hesse, W.: Uniform circuits for division: consequences and problems. In: Proceedings IEEE Conference Computational Complexity, pp. 150–159 (2001)
3. Chandra, A., Kozen, D., Stockmeyer, L.: Alternation. J. Assoc. Comput. Mach. **28**, 114–133 (1981)
4. Chang, R., Hartmanis, J., Ranjan, D.: Space bounded computations: review and new separation results. Theoret. Comput. Sci. **80**, 289–302 (1991)
5. Chiu, A.: Complexity of parallel arithmetic using the Chinese remainder representation. Master's thesis, Univ. Wisconsin-Milwaukee (1995). (G. Davida, supervisor)
6. Chiu, A., Davida, G., Litow, B.: Division in logspace-uniform NC^1. RAIRO Inform. Théor. Appl. **35**, 259–275 (2001)
7. Dietz, P., Macarie, I., Seiferas, J.: Bits and relative order from residues, space efficiently. Inform. Process. Lett. **50**, 123–127 (1994)
8. Emde Boas, P.: Machine models and simulations In: Leeuwen, J. (ed.) Handbook of Theoretical Computer Science. Elsevier Science (1989)
9. Geffert, V.: Nondeterministic computations in sublogarithmic space and space constructibility. SIAM J. Comput. **20**, 484–498 (1991)
10. Geffert, V.: Bridging across the log(n) space frontier. Inform. Comput. **142**, 127–158 (1998)
11. Geffert, V.: Alternating demon space is closed under complement and other simulations for sublogarithmic space. In: Brlek, S., Reutenauer, C. (eds.) DLT 2016. LNCS, vol. 9840, pp. 190–202. Springer, Heidelberg (2016). doi:10.1007/978-3-662-53132-7_16
12. Geffert, V., Pardubska, D.: Unary coded NP-complete languages in ASPACE(log log n). Internat. J. Found. Comput. Sci. **24**, 1167–1182 (2013)
13. Hartmanis, J., Immerman, N., Sewelson, W.: Sparse sets in NP–P: EXPTIME versus NEXPTIME. Inform. Control **65**, 158–181 (1985)
14. Hopcroft, J., Motwani, R., Ullman, J.: Introduction to Automata Theory, Languages, and Computation. Addison-Wesley, Reading (2001)
15. Ladner, B., Lipton, R., Stockmeyer, L.: Alternating pushdown and stack automata. SIAM J. Comput. **13**, 135–155 (1984)
16. Macarie, I.I.: Space-efficient deterministic simulation of probabilistic automata. In: Enjalbert, P., Mayr, E.W., Wagner, K.W. (eds.) STACS 1994. LNCS, vol. 775, pp. 109–122. Springer, Heidelberg (1994). doi:10.1007/3-540-57785-8_135
17. Meyer, A., Stockmeyer, L.: Word problems requiring exponential time. In: Proceeding of ACM Symposium Theory of Computation, pp. 1–9 (1973)
18. Savitch, W.: Relationships between nondeterministic and deterministic tape complexities. J. Comput. System Sci. **4**, 177–192 (1970)
19. Szepietowski, A.: Turing Machines with Sublogarithmic Space. LNCS, vol. 843. Springer, Heidelberg (1994)

Turing Degree Spectra of Minimal Subshifts

Michael Hochman[1] and Pascal Vanier[2]([⊠])

[1] Einstein Institute of Mathematics,
Hebrew University of Jerusalem, Jerusalem, Israel
[2] Laboratoire d'Algorithmique, Complexité et Logique,
Université de Paris-Est, LACL, UPEC, Créteil, France
pascal.vanier@lacl.fr

Abstract. Subshifts are shift invariant closed subsets of $\Sigma^{\mathbb{Z}^d}$, with Σ a finite alphabet. Minimal subshifts are subshifts in which all points contain the same patterns. It has been proved by Jeandel and Vanier that the Turing degree spectra of non-periodic minimal subshifts always contain the cone of Turing degrees above any of its degrees. It was however not known whether each minimal subshift's spectrum was formed of exactly one cone or not. We construct inductively a minimal subshift whose spectrum consists of an uncountable number of cones with incomparable bases.

A $\mathbb{Z}^d$-subshift is a closed shift invariant subset of $\Sigma^{\mathbb{Z}^d}$, with Σ finite. Subshifts may be seen as sets of colorings of $\mathbb{Z}^d$, with a finite number of colors, avoiding some set of forbidden patterns. The traditionally studied type of subshifts are Subshifts of Finite Type (SFTs), subshifts that may be defined using a finite family of forbidden patterns. For SFTs, there is a fundamental distinction between dimension one and higher dimensions: in dimension one, SFTs are essentially biinfinite walks on a graph while in higher dimensions SFTs become more complex and can embed Turing machine computations.

Subshift may also be seen as dynamical systems, and it has turned out that many dynamical properties of SFTs, sofic subshifts (letter by letter projections of SFTs) and effective subshifts (subshift that may be defined with a recursively enumerable family of forbidden patterns) can be characterized by means of computability theoretic objects. Examples of such characterizations include entropy, which can be characterized as the right recursively enumerable reals [HM10], slopes of periodicity [JV10], subactions [Hoc09, AS13, DRS10] and several other aspects [FS12, SS16].

From a computable point of view, effective subshifts, SFTs and sofic subshifts are all Π_1^0 *classes*. These are subsets of $\{0,1\}^{\mathbb{N}}$ for which there exists a Turing machine which, given a point of $\{0,1\}^{\mathbb{N}}$ as an oracle, halts if and only if it is not in the class. One measure of the computational power of classes of sets are Muchnik and Medvedev degrees: two classes are Muchnik equivalent if for each point of one of the classes there exists a computable function which maps it to some point of the other class. They are Medvedev equivalent if we have uniformity: if it is the same function for all points. Simpson [Sim11], building on

© Springer International Publishing AG 2017
P. Weil (Ed.): CSR 2017, LNCS 10304, pp. 154–161, 2017.
DOI: 10.1007/978-3-319-58747-9_15

the work of Hanf [Han74] and Myers [Mye74], proved that the Medvedev and Muchnik degrees of SFTs are the same as the Medvedev degrees of Π_1^0 classes in general.

However, this measure is not very fine-grained, for instance two subshifts can be Muchnik equivalent but not even have the same cardinality: one may be uncountable while the other is countable or even finite, in particular, two subshifts containing each a computable point will always be Muchnik/Medvedev equivalent. A more precise measure is the *Turing degree spectrum*: the set of Turing degrees of its points. These have first been studied on one dimensional subshifts Cenzer, Dashti, and King [CDK08] and Cenzer, Dashti, Toska, and Wyman [CDTW10, CDTW12].

Subsequently Jeandel and Vanier [JV13] focused on Turing degree spectra of the classical classes of multidimensional subshifts: SFTs, sofic and effective subshifts. They proved in particular that the Turing degree spectra of SFTs are almost the same as the spectra of Π_1^0 classes (see Kent and Lewis [KL10] for a survey of Π_1^0 spectra):

Theorem (Jeandel and Vanier [JV13], Theorem 4.1). *Let $S \subseteq \{0,1\}^{\mathbb{N}}$ be a Π_1^0 class, there exists an SFT X such that X and S are recursively homeomorphic up to a uniformly computable subset of X.*

This means that if we add $\mathbf{0}$ to the spectrum of S then we can construct an SFT with the same spectrum. In order to show that it is not possible to realize exactly some Π_1^0 classes, that is without adding a computable point, they studied the spectra of minimal subshifts.

Minimal subshifts are subshifts containing no proper subshift, or equivalently subshifts in which all configurations have the same finite subpatterns. They are fundamental in the sense that all subshifts contain at least one minimal subshift [Bir12].

It was proved in [JV13] that the spectrum of any non-periodic minimal subshift contains the cone above any of its degrees (a definition of cone of Turing degrees is given in Sect. 1.3):

Theorem (Jeandel and Vanier [JV13], Theorem 5.10). *Let X be a minimal non-finite subshift (i.e. non-periodic in at least one direction). For any point $x \in X$ and any degree $\mathbf{d} \geq_T \deg_T x$, there exists a point $y \in X$ such that $\mathbf{d} = \deg_T y$.*

Here we answer the followup question of whether a minimal subshift's spectrum always corresponds to a single cone or if there exists one containing at least two cones of incomparable bases. It is quite easy to prove the following theorem:

Theorem 1. *For any Turing degree $\deg_T d$, there exists a minimal subshift X whose spectrum of Turing degrees is a cone of base $\deg_T d$.*

For instance the spectrum of a Sturmian subshift [MH40] with an irrational angle is the cone whose base is the degree of the angle of the rotation. The theorem can also be seen as a corollary of Miller's proof [Mil12] [Proposition 3.1] of a result on Medvedev degrees.

Theorem 2. *There exist a minimal subshift $X \subset \{0,1\}^{\mathbb{Z}^d}$ and points $x_z \in X$ with $z \in \{0,1\}^{\mathbb{N}}$ such that for any $z \neq z' \in \{0,1\}^{\mathbb{N}}$, $\deg_T x_z$ and $\deg_T x_{z'}$ are incomparable and such that there exists no point $y \in X$ with $\deg_T y \leq_T \deg_T x_z$ and $\deg_T y \leq_T \deg_T x_{z'}$.*

That is to say, there exists a minimal subshift whose spectrum consists of $2^{\aleph_0}$ cones with incomparable bases.

The subshift constructed in this proof is not effective and cannot be "effectivized", since minimal effective subshifts always contain a computable point and thus their spectra are the whole set of Turing degrees when they are non-periodic.

1   Preliminary Definitions

1.1   Words and Languages

For a possibly infinite word $w = w_0 \ldots w_n$, we denote $w_{[i,j]} = w_i \ldots w_j$. For a language L, denote by L^* the set of finite words formed by concatenations of words of L and L^ω the set of infinite words formed by concatenations of words of L^*. We will say that a word w' extends a word w when w is a prefix of w'.

1.2   Subshifts

We give here some standard definitions and facts about subshifts, one may consult the book of Lind and Marcus [LM95] for more details.

Let Σ be a finite alphabet, its elements are called *symbols*, the d-dimensional full shift on Σ is the set $\Sigma^{\mathbb{Z}^d}$ of all maps (colorings) from $\mathbb{Z}^d$ to the Σ (the colors). For $v \in \mathbb{Z}^d$, the shift functions $\sigma_v : \Sigma^{\mathbb{Z}^d} \to \Sigma^{\mathbb{Z}^d}$, are defined locally by $\sigma_v(c_x) = c_{x+v}$. The full shift equipped with the distance $d(x,y) = 2^{-\min\left\{\|v\| \,\middle|\, v \in \mathbb{Z}^d, x_v \neq y_v\right\}}$ is a compact metric space on which the shift functions act as homeomorphisms. An element of $\Sigma^{\mathbb{Z}^d}$ is called a *configuration*.

Every closed shift-invariant (invariant by application of any σ_v) subset X of $\Sigma^{\mathbb{Z}^d}$ is called a *subshift*. An element of a subshift is called a *point* of this subshift.

Alternatively, subshifts can be defined with the help of forbidden patterns. A *pattern* is a function $p : P \to \Sigma$, where P, the *support*, is a finite subset of $\mathbb{Z}^d$. We say that a configuration x contains a pattern $p : P \to \Sigma$, or equivalently that the pattern p appears in x, if there exists $z \in \mathbb{Z}^d$ such that $x_{|z+P} = p$.

Let $\mathcal{F}$ be a collection of *forbidden* patterns, $X_{\mathcal{F}}$ is the subset of $\Sigma^{\mathbb{Z}^d}$ containing all configurations having nowhere a pattern of $\mathcal{F}$. More formally, $X_{\mathcal{F}}$ is defined by

$$X_{\mathcal{F}} = \left\{ x \in \Sigma^{\mathbb{Z}^d} \,\middle|\, \forall z \in \mathbb{Z}^d, \forall p \in F, x_{|z+P} \neq p \right\}.$$

In particular, a subshift is said to be a *subshift of finite type* (SFT) when it can be defined by a collection of forbidden patterns that is finite. Similarly, an *effective subshift* is a subshift which can be defined by a recursively enumerable collection of forbidden patterns. A subshift $Y \subseteq \Gamma^{\mathbb{Z}^d}$ is *sofic* if there exists an SFT X on some alphabet Σ and a letter by letter projection $\pi : \Gamma \to \Sigma$, such that $\pi(X) = Y$, where we extended π naturally on configurations.

Let us now come to the definition of the object of study of this article:

Definition 1 (Minimal subshift). *A subshift X is called* minimal *if it verifies one of the following equivalent conditions:*

- *There is no subshift Y such that $Y \subsetneq X$.*
- *All the points of X contain the same patterns.*
- *It is the closure of the orbit of any of its points.*

In the sequel, we will use the two latter conditions.

1.3   Computability

We give here some definitions and basic notations for computability theory, a detailed introduction may be found in Rogers [Rog87].

A set $A \subseteq \mathbb{N}$ is called *recursively enumerable* if there exists a Turing machine that enumerates each of its elements, or equivalently, if there exists a Turing machine that halts only when its input is an element of A.

For $x, y \subset \{0, 1\}^{\mathbb{N}}$, we say that $x \leq_T y$ if there exists a Turing machine M such that M with oracle y computes x, that is to say M has access to a supplementary tape on which y is written. Of course $x \equiv_T y$ when we have both $x \leq_T y$ and $y \leq_T x$. The Turing degree of x is the equivalence class of x with respect to $\equiv_T$. The spectrum of some subset $S \subseteq \{0, 1\}^{\mathbb{N}}$ is defined as $\mathbf{Sp}(S) = \{\mathbf{d} \mid \exists x \in S, \mathbf{d} = \deg_T x\}$.

A *cone* of Turing degrees is a set of degrees D for which there exists a degree $\mathbf{d}$ such that $D = \{d \mid d \geq_T \mathbf{d}\}$.

We call *recursive operator* a partial function $\phi : \{0, 1\}^{\mathbb{N}} \to \{0, 1\}^{\mathbb{N}}$ corresponding to a Turing machine whose input is its oracle and output is an infinite sequence of bits. We say that the function is undefined on the inputs on which the Turing machine does not output an infinite sequence of bits.

2   Minimal Subshifts with Several Cones

Lemma 1. *There exists a countable set $\mathcal{C} \subseteq \{0, 1\}^{\mathbb{N}}$ such that for any finite language L, any two recursive partial operators $\phi_1, \phi_2 : \{0, 1\}^{\mathbb{N}} \to \{0, 1\}^{\mathbb{N}}$ and two distinct words $w_1, w_2 \subseteq L^*$, there exist two words $w_1', w_2' \in L^*$ extending respectively w_1 and w_2 such that we have one of the following:*

- *(a) either for any pair $x, y \in L^{\omega}$, $\phi_1(w_1' x)$ differs from $\phi_2(w_2' y)$ when they are both defined,*
- *(b) or for any pair $x, y \in L^{\omega}$, $\phi_1(w_1 x) = \phi_2(w_2 y) \in \mathcal{C}$ when both defined.*

Proof. Let M_1, M_2 be the Turing machines computing the functions $x \mapsto \phi_1(w_1x)$, $x \mapsto \phi_2(w_2x)$ respectively. When restricting ourselves to inputs on which both operators are defined, by continuity of computable functions, it is quite clear that:

- either there exists some sequences $x, y \in L^*$ such that $M_1(x)$'s output differs from $M_2(y)$'s output at some step,
- or the outputs of both machines M_1, M_2 do not depend on their inputs on their respective domains and are equal, in this latter case, we are in case b. We define $\mathcal{C}$ to be the set of these outputs, for all finite languages L and all such pairs M_1, M_2 of operators.

In the former case, there exist prefixes w_1' and w_2' of w_1x and w_2y such that the partial outputs of M_1 once it has read w_1' already differs from the partial output of M_2 once it has read w_2'.

In the latter case, one may take $w_1' = w_1$ and $w_2' = w_2$. $\mathcal{C}$ is countable since there is a countable number of tuples $L, \phi_1, \phi_2, w_1, w_2$. □

Theorem 3. *There exists a minimal subshift $X \subseteq \{0,1\}^{\mathbb{N}}$ whose spectrum contains $2^{\aleph_0}$ disjoint cones of Turing degrees with incomparable bases with no degree below any two of them.*

Note that this proof is in no way effective. As a matter of fact, for minimal subshifts, it is equivalent to be effective and to contain a recursive point [BJ10]. The set of Turing degrees an effective subshift is thus always the cone of *all* degrees.

The following theorem establishes a variant of Theorem 2 for one-sided, one-dimensional subshifts. After the proof we comment on how to recover the full statement of Theorem 2, which applies to two-sided onedimensional, and to multidimensional, subshifts.

Proof. We construct a sequence of sofic subshifts $(X_i)_{i \in \mathbb{N}}$ such that $X_{i+1} \subseteq X_i$ and such that the limit $X = \bigcap_{i \in \mathbb{N}} X_i$ is minimal. In the process of constructing the X_i, which will be formed of concatenations of allowed words $x_1^i, \ldots, x_k^i$, we ensure that no extensions of two distinct words may compute an identical sequence with any of the first i Turing machines. At the same time, we make sure that all allowed words of level $i+1$ contain all words of level i, thus enforcing the minimality of the limit X. We also have to avoid that the limit X contains a computable point.

Let $(\mathcal{M}_i)_{i \in \mathbb{N}}$ be an enumeration of all minimal subshifts containing a point of the set $\mathcal{C}$ defined in Lemma 1. Such an enumeration exists since $\mathcal{C}$ is countable and minimal subshifts are the closure of any of their points. We will also need an enumeration $(\phi_i)_{i \in \mathbb{N}}$ of the partial recursive operators from $\{0,1\}^{\mathbb{N}}$ to $\{0,1\}^{\mathbb{N}}$.

Now let us define the sequence of sofic shifts $(X_i)_{i \in \mathbb{N}^*}$. These will actually be renewal systems, a subclass of sofic shifts, see Williams [Wil90]: each of them will be the shift invariant closure of the biinfinite words formed by concatenations of words of some language L_i which here will be finite languages.

We define $X_0 = \{0,1\}^{\mathbb{N}}$ that is to say X_0 is generated by $L_1 = \{w_0 = 0, w_1 = 1\}$. Let us now give the induction step. At each step, L_{i+1} will contain 2^{i+1} words $w_{0...0}, \ldots, w_{1...1}$, the indices being binary words of length $i+1$, which will verify the following conditions:

1. The words w_{b0}, w_{b1} of L_{i+1} start with the word w_b of L_i and consist only of concatenations of words of L_i.
2. The words w_b with $b \in \{0,1\}^{i+1}$ of L_{i+1} each contain all the words of $w_{b'}$ with $b' \in \{0,1\}^i$ of L_i.
3. For any two words $w_b \neq w_{b'}$ of L_{i+1} and for all $j, j' \leq i$:
   - Either for all $x, y \in L_i^{\omega}$, $\phi_j(w_b x) \neq \phi_{j'}(w_{b'} y)$ when both defined,
   - Or for all $x, y \in L_i^{\omega}$, $\phi_j(w_b x), \phi_{j'}(w_{b'} y)$ are in $\mathcal{C}$ when defined.
4. The words w_{b0}, w_{b1} do not appear in any configuration of $\mathcal{M}_j$, for all $j \leq i$.

Conditions 1 and 2 are easy to ensure: just make w_{ba}, $b \in \{0,1\}^i$, $a \in \{0,1\}$ start with w_b followed by all concatenations of words $w_{b'}$ with $b' \in \{0,1\}^i$. We then use Lemma 1 to extend any w previously constructed into a word w' verifying condition 3, this is done several times in a row, once for every quadruple $w, w', \phi_j, \phi_{j'}$, using language $L = L_i$. And finally, since X_i is not minimal, we can extend w' so that it contains a pattern appearing in none of the $\mathcal{M}_j$'s for $j \leq i$, to obtain condition 4. Now we can just extend w' with two different words thus obtaining w_{b0} and w_{b1}.

Now let's check that this leads to the desired result:

- $X = \bigcap X_i$ is a countable intersection of compact shift-invariant non-empty spaces, it is compact and shift-invariant and non-empty, thus a subshift.
- Any pattern p appearing in some point of X is contained in a pattern w_b, with $b \in \{0,1\}^i$ for some i, by construction (condition 2), all $w_{b'}$ with $b \in \{0,1\}^{i+1}$ contain w_b. Therefore, all points of X, since they are contained in X_{i+1}, contain w_b and hence p. So X is minimal.
- For all $z \in \{0,1\}^{\mathbb{N}}$, define the points $x_z = \lim_{i \to \infty} w_{z[0,i]}$, they are in X because they belong to each X_i. Condition 3 ensures that if from two of them one could compute the same sequence $y \in \{0,1\}^{\mathbb{N}}$, then this sequence would be in $\mathcal{C}$. But condition 4 ensures that no point of X belongs to a minimal subshift containing a point of $\mathcal{C}$.

  This means that for any two $z \neq z' \in \{0,1\}^{\mathbb{N}}$ no point in X is a common lower bound for x_z and $x_{z'}$, in particular this means that each of them is the base of the only cone it belongs to.   □

It is quite straightforward to transform this proof in order to get a subshift on $\{0,1\}^{\mathbb{Z}}$ instead of $\{0,1\}^{\mathbb{N}}$: it suffices to allow words to be extended in two directions and to now put w_b in the center of any w_{ba}, so that the limit sequence is well defined. Note that the words do not need to grow at the same rate on both sides, but need to be strictly increasing on both sides. One way to do this is for instance to put a copy of all concatenations of all words of the previous level on both sides instead of just one.

We thus obtain Theorem 2 by making the minimal subshift on $\mathbb{Z}$ periodic in all other directions of $\mathbb{Z}^d$:

Corollary 1. *For any dimension d, there exists a minimal subshift $X \subseteq \{0,1\}^{\mathbb{Z}^d}$ whose spectrum of Turing degrees contains $2^{\aleph_0}$ cones of incomparable bases with no degree below any two of them.*

References

[AS13]   Aubrun, N., Sablik, M.: Simulation of effective subshifts by two-dimensional subshifts of finite type. Acta Applicandae Mathematicae **126**(1), 35–63 (2013). ISSN: 1572-9036. 1007, doi:10.1007/s10440-013-9808-5, http://dx.doi.org/10.1007/s10440-013-9808-5

[Bir12]   Birkhoff, M.-G.D.: Quelques théorèmes sur le mouvement des systèmes dynamiques. Bulletin de la SMF **40**, 305–323 (1912)

[BJ10]   Ballier, A., Jeandel, E.: Computing (or not) quasi-periodicity functions of tilings. In: Second Symposium on Cellular Automata (JAC) (2010)

[CDK08]   Cenzer, D., Dashti, A., King, J.L.F.: Computable symbolic dynamics. Math. Logic Q. **54**(5), 460–469 (2008). doi:10.1002/malq.200710066

[CDTW10]   Cenzer, D., Dashti, A., Toska, F., Wyman, S.: Computability of countable subshifts. In: Ferreira, F., Löwe, B., Mayordomo, E., Mendes Gomes, L. (eds.) CiE 2010. LNCS, vol. 6158, pp. 88–97. Springer, Heidelberg (2010). doi:10.1007/978-3-642-13962-8_10

[CDTW12]   Cenzer, D., Dashti, A., Toska, F., Wyman, S.: Computability of countable subshifts in one dimension. Theory Comput. Syst. **51**, 352–371 (2012). doi:10.1007/s00224-011-9358-z

[DRS10]   Durand, B., Romashchenko, A., Shen, A.: Effective closed subshifts in 1D can be implemented in 2D. In: Blass, A., Dershowitz, N., Reisig, W. (eds.) Fields of Logic and Computation. LNCS, vol. 6300, pp. 208–226. Springer, Heidelberg (2010). doi:10.1007/978-3-642-15025-8_12

[FS12]   Fernique, T., Sablik, M.: Local rules for computable planar tilings. In: Formenti, E. (ed.) Proceedings 18th International Workshop on Cellular Automata and Discrete Complex Systems and 3rd International Symposium Journées Automates Cellulaires (AUTOMATA & JAC 2012), La Marana, Corsica, 19–21 September 2012, vol. 90, EPTCS, pp. 133–141. doi:10.4204/EPTCS.90.11, http://dx.doi.org/10.4204/EPTCS.90.11

[Han74]   Hanf, W.: Non recursive tilings of the plane I. J. Symb. Logic **39**(2), 283–285 (1974)

[HM10]   Hochman, M., Meyerovitch, T.: A characterization of the entropies of multidimensional shifts of finite type. Ann. Math. **171**(3), 2011–2038 (2010). doi:10.4007/annals.2010.171.2011

[Hoc09]   Hochman, M.: On the dynamics, recursive properties of multidimensional symbolic systems. Inventiones Mathematicae **176**, 131 (2009)

[JV10]   Jeandel, E., Vanier, P.: Slopes of tilings. In: Kari, J. (ed.) JAC, pp. 145–155. Turku Center for Computer Science (2010). ISBN: 978-952-12-2503-1

[JV13]   Jeandel, E., Vanier, P.: Turing degrees of multidimensional SFTs. In: Theoretical Computer Science 505.0. Theory and Applications of Models of Computation 2011, pp. 81–92 (2013). ISBN: 0304-3975. http://dx.doi.org/10.1016/j.tcs.2012.08.027, http://www.sciencedirect.com/science/article/pii/S0304397512008031

[KL10]   Kent, T., Lewis, A.E.M.: On the degree spectrum of a Π_1^0 class. Trans. Am. Math. Soc. **362**, 5283–5319 (2010)

[LM95]   Lind, D., Marcus, B.: An Introduction to Symbolic Dynamics and Coding. Cambridge University Press, New York (1995)

[MH40]   Morse, H.M., Hedlund, G.A.: Symbolic dynamics II. Sturmian trajectories. Am. J. Math. **62**(1), 1–42 (1940)

[Mil12]   Miller, J.S.: Two notes on subshifts. Proc. Am. Math. Soc. **140**(5), 1617–1622 (2012). doi:10.1090/S0002-9939-2011-11000-1

[Mye74]   Myers, D.: Non recursive tilings of the plane II. J. Symbol. Logic **39**(2), 286–294 (1974)

[Rog87]   Rogers Jr., H.: Theory of Recursive Functions and Effective Computability. MIT Press, Cambridge (1987)

[Sim11]   Simpson, S.G.: Medvedev degrees of 2-dimensional subshifts of finite type. In: Ergodic Theory and Dynamical Systems (2011)

[SS16]   Sablik, M., Schraudner, M.: Algorithmic complexity for the realization of an effective subshift by a sofic. In: Chatzigiannakis, I., Mitzenmacher, M., Rabani, Y., Sangiorgi, D. (eds.) 43rd International Colloquium on Automata, Languages, and Programming (ICALP 2016), Leibniz International Proceedings in Informatics (LIPIcs), vol. 55, pp. 110:1–110:14. Schloss Dagstuhl-Leibniz-Zentrum fuer Informatik, Dagstuhl (2016). ISBN: 978-3-95977-013-2, http://dx.doi.org/10.4230/LIPIcs.ICALP.2016. 110, http://drops.dagstuhl.de/opus/volltexte/2016/6245

[Wil90]   Williams, S.: Notes on renewal systems. Proc. Am. Math. Soc. **110**(3), 851–853 (1990)

Reordering Method and Hierarchies for Quantum and Classical Ordered Binary Decision Diagrams

Kamil Khadiev[1,2]([✉]) and Aliya Khadieva[2]

[1] University of Latvia, Riga, Latvia
kamilhadi@gmail.com
[2] Kazan Federal University, Kazan, Russia
aliya.khadi@gmail.com

Abstract. We consider Quantum OBDD model. It is restricted version of read-once Quantum Branching Programs, with respect to "width" complexity. It is known that maximal complexity gap between deterministic and quantum model is exponential. But there are few examples of such functions. We present method (called "reordering"), which allows to build Boolean function g from Boolean Function f, such that if for f we have gap between quantum and deterministic OBDD complexity for natural order of variables, then we have almost the same gap for function g, but for any order. Using it we construct the total function REQ which deterministic OBDD complexity is $2^{\Omega(n/log n)}$ and present quantum OBDD of width $O(n^2)$. It is bigger gap for explicit function that was known before for OBDD of width more than linear. Using this result we prove the width hierarchy for complexity classes of Boolean functions for quantum OBDDs.

Additionally, we prove the width hierarchy for complexity classes of Boolean functions for bounded error probabilistic OBDDs. And using "reordering" method we extend a hierarchy for k-OBDD of polynomial size, for $k = o(n/log^3 n)$. Moreover, we proved a similar hierarchy for bounded error probabilistic k-OBDD. And for deterministic and probabilistic k-OBDDs of superpolynomial and subexponential size.

Keywords: Quantum computing · Quantum OBDD · OBDD · Branching programs · Quantum vs classical · Quantum models · Hierarchy · Computational complexity · Probabilistic OBDD

1 Introduction

Branching programs are one of the well known models of computation. These models have been shown useful in a variety of domains, such as hardware verification, model checking, and other CAD applications (see for example the book

K. Khadiev—Partially supported by ERC Advanced Grant MQC. The work is performed according to the Russian Government Program of Competitive Growth of Kazan Federal University.

© Springer International Publishing AG 2017
P. Weil (Ed.): CSR 2017, LNCS 10304, pp. 162–175, 2017.
DOI: 10.1007/978-3-319-58747-9_16

by I. Wegener [Weg00]). It is known that the class of Boolean functions computed by polynomial size branching programs coincide with the class of functions computed by non-uniform log-space machines.

One of important restrictive branching programs are oblivious read once branching programs, also known as Ordered Binary Decision Diagrams (OBDD) [Weg00]. It is a good model of data streaming algorithms. These algorithms are actively used in industry, because of rapidly increasing of size of data which should be processed by programs. Since a length of an OBDD is at most linear (in the length of the input), the main complexity measure is "width", analog of size for automata. And it can be seen as nonuniform automata (see for example [AG05]). In the last decades quantum OBDDs came into play [AGK01, NHK00, SS05a, Sau06].

In 2005 F. Ablayev, A. Gainutdinova, M. Karpinski, C. Moore and C. Pollett [AGK+05] proved that the gap between width of quantum and deterministic OBDD is at most exponential. They showed that this bound can be reached for MOD_p function that takes the value 1 on input such that number of 1s by modulo p is 0. Authors presented quantum OBDD of width $O(\log p)$ for this function (another quantum OBDD of same width is presented in [AV08]) and proved that any deterministic OBDD has width at least p. However explicit function MOD_p presents a gap for OBDD of at most linear width. For bigger width it was shown that Boolean function $PERM$ has not deterministic OBDD of width less than $2^{\sqrt{n}/2}/(\sqrt{n}/2)^{3/2}$ [KMW91] and M. Sauerhoff and D. Sieling [SS05b] constructed quantum OBDD of width $O(n^2 \log n)$. F. Ablayev, A. Khasianov and A. Vasiliev [AKV08] presented the quantum OBDD of width $O(n \log n)$ for $PERM$. Let us note that for partial functions the gap between widths of quantum and deterministic OBDDs can be more than exponential [AGKY14, Gai15, AGKY16].

Better difference between quantum and deterministic complexity was proven in [AKV08] for Equality function. But there authors had exponential gap only for natural order of variables. They presented the quantum OBDD of width $O(n)$, which is based on quantum fingerprinting technique, at the same time any deterministic OBDD has width at least $2^{n/2}$ for natural order. But if we consider any order, then we can construct deterministic OBDD of constant width.

Changing the order is one of the main issues for proving lower bound on width for OBDD. We present a technique that allows to build Boolean Function g from Boolean function f. We consider f such that any deterministic OBDD with natural order for the function has width at least $d(n)$ and we can construct quantum OBDD of width $w(n)$. In that case we present quantum OBDD of width $O(w(n/logn) \cdot n/logn)$ for function g and any deterministic OBDD has width at least $d(O(n/logn))$. It means that if difference between quantum OBDD complexity of function f and deterministic OBDD complexity for natural order is exponential, then we have almost exponential difference for function g. We called this method "reordering". And idea is based on adding addresses of variables to input. Similar idea was used in [Kha15].

Then, we present Boolean function *Reordered Equality* (*REQ*), it is modification of Equality function [AKV08]. We apply the main ideas of *reordering* and prove that for REQ deterministic OBDD has width $2^{\Omega(n/\log n)}$ and bounded error quantum OBDD has width $O(n^2/\log^2 n)$. This gap between deterministic and quantum complexity is better than for $PERM$ function. And it has advantage over results on EQ, because we prove distance for any order of input variables. And in comparing to MOD_p, we can show a distance for bigger width.

Using complexity properties of MOD_p function, REQ function and *Mixed weighted sum function* (MWS_n) [Sau05] we prove the width hierarchy (not tight) for classes of Boolean functions computed by bounded error quantum OBDD. The hierarchy is separated to three cases: first one is for width less than $\log n$, and for this case we prove hierarchy with small gap between width-parameters of classes. Second one is for width less than n and bigger gap. And third one is for width less than $2^{O(n)}$ and here gap is the biggest. Similar hierarchy is already known for deterministic and nondeterministic OBDD [AGKY14, AGKY16], for deterministic k-OBDD [Kha15]. And we present not tight width hierarchy for bounded error probabilistic OBDD in the paper.

Forth group of results is extending the hierarchy for deterministic and bounded error probabilistic k-OBDD of polynomial size. The known tight hierarchy for classes of Boolean functions that computed by a deterministic k-OBDD of polynomial size is result of B. Bollig, M. Sauerhoff, D. Sieling, and I. Wegener [BSSW98]. They proved that P-$(k-1)$OBDD $\subsetneq$ P-kOBDD for $k = o(\sqrt{n} \log^{3/2} n)$. It is known extension of this hierarchy for $k = o(n/\log^2 n)$ in papers [Kha16, AK13]. But this hierarchy is not tight with the gap between classes in hierarchy at least not constant. We prove almost tight hierarchy P-kOBDD $\subsetneq$ P-$2k$OBDD for $k = o(n/\log^3 n)$. Our result is better than both of them. It is better than first one, because k is bigger, but at the same time it is not enough tight. And it is better than second one because proper inclusion of classes is proven if k is 2 times increased. Additionally, we prove almost tight hierarchy for $k - OBDD$ of superpolynomial and subexponential size. These hierarchies improve known not tight hierarchy from [Kha16]. Our hierarchy is almost tight (with small gap), but for little bit smaller k. The proof of hierarchies is based on complexity properties of Boolean function *Reordered Pointer Jumping*, it is modification of *Pointer Jumping* function from [NW91, BSSW98], is based on ideas of *reordering* method. For probabilistic case it is not known tight hierarchy for polynomial size only for sublinear width [Kha16]. Additionally, for more general model Probabilistic k-BP Hromkovich and Sauerhoff in 2003 [HS03] proved the tight hierarchy for $k \leq \log n/3$. We proved similar almost tight hierarchy for polynomial size bounded error probabilistic k-OBDD with error at most $1/3$ for $k = o(n^{1/3}/\log n)$. And almost tight hierarchies for superpolynomial and subexponential size, these results improve results from [Kha16]. Note that, for example for nondeterministic k-OBDD we cannot get result better than [Kha16], because for constant k 1-OBDD of polynomial size and k-OBDD compute the same Boolean functions [BHW06].

Structure of the paper is following. Section 2 contains description of models, classes and other necessary definitions. Discussion reordering method and applications for quantum OBDD located in Sect. 3. The width hierarchies for quantum and probabilistic OBDDs are proved in Sect. 4. Finally, Sect. 5 contains applying reordering method and hierarchy results to deterministic and probabilistic k-OBDD.

2   Preliminaries

Ordered read ones Branching Programs (OBDD) are well known model for Boolean functions computation. A good source for different models of branching programs is the book by I. Wegener [Weg00].

A branching program over a set X of n Boolean variables is a directed acyclic graph with two distinguished nodes s (a source node) and t (a sink node). We denote such program $P_{s,t}$ or just P. Each inner node v of P is associated with a variable $x \in X$. *Deterministic* P has exactly two outgoing edges labeled $x = 0$ and $x = 1$ respectively for such node v.

The program P computes the Boolean function $f(X)$ $(f : \{0,1\}^n \to \{0,1\})$ as follows: for each $\sigma \in \{0,1\}^n$ we let $f(\sigma) = 1$ if and only if there exists at least one $s - t$ path (called *accepting* path for σ) such that all edges along this path are consistent with σ.

A branching program is *leveled* if the nodes can be partitioned into levels $V_1, \ldots, V_\ell$ and a level $V_{\ell+1}$ such that the nodes in $V_{\ell+1}$ are the sink nodes, nodes in each level V_j with $j \leq \ell$ have outgoing edges only to nodes in the next level V_{j+1}. For a leveled $P_{s,t}$ the source node s is a node from the first level V_1 of nodes and the sink node t is a node from the last level $V_{\ell+1}$.

The *width* $w(P)$ of a leveled branching program P is the maximum of number of nodes in levels of P. $w(P) = \max_{1 \leq j \leq \ell} |V_j|$. The *size* of branching program P is a number of nodes of program P.

A leveled branching program is called *oblivious* if all inner nodes of one level are labeled by the same variable. A branching program is called *read once* if each variable is tested on each path only once. An oblivious leveled read once branching program is also called Ordered Binary Decision Diagram (OBDD). OBDD P reads variables in its individual order $\pi = (j_1, \ldots, j_n)$, $\pi(i) = j_i$, $\pi^{-1}(j)$ is position of j in permutation π. We call $\pi(P)$ the order of P. Let us denote natural order as $id = (1, \ldots, n)$. Sometimes we will use notation id-OBDD P, it means that $\pi(P) = id$. Let $width(f) = \min_P w(P)$ for OBDD P which computes f and $id-width(f)$ is the same but for id-OBDD.

The Branching program P is called k-OBDD if it consists from k layers, where i-th $(1 \leq i \leq k)$ layer P^i of P is an OBDD. Let π_i be an order of P^i, $1 \leq i \leq k$ and $\pi_1 = \cdots = \pi_k = \pi$. We call order $\pi(P) = \pi$ the order of P.

Let $tr_P : \{1, \ldots, n\} \times \{1, \ldots, w(P)\} \times \{0,1\} \to \{1, \ldots, w(P)\}$ be transition function of OBDD P on level i. OBDD P is called *commutative* if for any permutation π' we can construct OBDD P' by just reordering transition functions and P' still computes the same function. Formally, it means $tr_{P'}(i, s, x_{\pi'(i)}) = tr_P(\pi^{-1}(\pi'(i)), s, x_{\pi'(i)})$, for π is order of P, $i \in \{1, \ldots, n\}$, $s \in \{1, \ldots, w(P)\}$. k-OBDD P is commutative if each layer is commutative OBDD.

Nondeterministic OBDD (NOBDD) is nondeterministic counterpart of OBDD. Probabilistic OBDD (POBDD) can have more than two edges for node, and choose one of them using probabilistic mechanism. POBDD P computes Boolean function f with bounded error $0.5 - \varepsilon$ if probability of right answer is at least $0.5 + \varepsilon$.

Let us discuss a definition of quantum OBDD (QOBDD). It is given in different terms, but you can see that it is equivalent. You can see [AGK+05], [AGK01] for more details.

For a given $n > 0$, a quantum OBDD P of width w, defined on $\{0,1\}^n$, is a 4-tuple $P = (T, |\psi\rangle_0, Accept, \pi)$, where

- $T = \{T_j : 1 \leq j \leq n \text{ and } T_j = (G_j^0, G_j^1)\}$ are ordered pairs of (left) unitary matrices representing the transitions is applied at the j-th step, where G_j^0 or G_j^1, determined by the corresponding input bit, is applied.
- $|\psi\rangle_0$ is initial vector from w-dimensional Hilbert space over field of complex numbers. $|\psi\rangle_0 = |q_0\rangle$ where q_0 corresponds to the initial node.
- $Accept \subset \{1, \ldots, w\}$ is accepting nodes.
- π is a permutation of $\{1, \ldots, n\}$ defining the order of testing the input bits.

For any given input $\sigma \in \{0,1\}^n$, the computation of P on σ can be traced by a vector from w-dimensional Hilbert space over field of complex numbers. The initial one is $|\psi\rangle_0$. In each step j, $1 \leq j \leq n$, the input bit $x_{\pi(j)}$ is tested and then the corresponding unitary operator is applied: $|\psi\rangle_j = G_j^{x_{\pi(j)}}(|\psi\rangle_{j-1})$, where $|\psi\rangle_{j-1}$ and $|\psi\rangle_j$ represent the state of the system after the $(j-1)$-th and j-th steps, respectively, where $1 \leq j \leq n$.

In the end of computation program P measure qubits. The accepting (return 1) probability $Pr_{accept}(\sigma)$ of P_n on input σ is $Pr_{accept}(\nu) = \sum_{i \in Accept} v_i^2$, for $|\psi\rangle_n = (v_1, \ldots, v_w)$. We say that a function f is computed by P with bounded error if there exists an $\varepsilon \in (0, \frac{1}{2}]$ such that P accepts all inputs from $f^{-1}(1)$ with a probability at least $\frac{1}{2} + \varepsilon$ and P_n accepts all inputs from $f^{-1}(0)$ with a probability at most $\frac{1}{2} - \varepsilon$. We can say that error of answer is $\frac{1}{2} - \varepsilon$.

3 Reordering Method and Exponential Gap Between Quantum and Classical OBDD

Let us introduce some helpful definitions. Let $\theta = (\{x_{j_1}, \ldots, x_{j_u}\}, \{x_{i_1}, \ldots, x_{i_{n-u}}\}) = (X_A, X_B)$ be a partition of set X into two parts. Below we will use equivalent notations $f(X)$ and $f(X_A, X_B)$. Let $f|_\rho$ be a subfunction of f, where ρ is a mapping $\rho : X_A \to \{0,1\}^{|X_A|}$. Function $f|_\rho$ is obtained from f by applying ρ, so if $\rho : X_A \to \nu$, then $f|_\rho(X_B) = f(\nu, X_B)$. Let $N^\theta(f)$ be number of different subfunctions with respect to partition θ. Let $\Pi(n)$ be the set of all permutations of $\{1, \ldots, n\}$. We say, that partition θ agrees with permutation $\pi = (j_1, \ldots, j_n) \in \Pi(n)$, if for some u, $1 < u < n$ the following is right: $\theta = (\{x_{j_1}, \ldots, x_{j_u}\}, \{x_{j_{u+1}}, \ldots, x_{j_n}\})$. We denote $\Theta(\pi)$ a set of all partitions which agrees with π. Let $N^\pi(f) = \max_{\theta \in \Theta(\pi)} N^\theta(f)$, $N(f) = \min_{\pi \in \Pi(n)} N^\pi(f)$.

It is known that the difference between quantum and deterministic OBDD complexity is at most exponential [AGK+05]. But one of the main issues in proof of complexity of OBDD is different orders of input variables. We suggest a method, called "reordering", which allows to construct partial function f' from Boolean function f such that $N^{id}(f) = d(n), N(f') \geq d(q), n = q(\lceil \log q \rceil + 1)$. Note that $N(f') = width(f')$ and $N^{id}(f) = id\text{-}width(f)$, due to [Weg00]. At the same time, if commutative QOBDD P of width $g(n)$ computes f, then we can construct QOBDD P' of width $g(q) \cdot q$ which computes f'. If $g(n) = O(n)$ and $d(n) = O(2^n)$, then we can say that $d(q/\lceil \log q + 1 \rceil)$ is almost exponential great than $g(q) \cdot q$. And total boolean function f'' with same properties can be built using result of computation of P' for unspecified inputs. And for some functions we can give explicit definition of such total *reordered* function.

Reordering Method. Let us shuffle input bits for solving "order issues". It means that order of *value* bits is determined by input. Let us consider input $X = (x_1, \ldots, x_n)$, among the variables we have q *value* bits $Z = \{z_1, \ldots, z_q\}$, where q is such that $n = q(\lceil \log q \rceil + 1)$. And any *value* bit has $\lceil \log q \rceil$ bits as *address*, that is binary representation of number of real position of *value* bit in input. We call this process as *reordering* of input or *reordering* of Boolean function $f(X)$. Now from $f(X)$ we obtain a new partial Boolean function $f'(X)$ on reordered input, such that any *value* bit has unique address and all addresses from $\{1, \ldots, q\}$ are occurred in input. In a case of *xor-reordering* address of *value* bit can be obtain as parity of current and previous *address* bits (Fig. 1).

Let us formally describe a partial function $f'(X)$:

- X consists of q blocks, for $n = q(\lceil \log q \rceil + 1)$ or $q = O(n/log n)$.
- Block i consists of $p = \lceil \log q \rceil$ *address* bits $y_1^i, \ldots, y_p^i$ and one *value* bit z^i. Formally, $(x_{(i-1)(p+1)+1}, \ldots, x_{i(p+1)}) = (y_1^i, \ldots, y_p^i, z^i)$.
- Function $Adr : \{0,1\}^n \times \{1, \ldots, q\} \to \{1, \ldots, q\}$, $Adr(X, i)$ is the address of i-th *value* bit. Let $bin(y_1, \ldots, y_p)$ is a number, which binary representation is $(y_1, \ldots, y_p)$, then in a case of *reordering* $Adr(X, i) = bin(y_1^i, \ldots, y_p^i) + 1$. If we consider *xor-reordering* then $Adr(X, i) = Adr'(X, i) + 1$, $Adr'(X, i) = Adr'(X, i - 1) \oplus bin(y_1^i, \ldots, y_p^i)$ for $i \geq 1$ and $Adr'(X, 0) = 0$. Here when we apply parity function to integers, we mean parity of their bits in binary representation.
- We consider only such inputs $\sigma \in \{0,1\}^n$ that addresses of the blocks are different and all addresses are occurred. Formally, $\{1, \ldots, q\} = \{Adr(\sigma, 1), \ldots, Adr(\sigma, q)\}$.
- Let a permutation $\pi = (Adr(\sigma, 1), \ldots, Adr(\sigma, q))$, and γ is string of *value* bits of σ then $f'(\sigma) = f(\gamma_{\pi^{-1}(1)}, \ldots, \gamma_{\pi^{-1}(q)})$.

Theorem 1. *Let Boolean function f over $X = (x_1, \cdots, x_n)$, such that $N^{id}(f) \geq d(n)$. Then partial Boolean function f', reordered or xor-reordered version of f, such that $N(f') \geq d(q)$, where $n = q(\lceil \log q \rceil + 1)$.*

Proof. Let us consider function $f'(X)$ and any order $\pi = (j_1, \ldots, j_n)$ and $\pi' = (i_1, \ldots, i_q)$ is the order of *value* bits according to order π. Let Σ be the set of

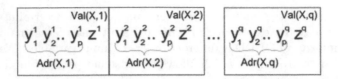

Fig. 1. Input. Blocks of *address* and *value* bits.

inputs with natural order of blocks (*value* bits) with respect to π, that is $\Sigma = \{\sigma \in \{0,1\}^n : Adr(\sigma, i_r) = r,$ for $1 \leq r \leq q\}$. Let partition $\theta \in \Theta((1,\dots,q)), \theta = (\{x_1,\dots,x_u\},\{x_{u+1},\dots,x_q\})$ be such that $N^\theta(f) = N^{id}(f)$. And let partition $\theta' = (X_A, X_B) = (\{x_1,\dots,x_{u'}\},\{x_{u'+1},\dots,x_n\})$, for $\theta' \in \Theta(\pi)$, be such that exactly u *value* bits belongs to X_A and others to X_B. Let $\Gamma = \{\gamma \in \{0,1\}^u :$ for different γ, γ' and corresponding subfunctions holds $f|_\rho \neq f|_{\rho'}\}$. And $\Xi = \{\xi \in \{0,1\}^{u'} :$ there are $\nu \in \{0,1\}^{n-u'}$ such that $(\xi, \nu) \in \Sigma$ and string of *value* bits of ξ belongs to $\Gamma\}$. It is easy to see that $|\Gamma| = |\Xi|$ and each $\xi \in \Xi$ produce own subfunction of function f'. Therefore $N^{\theta'}(f') \geq N^\theta(f)$. Due to definition, $N^\pi(f') \geq N^{\theta'}(f') \geq N^{id}(f)$. It is right for any order, hence $N(f') \geq N^{id}(f)$. $\square$

Theorem 2. *Let commutative QOBDD P of width $g = g(n)$ computes a Boolean function $f(X)$ over $X = (x_1,\dots,x_n)$. Then there is id-QOBDD P' of width $g(q) \cdot q$ which computes partial Boolean function f', xor-reordered version of f, where q is such that $n = q(\lceil \log q \rceil + 1)$.*

Proof. Because of P is commutative, we can consider *id*-QOBDD P_{id} of the same width $g(n)$ for function f. For description of computation of P' we use quantum register $|\psi\rangle = |\psi_1\psi_2\dots\psi_t\rangle$, where $t = \lceil \log g \rceil$ and $g \times g$ matrices for unitary operators (G_i^0, G_i^1), $i \in \{1, 2 \dots q\}$.

Then we consider partial function $f'(X)$ described above. In this case we have q *value* bits and $p = \lceil \log q \rceil$ *address* bits for any *value* bit. Let us construct QOBDD P' for computing f'.

Program P' has quantum register of $\lceil \log g \rceil + \lceil \log q \rceil$ qubits, having $g \cdot q$ states. Let us denote it as $|\phi\rangle = |\phi_1\phi_2\dots\phi_p\psi_1\psi_2\dots\psi_t\rangle$, where $t = \lceil \log g \rceil$, $p = \lceil \log q \rceil$.

Part of register $|\phi\rangle$ consisting of $|\psi_1\psi_2\dots\psi_t\rangle$ qubits (we note it as a *computing* part) is modified on reading *value* bit. In other hand, we added qubits $|\phi_1\phi_2\dots\phi_p\rangle$ (let this part be an *address* part) to determine address of *value* bit. And superposition of the states of these two parts will give us a right computation of function. Program P' consists of q parts, because input contains q blocks, for $n = (\lceil \log q \rceil + 1)q$ or $q = O(n/\log n)$. For any $i \in \{1,\dots,q\}$ block i handles *value* bit z^i.

Informally, when P' processes a block, it stores address in *address* part by applying parity function. After that some modifications are produced on the *computation* part, with respect to *value* bit.

Let us describe i-th block of levels formally, for $i \in \{1,\dots,q\}$. In the first $\lceil \log q \rceil$ levels the program computes address $Adr'(X, i)$, it reads bits one by one, and for bit y_j^i we use unitary operator $U_j^{y_j^i}$ on the *address* part of register $|\phi\rangle$, for $j \in \{1,2\ldots,p\}$ (see Picture 1). $U_j^{y_j^i} = I \otimes I \otimes \ldots \otimes I \otimes A^{y_j^i} \otimes I \ldots \otimes I$, where $A^0 = I$ and $A^1 = NOT$, I and NOT are 2×2 matrices, such that I is diagonal 1-matrix and NOT is anti-diagonal 1-matrix. And we do not modify computation part.

After these operations *address* part of register in binary notation equals to the address $Adr'(X,i)$. In the vector of the states all elements are equals to zero except elements of block where address part of qubits corresponds to $Adr(X,i)$.

After reading z^i we transform system $|\phi\rangle$ by unitary $(g \cdot q \times g \cdot q)$-matrix D^{z^i}.

$$D^0 = \begin{pmatrix} G_1^0 & 0 & \cdots & 0 \\ 0 & G_2^0 & \cdots & 0 \\ \vdots & \vdots & \ddots & \vdots \\ 0 & 0 & \cdots & G_q^0 \end{pmatrix} \text{ and } D^1 = \begin{pmatrix} G_1^1 & 0 & \cdots & 0 \\ 0 & G_2^1 & \cdots & 0 \\ \vdots & \vdots & \ddots & \vdots \\ 0 & 0 & \cdots & G_q^1 \end{pmatrix},$$

where matrices $\{(G_i^0, G_i^1), 1 \leq i \leq q\}$ are unitary matrices transforming quantum system in *id*-QOBDD P_{id}.

Because of size of register, QOBDD P' has width $g(q) \cdot q$. Let us prove that P' computes f'.

Let us consider an input $\sigma \in \{0,1\}^n$. Let a permutation $\pi = (j_1,\ldots,j_q) = (Adr(\sigma,1),\ldots,Adr(\sigma,q))$ be an order of *value* variables with respect to input σ.

Due to *id*-QOBDD P_{id} is commutative, we can reorder unitary operators $\{(G_i^0, G_i^1), 1 \leq i \leq q\}$ according to order π and get a QOBDD P_π computing f as well.

It is easy to see that P' exactly emulates computation of P_π, therefore P' on σ gives us the same result as P_π on corresponding *value* bits. So, by definition of f' we have P' computes f'. $\square$

Corollary 1. *Let commutative k-QOBDD P of width $g = g(n)$ computes a Boolean function $f(X)$ over $X = (x_1,\ldots,x_n)$. Then there is k-QOBDD P' of width $g(q) \cdot q$ which computes partial Boolean function f', xor-reordered version of f, where q is such that $n = q(\lceil \log q \rceil + 1)$.*

It is proved exactly by the same way as Theorem 2.

Theorem 3. *If for some Boolean function $f : \{0,1\}^n \to \{0,1\}$ there are commutative k-OBDD P_1, k-NOBDD P_2 and k-POBDD P_3 that computes f and width of P_i is d_i for $i \in \{1,2,3\}$. Then there are k-OBDDs D_1, D_4, k-NOBDDs D_2, D_5 and k-POBDDs D_3, D_6 such that width of D_i is $d_i(q) \cdot q$ for $i \in \{1,2,3\}$ and $d_{i-3}(q) \cdot q$ for $i \in \{4,5,6\}$, and D_i computes f', reordered version of f, and D_j computes f'', xor-reordered version of f, for $i \in \{1,2,3\}$, $j \in \{4,5,6\}$, $n = q(\lceil \log q \rceil + 1)$, f' and f'' are partial Boolean function.*

Proof. Let P_1 be commutative deterministic k-OBDD of width $d_1(n)$ which computes Boolean function f. Let Boolean function f' is reordered f and f'' is xor-reordered f, note that f' and f'' are partial Boolean functions.

We want to construct deterministic k-OBDDs D_1 and D_4 of width $q \cdot d_1(q)$, for $n = q(\lceil \log_2 q \rceil + 1)$. D_1 is for reordering case and D_4 is for xor-reordering one. D_1

and D_4 read variables in natural order. D_1 and D_4 have $q \cdot d_1(q)$ nodes on level, each of them corresponds to pair (i, s), where $i \in \{1, \ldots, q\}$, $s \in \{1, \ldots, d(q)\}$. Let us describe computation on block j.

- A case of reordering and program D_1. In the begin of the block D_1 situated in one of the nodes $(1, s)$. After reading first $\lceil \log q \rceil$ bits of the block D' just store a number $Adr(X, j) = a$ in states and program reaches node corresponding to (a, s). Then if transition function of P is such that $s' = tr_D(\pi^{-1}(a), s, z^j)$ then D' reaches $(1, s')$.
- A case of xor-reordering and program D_4. In the begin of the block D_4 is situated in one of the nodes (b, s). After reading first $\lceil \log q \rceil$ bits of the block D_4 just computes parity of $b - 1$ and *address* bits, so computes a number $Adr'(X, j) = a'$ and program reaches node, which is correspond to (a, s), for $a = a' + 1$.

In the case when all addresses are different, D_1 and D_4 just emulate work of D_π which is constructed from D by permutation of transition function with respect to order $(Adr(X, 1), \ldots, Adr(X, q))$. By the definition of commutative k-OBDD the D_π computes the same function f. Therefore D_1 and D_4 also return the same result. And by the definition of functions f' and f'' programs D_1 computes f' and D_4 computes f''.

We can construct nondeterministic k-OBDDs D_2, D_5 and probabilistic k-OBDDs D_3, D_6 by the similar way.   □

Corollary 2. *Let Boolean function f over $X = (x_1, \cdots, x_n)$, such that $N^{id}(f) \geq d(n)$ and commutative k-QOBDD P, k-OBDD D, k-NOBDD H and k-POBDD R of width $g(n), d(n), h(n)$ and $w(n)$, respectively, computes f. Then there are total Boolean functions $f^{(i)}$, total xor-reordered version of f, for $i \in \{1, \ldots, 4\}$ and $f^{(j)}$, total reordered version of f, for $j \in \{5, \ldots, 7\}$, such that $N(f^{(i)}), N(f^{(j)}), \geq d(q)$, where $n = q(\lceil \log q \rceil + 1)$. And there are k-QOBDD P' of width $g(q) \cdot q$ which computes $f^{(1)}$ and k-OBDD D', k-NOBDD H' and k-POBDD R' of width $d(q) \cdot q$, $h(q) \cdot q$ and $w(q) \cdot q$, respectively, such that D' computes computing $f^{(2)}, f^{(5)}$, H' computes computing $f^{(3)}, f^{(6)}$, R' computes computing $f^{(4)}, f^{(7)}$.*

Proof. Let partial Boolean function f' be xor-reordered version of f. Due to Theorems 1 and 2, $N(f') \geq d(q)$, id-QOBDD P' of width $g(q) \cdot q$ computes f'. Let total function $f^{(1)}$ be such that $f^{(1)}(\sigma) = f'(\sigma)$ for input σ allowed for f'. And for input σ', not allowed for f', $f^{(1)}(\sigma)$ equals to result of P'. It is easy to see that $N(f^{(1)}) \geq N(f')$. Similar prove for $f^{(i)}$ and $f^{(j)}$.   □

Exponential Gap Between Quantum and Classical OBDDs. Let us apply reordering method to *Equality function $EQ_n : \{0, 1\}^n \to \{0, 1\}$*. $EQ_n(\sigma) = 1$, iff $(\sigma_1, \sigma_2 \ldots \sigma_{n/2}) = (\sigma_{n/2+1}, \sigma_{n/2+2} \ldots \sigma_n)$.

From [AKV08] we know, that there is commutative id-QOBDD P of width $O(n)$, which computes this function. After xor-reordering we get partial function $EQ'_n(X)$ computed by $QOBDD$ of width $O(q) \cdot q = O(q^2)$, where $q = O(n/\log n)$.

It is known that $N^{id}(EQ_n) = 2^{n/2}$. Due to Theorem 1 we have $N(EQ'_n) \geq 2^{q/2}$, therefore deterministic OBDD has width at least $2^{q/2}$.

So we have following Theorem for EQ'_n:

Theorem 4. *Let partial Boolean function EQ'_n is xor-reordered EQ_n. Then there is quantum OBDD P of width $O(n^2/\log^2 n)$ which computes EQ'_n and any deterministic OBDD D which computes EQ'_n has width $2^{\Omega(n/\log n)}$.*

Let us consider *Reordered Equality function* $REQ_n : \{0,1\}^n \to \{0,1\}$. This is total version of EQ'_n and on inputs which is not allowed for EQ'_n the result of function is exactly result of $QOBDD$ P' which was contracted for EQ'_n by the method from the proof of Theorem 2. Due to fingerprinting algorithm for EQ_n from [AKV08], we can see that $REQ_n(\sigma) = 1$ iff $\sum_{i=1}^{q/2} 2^{Adr'(\sigma,i)} Val(\sigma, i) = \sum_{i=q/2+1}^{q} 2^{Adr'(\sigma,i)} Val(\sigma, i)$. We can prove the following lemma for this function:

Lemma 1. $N(REQ_n) \geq 2^{n/(2\lceil \log_2 n + 1 \rceil)}$.

It means that any deterministic OBDD P of width w computing REQ_n is such that $w \geq 2^{n/(2\lceil \log_2 n + 1 \rceil)}$. Therefore function REQ such that:

Theorem 5. *There is quantum OBDD P of width $O(n^2/\log^2 n)$ which computes total Boolean function REQ_n and any deterministic OBDD D which computes REQ has width $2^{\Omega(n/\log n)}$.*

Proof. By the definition of the function we can construct $QOBDD$ P and Lemma 1 shows a bound for deterministic case. $\qquad\square$

So, REQ_n is explicit function which shows the following distance between quantum and deterministic ODDD complexity: $O(n^2/\log^2 n)$ and $2^{\Omega(n/\log n)}$.

4   Hierarchy for Probabilistic and Quantum OBDDs

Let us consider classes **BPOBDD**$_d$ and **BQOBDD**$_d$ of Boolean functions that will be computed by probabilistic and quantum OBDDs with bounded error of width d, respectively. We want to prove a width hierarchy for these classes.

Hierarchy for Probabilistic OBDDs. Before proof of hierarchy let us consider the Boolean function $WS_n(X)$ due to Savický and Žák [SŽ00]. For a positive integer n and $X = (x_1, \ldots, x_n) \in \{0,1\}^n$, let $p(n)$ be the smallest prime larger than n and let $s_n(X) = (\sum_{i=1}^{n} i \cdot x_i)\ mod\ p(n)$. Define the weighted sum function by $WS_n(x) = x_{s_n(X)}$. For this function it is known that for every n large enough it holds that any bounded error probabilistic OBDD P which computes $WS_n(X)$ has size no less than $2^{\Omega(n)}$. Let us modify Boolean function $WS_n(X)$ using pending bits. We will denote it $WS_n^b(X)$. For a positive integers n and b, $b \leq n/3$ and $X = (x_1, \ldots, x_n) \in \{0,1\}^n$, let $p(b)$ be the smallest prime larger than b, $s_b(X) = (\sum_{i=1}^{b} i \cdot x_i)\ mod\ p(b)$. Define the weighted sum function by $WS_n^b(x) = x_{s_b(X)}$. We can prove the following lemma by the way as in [SŽ00].

Lemma 2. *For large enough n and $const = o(b)$, any bounded error probabilistic OBDD P computing $WS_n^b(X)$ has width no less than $2^{\Omega(b)}$. There is bounded error probabilistic OBDD P of width 2^b which computes $WS_n^b(X)$.*

The second claim of the Lemma follows form the fact that any Boolean function over $X \in \{0,1\}^n$ can be computed by deterministic OBDD of width 2^n, just by building full binary tree.

Let us prove hierarchy for $\mathbf{BPOBDD}_d$ classes using these properties of Boolean function $WS_n^b(X)$.

Theorem 6. *For integer $d = o(2^n), const = o(d)$, the following statement holds:*
$\mathbf{BPOBDD}_{d^{1/\delta}} \subsetneq \mathbf{BPOBDD}_d$, *for $const = o(\delta)$.*

Proof. It is easy to see that $\mathbf{BPOBDD}_{d^{1/\delta}} \subseteq \mathbf{BPOBDD}_d$. Let us prove inequality of these classes. Due to Lemma 2, Boolean function $WS_n^{\log d} \in \mathbf{BPOBDD}_d$, at the same time for any bounded error probabilistic OBDD P we have $w(P) = 2^{\Omega(\log d)} > 2^{(\log d)/\delta} = d^{1/\delta}$. Therefore $WS_n^{\log d} \notin \mathbf{BPOBDD}_{d^{1/\delta}}$. □

Hierarchy for Quantum OBDDs. Let us modify Boolean function $REQ_n(X)$ using pending bits as for $WS_n^b(X)$. We will denote it $REQ_n^b(X)$. Also let us consider complexity property of MOD_p function (number of 1 s by modulo p is 0). And Boolean function $MSW_n^b(X)$, it is similar modification of $MSW_n(X)$ function [Sau06] using pending bits. $MSW_n^b(X) = x_z \oplus x_{r+n/2}$, where $z = s_{b/2}(x_1,\ldots,x_{b/2}), r = s_{b/2}(x_{b/2+1},\ldots,x_b)$, if $r = z$ and $MSW_n^b(X) = 0$ otherwise. Complexity properties of functions are described in the following lemma.

Lemma 3. *Claim 1. Any bounded error quantum OBDD P which computes $REQ_n^b(X)$ has width at least $\lfloor b/\lceil \log b + 1 \rceil \rfloor$, for $\lfloor b/\lceil \log b + 1 \rceil \rfloor \geq 1$. There is bounded error quantum OBDD P of width b^2 which computes $REQ_n^b(X)$.*

Claim 2. Any bounded error quantum OBDD P which computes $MOD_p(X)$ has width no less than $\lfloor \log p \rfloor$, for $2 \leq p \leq n$. There is bounded error quantum OBDD P of width $O(\log p)$ which computes $MOD_p(X)$.

Claim 3. Any bounded error quantum OBDD P which computes $MSW_n^b(X)$ has width no less than $2^{\Omega(b)}$, for $const = o(b)$. There is bounded error quantum OBDD P of width 2^b which computes $MSW_n^b(X)$.

A proof of Claim 1 is similar to Theorem 5, a proof of Claim 2 is presented in [AGK+05, AV08] and a proof of Claim 3 is based on result from [Sau06].

Let us prove hierarchies for $\mathbf{BQOBDD}_d$ classes using presented above lemma.

Theorem 7. *For a integer d following statements are right:*
$\mathbf{BQOBDD}_{d/\delta^2} \subsetneq \mathbf{BPOBDD}_{d^2}$, *for $d < \log n, d > 2, const = o(\delta)$.*
$\mathbf{BQOBDD}_{d/\log_2^2 d} \subsetneq \mathbf{BPOBDD}_{d^2}$, *for $d < nd > 2$.*
$\mathbf{BQOBDD}_{d^{1/\delta}} \subsetneq \mathbf{BQOBDD}_d$, *for $d = o(2^n), const = o(d), const = o(\delta)$.*

A proof is based on Lemma 3.

5   Extension of Hierarchy for Deterministic and Probabilistic k-OBDD

Let us apply the reordering method to k-OBDD model. We will prove almost tight hierarchy for Deterministic and Probabilistic k-OBDDs using complexity properties of *Pointer jumping* function (PJ) [NW91,BSSW98]. These hierarchies are extention of existing ones. At first, let us present version of function which works with integer numbers.

Let V_A, V_B be two disjoint sets (of vertices) with $|V_A| = |V_B| = m$ and $V = V_A \cup V_B$. Let $F_A = \{f_A : V_A \to V_B\}$, $F_B = \{f_B : V_B \to V_A\}$ and $f = (f_A, f_B) : V \to V$ defined by $f(v) = f_A(v)$, if $v \in V_A$ and $f = f_B(v)$, $v \in V_B$. For each $k \geq 0$ define $f^{(k)}(v)$ by $f^{(0)}(v) = v$, $f^{(k+1)}(v) = f(f^{(k)}(v))$. Let $v_0 \in V_A$. The function we will be interested in computing is $g_{k,m} : F_A \times F_B \to V$ defined by $g_{k,m}(f_A, f_B) = f^{(k)}(v_0)$. Boolean function $PJt_{,n} : \{0,1\}^n \to \{0,1\}$ is boolean version of $g_{k,m}$, where we encode f_A in a binary string using $m \log m$ bits and do it with f_B as well. The result of function is parity of binary representation of result vertex.

Let us apply reordering method to $PJ_{k,n}$ function. $RPJ_{k,n}$ is total version of reordered $PJ_{k,n}$. Formally: Boolean function $RPJ_{k,n} : \{0,1\}^n \to \{0,1\}$ is following. Let us separate whole input $X = (x_1, \ldots, x_n)$ to b blocks, such that $b\lceil \log_2 b + 1 \rceil = n$, therefore $b = U(n/\log n)$. And let $Adr(X, i)$ be integer, which binary representation is first $\lceil \log_2 b \rceil$ bits of i-th block and $Val(X, i)$ be a value of bit number $\lceil \log_2 b + 1 \rceil$ of block i, for $i \in \{0, \ldots, b-1\}$. Let a be such that $b = 2a\lceil \log_2 a \rceil$ and $V_A = \{0, \ldots, a-1\}$, $V_B = \{a, \ldots, 2a-1\}$.

Let function $BV : \{0,1\}^n \times \{0, \ldots, 2a-1\} \to \{0, \ldots, a-1\}$ be the following:

$$BV(X, v) = \sum_{i : (v-1)\lceil \log_2 b \rceil < Adr(X,i) \leq v\lceil \log_2 b \rceil} 2^{Adr(X,i)-(v-1)\lceil \log_2 b \rceil} \cdot Val(X, i) \ (mod \ a)$$

Then $f_A(v) = BV(X, v) + a$, $f_B(v) = BV(X, v)$.
Let $r = g_{k,a}(f_A, f_B)$, then

$$RPJ_{k,n}(X) = \bigoplus_{i : (r-1)\lceil \log_2 b \rceil < Adr(X,i) \leq r\lceil \log_2 b \rceil} Val(X, i).$$

Let us prove lower bound for this function:

Lemma 4. *Claim 1. The functions $RPJ_{2k-1,n}$ can be computed by $2k$-OBDD of size $O(n^3)$.*

Claim 2. Each k-OBDD for $RPJ_{2k-1,n}$, has size $2^{\Omega(n/(k \log n) - \log(n/\log n))}$. Each k-POBDD for $RPJ_{2k-1,n}$ which computed with bounded error at least $1/3$, has size $2^{\Omega(n/(k^3 \log n) - \log(n/\log n))}$.

A proof of lower bound is based on communication complexity properties of the function $PJ_{k,n}$ from [NW91]. And a proof of upper bound is based on Theorem 3 and Corollary 2.

Using this lemma we extend hierarchy for following classes: P-kOBDD, BP$_{1/3}$-kOBDD, SUPERPOLY-kOBDD, BSUPERPOLY$_{1/3}$-kOBDD, SUBEXP$_\alpha$ -kOBDD and BSUBEXP$_{\alpha,1/3}$-kOBDD. These are classes of Boolean functions computed by following models:

- P-kOBDD and BP$_\delta$-kOBDD are for polynomial size k-OBDD, the first one is for deterministic case and the second one is for bounded error probabilistic k-OBDD with error at least δ.
- SUPERPOLY-kOBDD and BSUPERPOLY$_{1/3}$-kOBDD are similar classes for superpolynomial size models.
- SUBEXP$_\alpha$-kOBDD and BSUBEXP$_{\alpha,1/3}$-kOBDD are similar classes for size at most $2^{O(n^\alpha)}$, for $0 < \alpha < 1$.

Theorem 8. *Claim 1. P-kOBDD $\subsetneq$ P-$2k$OBDD, for $k = o(n/\log^3 n)$. BP$_{1/3}$-$kOBDD \subsetneq BP_{1/3}$-$2kOBDD$, for $k = o(n^{1/3}/\log n)$.*

Claim 2. SUPERPOLY-$kOBDD \subsetneq SUPERPOLY$-$2kOBDD$, for $k = o(n^{1-\delta})$, $\delta > 0$. BSUPERPOLY$_{1/3}$-$kOBDD \subsetneq BSUPERPOLY_{1/3}$-$2kOBDD$, for $k = o(n^{1/3-\delta})$, $\delta > 0$.

Claim 3. SUBEXP$_\alpha$-$kOBDD \subsetneq SUBEXP_\alpha$-$2kOBDD$, for $k = o(n^{1-\delta})$, $1 > \delta > \alpha + \varepsilon$, $\varepsilon > 0$. BSUBEXP$_{\alpha,1/3}$-$kOBDD \subsetneq BSUBEXP_{\alpha,1/3}$-$2kOBDD$, for $k = o(n^{1/3-\delta/3})$, $1/3 > \delta > \alpha + \varepsilon$, $\varepsilon > 0$.

A proof is based on Lemma 4.

Acknowledgements. We thank Alexander Vasiliev and Aida Gainutdinova from Kazan Federal University and Andris Ambainis from University of Latvia for their helpful comments and discussions.

References

[AG05] Ablayev, F., Gainutdinova, A.: Complexity of quantum uniform and nonuniform automata. In: Felice, C., Restivo, A. (eds.) DLT 2005. LNCS, vol. 3572, pp. 78–87. Springer, Heidelberg (2005). doi:10.1007/11505877_7

[AGK01] Ablayev, F., Gainutdinova, A., Karpinski, M.: On computational power of quantum branching programs. In: Freivalds, R. (ed.) FCT 2001. LNCS, vol. 2138, pp. 59–70. Springer, Heidelberg (2001). doi:10.1007/3-540-44669-9_8

[AGK+05] Ablayev, F., Gainutdinova, A., Karpinski, M., Moore, C., Pollett, C.: On the computational power of probabilistic and quantum branching program. Inf. Comput. **203**(2), 145–162 (2005)

[AGKY14] Ablayev, F., Gainutdinova, A., Khadiev, K., Yakaryılmaz, A.: Very narrow quantum OBDDs and width hierarchies for classical OBDDs. In: Jürgensen, H., Karhumäki, J., Okhotin, A. (eds.) DCFS 2014. LNCS, vol. 8614, pp. 53–64. Springer, Cham (2014). doi:10.1007/978-3-319-09704-6_6

[AGKY16] Ablayev, F., Gainutdinova, A., Khadiev, K., Yakaryılmaz, A.: Very narrow quantum obdds and width hierarchies for classical obdds. Lobachevskii J. Math. **37**(6), 670–682 (2016)

[AK13] Ablayev, F., Khadiev, K.: Extension of the hierarchy for k-OBDDs of small width. Russ. Math. **53**(3), 46–50 (2013)

[AKV08] Ablayev, F., Khasianov, A., Vasiliev, A.: On complexity of quantum branching programs computing equality-like boolean functions. In: ECCC (2008)

[AV08] Ablayev, F., Vasiliev, A.: On the computation of boolean functions by quantum branching programs via fingerprinting. In: Electronic Colloquium on Computational Complexity (ECCC), vol. 15 (2008)

[BHW06] Brosenne, H., Homeister, M., Waack, S.: Nondeterministic ordered binary decision diagrams with repeated tests and various modes of acceptance. Inf. Process. Lett. **98**(1), 6–10 (2006)

[BSSW98] Bollig, B., Sauerhoff, M., Sieling, D., Wegener, I.: Hierarchy theorems for kobdds and kibdds. Theor. Comput. Sci. **205**(1), 45–60 (1998)

[Gai15] Gainutdinova, A.F.: Comparative complexity of quantum and classical obdds for total and partial functions. Russ. Math. **59**(11), 26–35 (2015)

[HS03] Hromkovič, J., Sauerhoff, M.: The power of nondeterminism and randomness for oblivious branching programs. Theory Comput. Syst. **36**(2), 159–182 (2003)

[Kha15] Khadiev, K.: Width hierarchy for k-obdd of small width. Lobachevskii J. Math. **36**(2), 178–183 (2015)

[Kha16] Khadiev, K.: On the hierarchies for deterministic, nondeterministic and probabilistic ordered read-k-times branching programs. Lobachevskii J. Math. **37**(6), 682–703 (2016)

[KMW91] Krause, M., Meinel, C., Waack, S.: Separating the eraser turing machine classes Le, NLe, co-NLe and Pe. Theor. Comput. Sci. **86**(2), 267–275 (1991)

[NHK00] Nakanishi, M., Hamaguchi, K., Kashiwabara, T.: Ordered quantum branching programs are more powerful than ordered probabilistic branching programs under a bounded-width restriction. In: Du, D.-Z.-Z., Eades, P., Estivill-Castro, V., Lin, X., Sharma, A. (eds.) COCOON 2000. LNCS, vol. 1858, pp. 467–476. Springer, Heidelberg (2000). doi:10.1007/3-540-44968-X_46

[NW91] Nisan, N., Widgerson, A.: Rounds in communication complexity revisited. In: Proceedings of the Twenty-Third Annual ACM Symposium on Theory of Computing, pp. 419–429. ACM (1991)

[Sau05] Sauerhoff, M.: Quantum vs. classical read-once branching programs. arXiv preprint quant-ph/0504198 (2005)

[Sau06] Sauerhoff, M.: Quantum vs. classical read-once branching programs. In: Krause, M., Pudlák, P., Reischuk, R., van Melkebeek, D., (eds.) Complexity of Boolean Functions, number 06111 in Dagstuhl Seminar Proceedings, Dagstuhl, Germany, Internationales Begegnungs- und Forschungszentrum für Informatik (IBFI), Schloss Dagstuhl, Germany (2006)

[SS05a] Sauerhoff, M., Sieling, D.: Quantum branching programs and space-bounded nonuniform quantum complexity. Theor. Comput. Sci. **334**(1–3), 177–225 (2005)

[SS05b] Sauerhoff, M., Sieling, D.: Quantum branching programs and space-bounded nonuniform quantum complexity. Theor. Comput. Sci. **334**(1), 177–225 (2005)

[SŽ00] SŽák, P., Žák, S.: A read-once lower bound and a (1,+ k)-hierarchy for branching programs. Theor. Comput. Sci. **238**(1), 347–362 (2000)

[Weg00] Wegener, I.: Branching Programs and Binary Decision Diagrams: Theory and Applications. SIAM (2000)

Dynamic Stabbing Queries with Sub-logarithmic Local Updates for Overlapping Intervals

Elena Khramtcova[1(✉)] and Maarten Löffler[2]

[1] Computer Science Department, Université Libre de Bruxelles, Brussels, Belgium
elena.khramtsova@gmail.com
[2] Department of Information and Computing Sciences,
Utrecht University, Utrecht, The Netherlands
m.loffler@uu.nl

Abstract. We present a data structure to maintain a set of intervals on the real line subject to fast insertions and deletions of the intervals, stabbing queries, and *local updates*. Intuitively, a local update replaces an interval by another one of roughly the same size and location. We investigate whether local updates can be implemented faster than a deletion followed by an insertion.

We present the first results for this problem for sets of possibly overlapping intervals. If the maximum depth of the overlap (a.k.a. *ply*) is bounded by a constant, our data structure performs insertions, deletions and stabbing queries in time $O(\log n)$, and local updates in time $O(\log n/\log\log n)$, where n is the number of intervals. We also analyze the dependence on the ply when it is not constant. Our results are adaptive: the times depend on the current ply at the time of each operation.

1 Introduction

Preprocessing a set of objects for fast containment queries is a classic data structure problem. One of the most basic variants is maintaining a set S of one-dimensional intervals on the real line $\mathbb{R}^1$, subject to *stabbing queries*. Given a point $q \in \mathbb{R}^1$, the stabbing query for q aims to find all the intervals in S that contain q. Maintaining a set of intervals subject to stabbing queries is well understood both in the static [9] and in the dynamic setting [6,10]. In particular, it is well known how to maintain a set of n intervals subject to insertions and deletions in time $O(\log n)$, and stabbing queries in time $O(\log n + k)$, where k is the size of the output [10]. Data structures for stabbing queries remain an active research area and many variations of the problem have been studied: reporting the number of the stabbed intervals [1], finding the maximum priority stabbed interval [7,12], or considering different computational models and tradeoffs between memory requirements and query time [3,13].

E. Khramtcova was partially supported by F.R.S.-FNRS, and by the SNF grant P2TIP2-168563 under the Early PostDoc Mobility program.

M. Löffler was partially supported by the Netherlands Organisation for Scientific Research (NWO) under project no. 639.021.123 and 614.001.504.

P. Weil (Ed.): CSR 2017, LNCS 10304, pp. 176–190, 2017.
DOI: 10.1007/978-3-319-58747-9_17

In certain applications, for example involving moving or uncertain data, a special kind of update is frequently performed, called *local update* by Nekrich [11]; see also Löffler *et al.* [8] and references therein. Intuitively, a local update replaces an interval by another interval *similar* to it: the new interval has roughly the same size and location as the old interval (we make this definition precise in the next section). The particular nature of the local update suggests that it should be possible to perform such updates strictly faster than in logarithmic time, as in that much time one could delete an interval and insert another interval that would not need to be similar to the deleted one. In this paper we show that this is indeed the case for stabbing data structures in $\mathbb{R}^1$.

Sub-logarithmic local updates for containment queries in a set of *disjoint* intervals have already been studied in Löffler *et al.* [8]. However, the condition that intervals are and remain pairwise disjoint at all times is unrealistic in many applications. The method in [8] is hard to generalize to overlapping intervals even if the depth of the overlap is constant (see also Fig. 4). Therefore designing a new data structure that would handle overlapping intervals is posed as an open problem in [8]. Here we address this problem.

In this paper, we present a data structure to store a set of possibly overlapping intervals, allowing fast insertions, deletions, and local updates. The performance of the data structure is measured in terms of the number of intervals in the set and the *ply* of the set: the maximum number of intervals containing any point in $\mathbb{R}^1$. If the ply is bounded by a constant, then the operations of insertion, deletion, and answering a stabbing query require $O(\log n)$ time each, and a local update requires $O(\log n / \log \log n)$ time.

We conclude this section with the directions for further research. First direction is extending our data structure to operate with two-dimensional objects, which are very important in many applications. Another open question concerns the fact that the performance of our data structure depends linearly on the current ply of the interval set, see Theorem 3. Thus one would desire to be able, if ply is too large, to quickly transit from our data structure to a more efficient one, e.g., the interval tree [4,6]. Clearly, such transition should be made in $o(n \log n)$ time, i.e., faster than building an interval tree from the scratch.

1.1 The Problem Statement and Our Result

Given a set S of intervals in $\mathbb{R}^1$, we aim to maintain S subject to fast *stabbing queries* and *local updates*, as well as fast insertions and deletions of the intervals. We assume that at all times the intervals in S are contained in a bounding box,[1] i.e., they are contained in a large interval $B \subset \mathbb{R}^1$.

To state the problem formally, we need some definitions.

Definition 1 (Stabbing query). *Given a query point q, return all the intervals in S that contain q, or report that there is no such interval.*

---

[1] If an update of S violates this bounding box condition, B can easily be enlarged. Thus our assumption does not restrict the setting, but rather simplifies the description.

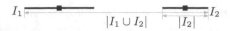

Fig. 1. 4-similar intervals I_1, I_2: the diameter $|I_2|$ is less than $|I_1|$ and $|I_1 \cup I_2| = 4|I_2|$.

For a closed, bounded, possibly disconnected region $R \subset \mathbb{R}^1$, the diameter of R is $|R| = \max_{p,q \in R} |q-p|$ (The diameter of an interval $I = [a, b]$ is $|I| = |b-a|$.)

Definition 2 ([8]). *For a pair I_1, I_2 of intervals in $\mathbb{R}^1$, and a real number $\rho > 0$, intervals I_1, I_2 are called ρ-similar, if $|I_1 \cup I_2| \leq \rho \min\{|I_1|, |I_2|\}$. See Fig. 1.*

Definition 3 (Local update [8]). *Given an interval $I \in S$ and a pointer to an interval $I' \subset B$ that is ρ-similar to I for some constant ρ, replace I by I'.*

Definition 4 (Ply). *For a point $p \in \mathbb{R}^1$, the ply of S at p is the number of intervals in S that contain p. The ply of S is the maximum ply of S at any point in $\mathbb{R}^1$.*

Now we are ready to formulate our main problem.

Problem 1. Given a set S of n intervals in $\mathbb{R}^1$ that can possibly overlap, a bounding interval $B \in \mathbb{R}^1$ that contains each interval in S, and a real constant $\rho > 0$, preprocess S subject to fast stabbing queries, insertions and deletions of intervals, and local updates with parameter ρ.

We show how to solve Problem 1, such that the resulting data structure requires $O(n)$ space, and the following holds:

- If the ply of S is always at most some constant number, then stabbing queries, insertions and deletions of intervals require $O(\log n)$ time; local updates require $O(\log n / \log \log n)$ time.
- Otherwise, stabbing queries, insertions and deletions of intervals require $O(\log n + k \log n / \log \log n)$ time; local updates require $O(k \log n / \log \log n)$ time. Here k is the ply of S at the moment when the operation is performed.

In both cases, the time complexity bound for insertion of an interval is amortized; all other time bounds are worst-case.

We begin by reviewing an existing solution to Problem 1 for disjoint intervals [8], see Sect. 1.2. In Sect. 2 we give an alternative solution for disjoint intervals, that contains the ideas important for our data structure for overlapping intervals, which we present in Sect. 3. Section 4 discusses compression of the quadtree, an additional detail deferred to the end of the paper to ease the exposition.

1.2   A Data Structure for Disjoint Intervals (Ply = 1) [8]

Below we review the data structure for disjoint intervals by Löffler *et al.* [8]. It consists of two trees: one for performing updates, and one for performing queries.

The first tree, further referred to as the *quadtree*, is a one-dimensional compressed balanced quadtree on the center points of the intervals in S. In particular, an interval I is stored in the largest quadtree cell C such that C contains the center point of I and does not contain the center point of any other interval in S. Such cell C is a quadtree leaf. Further, since the intervals in S are disjoint, $|I| \leq 4|C|$. The quadtree is additionally augmented with level links, i.e., each quadtree cell has a pointer to its adjacent cells of the same size (if they exist). The quadtree is compressed, i.e., it contains a-compressed cells for some large constant a. An a-compressed cell C has only one child C', the size of C' is at most $|C|/a$, and $C \backslash C'$ contains no central points of intervals in S. Each non-compressed cell has zero or two children (in the former case it is a leaf).

The second tree, referred to as the *query tree*, is a balanced binary tree over the subdivision of $\mathbb{R}^1$ induced by the leaves of the quadtree. The leaves of the query tree store pointers to the corresponding leaves of the quadtree.

Stabbing Queries. Given a query point $q \in \mathbb{R}^1$, we must return the interval in S that contains q (if it exists), see Definition 1. To do this, we use the following:

Property 1 ([8]). For a quadtree leaf C, any interval that intersects C is either stored in C, or it is stored in the closest to C non-empty quadtree leaf[2] either to the left, or to the right of C.

The stabbing query is performed by checking whether q is contained in one of the intervals stored in the three candidate cells from Property 1. The leaf C is found in $O(\log n)$ time by a binary search for q in the query tree. To find the other two leaves in $O(1)$ time, we store with each leaf the pointers to the two closest non-empty leaves from both sides. Checking if a given interval contains q takes $O(1)$ time. Thus the stabbing query requires $O(\log n)$ time in total.

Local Updates. For an interval $I \in S$, we need to replace I with a new interval I' that is $O(1)$-similar to I, see Definitions 2 and 3. To do this, we follow pointers in the quadtree to find the cell that must store I'. Since I and I' are $O(1)$-similar, such cell is at most a constant number of cells away from the one that stores I, and thus can be determined in $O(1)$ time. We remove I from the old cell and insert I' into the new cell, performing the necessary compression, decompression, and balancing in the quadtree. This requires $O(1)$ worst-case time. The corresponding deletion and insertion in the query tree is done in $O(1)$ time, since the pointers are given and no search is needed. After each insertion or deletion the balance in the search tree is restored in worst-case $O(1)$ time [5]. Thus, a local update operation can be completed in $O(1)$ worst-case total time.

---

[2] Such leaves are not necessarily adjacent to C, as the adjacent ones might be empty.

Classic Updates. Intervals can be inserted or deleted from the data structure in $O(\log n)$ time: First, the insertion/deletion in the query tree is performed. This provides a pointer to the place in the quadtree where the insertion/deletion should be done.

We conclude this overview with stating the result.

Theorem 1 ([8]). *A set of n non-overlapping intervals can be stored in a data structure of size $O(n)$, subject to stabbing queries, insertion and deletion of intervals in $O(\log n)$ worst-case time, and local updates in $O(1)$ worst-case time.*

2   An Alternative Data Structure for Disjoint Intervals

Below we describe another solution to Problem 1 for disjoint intervals. This solution is less efficient than the one summarized in Sect. 1.2: it ignores Property 1, and it rather can be seen as a version of the data structure for 2-dimensional disjoint fat regions [8]. The main goal of this section is to simplify the latter data structure as much as possible, still making sure it contains the ideas, useful for our solution for overlapping intervals.

The data structure, similarly to the one of Sect. 1.2, contains the *quadtree* and the *query tree* built on its leaves, but they are now defined differently. Moreover, we use an additional type of structure: a *marked-ancestor tree* built on top of the quadtree cells.

The Quadtree. We maintain a compressed (but not balanced) quadtree that stores the intervals in S. Notice that sometimes we need to create the cells that would automatically exist, should the quadtree be balanced or non-compressed. We defer the discussion on handling this to Sect. 4, and until then we assume that the quadtree does not contain compressed nodes. The quadtree cells store intervals according to the following:

Condition 1. *An interval $I \in S$ is stored in a cell C if and only if C is the largest cell that contains I's center point and is entirely covered by I. See Fig. 2.*

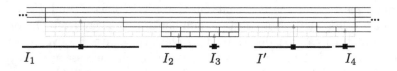

Fig. 2. A quadtree storing a set of disjoint intervals according to Condition 1

The above condition implies that if an interval I is stored in a quadtree cell C, then $|C| \le |I| < 4|C|$. In Fig. 2, the diameter of interval I_3 equals the size of its cell, and the diameter of interval I_2 is almost four times the size of its cell.

For the purpose which will be evident soon, we add more cells to the quadtree: For each interval I stored in a cell C according to Condition 1, we make sure that all (at most four) the cells of size $|C|$ intersected by I exist in the quadtree.

Figure 2 shows the quadtree for a set of five intervals. Black lines indicate the existing quadtree cells, and the light-gray lines indicate their further subdivision.

Marked-Ancestor Trees. We maintain three marked-ancestor trees built on cells of the quadtree. We denote these trees *L-MAT*, *R-MAT*, and *C-MAT*, standing for the left, the right, and the center marked-ancestor tree.

The quadtree cells are marked according to the following criteria:

Condition 2. *A quadtree cell C is marked in one of the marked-ancestor trees, if there is an interval $I \in S$ such that I intersects C, and the cell that stores I has size $|C|$. Specifically:*

 (i) *If C contains I's right endpoint (and thus I's center is to the left of C), then C is marked in L-MAT.*
 (ii) *If C contains I's left endpoint then C is marked in R-MAT.*
 (iii) *If I covers C entirely, then C is marked in C-MAT.*

We say that C is marked by I, or that I marks C. See Fig. 3.

Each interval I marks either three or four quadtree cells: the cell C that stores I, the cells of size $|C|$ that contain respectively the left and the right endpoint of C, and possibly C's neighbor of size $|C|$ entirely covered by I. Note that, by the way we have defined the quadtree in the beginning of this section, such quadtree cells are always present in it.

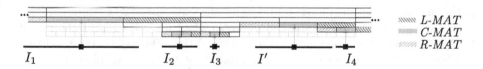

Fig. 3. Marking the cells of the quadtree from Fig. 2 in *L-MAT*, *C-MAT*, and *R-MAT*

Below Lemmas 1 and 2 are useful properties of the marked-ancestor trees, which are not hard to see. Lemma 3 provides an implementation of these structures.

Lemma 1. *Let C and C_a be two quadtree cells, such that C_a is an ancestor of C, and C_a is marked in C-MAT by some interval I_a. Then I_a entirely covers C.*

Lemma 2. *Let C and C_a be two quadtree cells that are both marked in L-MAT by an interval I and by an interval I_a, respectively. If C_a is an ancestor of C, then I_a lies to the left of I. A symmetric property holds for R-MAT.*

Lemma 3 ([2]). *For any rooted tree T, a data structure over the nodes of T can be maintained, supporting insertions and deletions of leaves in $O(1)$ time, marking and unmarking nodes in $O(\log \log n)$ time, and $O(\log n / \log \log n)$-time lowest marked-ancestor queries. The time bound for insertions is amortized; other bounds are worst-case.*

If T is a path, then marking, unmarking, and answering a marked-successor (or a marked-predecessor) query require $O(\log \log n)$ time each.

The Query Tree. The query tree is a balanced binary tree whose leaves correspond to leaves of the quadtree, ordered as they appear on $\mathbb{R}^1$. Unlike Sect. 1.2, the leaves of the query tree do not have pointers to their non-empty neighbors.

Stabbing Queries. Given a query point q, we need to return an interval in S that contains q, if such interval exists. By locating q in the query tree, we find the quadtree leaf C that contains q. If C is marked in C-MAT by some interval I (C may or may not store I), then we report I. Otherwise, we find the lowest marked ancestor C_a of C in L-MAT, and check whether the interval that marks C_a contains q. We do the same in R-MAT.

Lemma 4. *The above procedure is correct and requires $O(\log n)$ time.*

Proof. Let C be the quadtree leaf that contains q. If C is marked by an interval I in C-MAT then by definition I covers C, and thus it contains q. Let I be an interval in S that intersects C, but does not cover it. Then I intersects one of the borders of C, say, the left border. Thus I marks an ancestor C_a of C in L-MAT. Moreover, C_a is the lowest marked ancestor of C in L-MAT: suppose that some $C_a' \neq C_a$ is the lowest marked ancestor of C; let I' be the interval that marks C_a'. Then both I and I' contain the left border of C_a'. We obtain a contradiction to the disjointness of the intervals in S. The argument for R-MAT is symmetric.

By Lemma 3, the lowest marked ancestor of C in L-MAT and the one in R-MAT can be determined in $O(\log n / \log \log n)$ time. Therefore, the total time required for the stabbing query is dominated by the time required for point location in the query tree, and thus it is $O(\log n)$.   □

Local Updates. Given a pointer to an interval I stored in cell C, we need to replace it with a $O(1)$-similar interval I'. Let C' be the quadtree cell that needs to store I', see Condition 1. Assume that the pointer to C' is available.

Deleting I and inserting I' then reduces to repairing the data structures. Modifying the quadtree and the query tree is done as in Sect. 1.2. Deleting I requires unmarking the quadtree cells marked by I. The latter cells are easy to find in constant time as they are the cells of size $|C|$ that intersect I. If these unmarked cells are leaves, they may get deleted from the quadtree. Inserting I' requires the reverse manipulations: marking the cells of size $|C'|$ intersected by I, and adding them if they do not yet exist.

Lemma 5. *A local update requires $O(\log n / \log \log n)$ time.*

Proof. Intervals I and I' mark at most four quadtree cells each; such cells can be found in constant time. To insert and to delete a quadtree leaf is a constant-time operation. Thus the time required by the local update (after C' is available) is dominated by the time for a constant number of marking and unmarking operations, each of which by Lemma 3 requires $O(\log \log n)$ time. The overhead due to compression of the quadtree for finding C' is $O(\log n / \log \log n)$, see Lemma 8. The claim follows.   □

Classic Updates. Insertion (resp., deletion) of an interval requires an insertion (resp., deletion) in the query tree, and an update to marked-ancestor structures. The former two operations require worst-case $O(\log n)$ time (including the overhead due to compression, see Lemma 8). The latter operation requires $O(1)$ time, amortized for insertions and worst-case for deletions. Therefore we say that the classic updates require $O(\log n)$ time per operation, amortized for insertion and worst-case for deletions.

3   A Data Structure for Overlapping Intervals

In this section we present our solution to Problem 1 for sets of intervals that may overlap. Section 3.1 considers the case when the ply of the interval set is always at most two. Generalization of the data structure to the case of higher ply is quite intuitive, and we sketch it briefly due to the space constraints.

3.1   Intervals with Ply ≤ 2

Now we are given a set S of intervals that may overlap, such that the ply of S is guaranteed to be at most two at any moment. We solve Problem 1 for such S.

We remark that partitioning S into a constant number of layers, such that at each layer the intervals are disjoint, does not seem to work.

Should we do this, we would need to restore the properties of the layers when after a local update two intervals of the same layer start overlapping. A natural way to handle this, i.e., to assign the interval that just has been updated the next possible layer, does not work: Fig. 4 shows three intervals and a sequence of local updates, where every local update causes a change of the layer for the updated interval. Thus we need a more involved data structure for solving our problem.

Our starting point is the data structure of Sect. 2. Note that a single *L-MAT* (and a single *R-MAT*) is not enough for our setting: It can happen that the quadtree leaf that contains a query point q has linearly many marked ancestors in *L-MAT* before the one marked by the interval that actually contains q. Figure 5 shows such situation with three marked ancestors; the construction can be continued to increase this sequence arbitrarily, using intervals and cells of smaller size. To overcome this issue, we use marked-ancestor trees of two levels. Below we discuss the modifications to the data structure of Sect. 2 in detail.

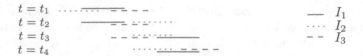

Fig. 4. Subdividing intervals in two layers of disjoint intervals is not efficient: A set of three intervals $\{I_1, I_2, I_3\}$, and a sequence of local updates that would cause linear number of layer changes. We illustrate here the first three updates in this sequence, i.e., the segment set in each of the first four time moments $t_1, \ldots, t_4$.

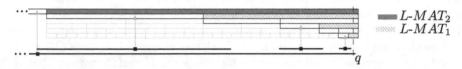

Fig. 5. A set of intervals with ply two, and a query point q, such that the cell of q has many marked ancestors in $L\text{-}MAT$ (the shaded cells), but only the largest one is marked by the interval that contains q. This cell must be marked in $L\text{-}MAT_2$.

The Quadtree. The rule to store intervals in the quadtree cells is still provided by Condition 1. Since intervals are now allowed to overlap, they may be stored in intermediate cells of the quadtree. If an interval $I \in S$ does not overlap any other intervals, then I is stored in a quadtree leaf. The inverse is not true. Notice that since ply of S is at most two, one cell C stores at most two intervals. If C stores two intervals, then C is a leaf.

Marked-Ancestor Trees. We maintain two levels of the left and the right marked-ancestor trees, denoted by $L\text{-}MAT_i, R\text{-}MAT_i, i \in \{1, 2\}$. The center marked-ancestor tree ($C\text{-}MAT$) is unique, and is defined in the same way as previously.

Marking the quadtree cells in the left marked-ancestor trees is now done according to the following (marking in the right trees is symmetric):

Condition 3. *For a cell C, if there is an interval I such that Condition 2(i) holds for C and I, then C is marked in one of the left marked-ancestor trees. Specifically (See Figs. 5 and 6):*

- *C is marked in $L\text{-}MAT_2$ if there is a descendant C' of C and an interval I' such that Condition 2(i) holds for C' and I', and I entirely covers I'.*
- *C is marked in $L\text{-}MAT_1$, otherwise.*

Since $C\text{-}MAT$ is defined in the same way as in Sect. 2, Lemma 1 still holds. The following lemma generalizes Lemma 2.

Lemma 6. *Let C and C_a be two quadtree cells that are both marked in $L\text{-}MAT_i$ for some level $i \in \{1, 2\}$ respectively by interval I and by interval I_a. If C_a is an ancestor of C, then the right endpoint of I_a lies to the left of the right endpoint of I. A symmetric property holds for $R\text{-}MAT_i$.*

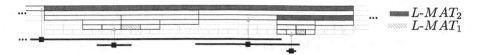

Fig. 6. A set of four intervals with ply 2; marking in $L\text{-}MAT_1$ and $L\text{-}MAT_2$ induced by them

Proof. Suppose for the sake of contradiction, that there are two quadtree cells C and C_a that are both marked in $L\text{-}MAT_1$ or both in $L\text{-}MAT_2$ by intervals I and I_a respectively, such that C_a is an ancestor of C, and the right endpoint of I_a is to the right of the right endpoint of I. Then I is covered entirely by I_a. It is not possible that both C and C_a are marked in $L\text{-}MAT_1$: by definition C_a must be marked in $L\text{-}MAT_2$, since I_a covers I. If both C and C_a were marked in $L\text{-}MAT_2$, then there would be an interval I' entirely covered by I. This would contradict the ply ≤ 2 condition for S. □

Marked-Descendant Trees. To update the marked-ancestor trees efficiently, we need a data structure built on the quadtree, that would quickly answer the following queries.

Definition 5 (Leftmost/rightmost marked-descendant query). *The left-most marked-descendant query for an intermediate quadtree cell C is as follows. If the subtree of C contains cells marked in $L\text{-}MAT_1$, among these cells, return the one that comes first after C in the pre-order traversal of the quadtree.[3] Otherwise return nil. The rightmost marked-descendant query for C is: If the subtree of C contains cells marked in $R\text{-}MAT_1$, among these cells, return the one that precedes C in the post-order traversal of the quadtree.[4] Otherwise return nil.*

The following lemma justifies the use of marked-descendant queries to manipulate the marked-ancestor trees.

Lemma 7. *Suppose a quadtree cell C and an interval I satisfy Condition 2(i). Cell C is marked in $L\text{-}MAT_2$ if and only if (1) the leftmost marked descendant C_d of C exists, and (2) $I_d \subset I$, where I_d is the interval that marks C_d.*

Same holds for a cell marked in $R\text{-}MAT_2$ and its rightmost marked descendant.

Proof. If items (1) and (2) hold, then C is marked in $L\text{-}MAT_2$ by Condition 3.

Suppose C is marked in $L\text{-}MAT_2$. Then there is a descendant C' of C marked in $L\text{-}MAT_1$ by an interval I' such that $I' \subset I$. First note that C' is a marked descendant of C, thus the leftmost marked descendant C_d of C exists. Suppose $C_d \neq C'$. Since C_d appears before C' in the pre-order traversal of the quadtree,

---

[3] The pre-order traversal of a binary tree first visits the root, then it recursively visits the left subtree, and finally it recursively visits the right subtree.

[4] The post-order traversal of a binary tree first recursively visits the left subtree, then it recursively visits the right subtree, and finally it visits the root.

the left border of C_d either coincides with left border of C', or lies to the left of it. Thus I_d is completely to the left of I'. Since I_d is stored in a cell whose size is less than $|C|$, the left border of I_d cannot be to the left of the left border of I, and thus $I_d \subset I$.   □

To efficiently answer the above queries, we maintain a pair of data structures, which we call the *left* and the *right marked-descendant tree*, respectively. The left marked-descendant tree is implemented by maintaining the path induced by the pre-order traversal of the quadtree, and the marked-successor data structure of Lemma 3 on top of this path for the cells marked in L-MAT_1. Maintaining the path can be done by augmenting each node with a pointer to its successor in the traversal. The leftmost marked-descendant query is then performed by querying the marked successor of C in the pre-order, and checking whether the returned cell is in the subtree of C. If no cell is returned, or if the returned cell is not a descendant of C, then we return *nil*. Otherwise we return that cell. The marked-successor queries require worst-case $O(\log \log n)$ time per query by Lemma 3.

The right marked-descendant tree is symmetric: we maintain the reversed post order, (i.e., determine the pointers for the post-order traversal and reverse all of them), and the marked-successor data structure on top of this path.

Stabbing Queries. Given a query point q, we need to report all intervals in S that contain q. To do that, we should first find the quadtree leaf C that contains q. If C or C's ancestor(s) are marked in C-MAT, we report the interval(s) that mark them, similar to Sect. 2. After that we query left and right marked-ancestor trees: We check the intervals that mark the lowest and the second lowest marked ancestor of C in L-MAT_1 (in case C is marked in L-MAT_1, we check its marking interval too). We report those of the above intervals that contain q. We repeat the procedure for L-MAT_2, R-MAT_1, and R-MAT_2.

The above procedure requires $O(\log n)$ time, as it is dominated by the time for point location in the query tree.

Local Updates. We need to move an interval I stored in a cell C so that it becomes an interval I', $O(1)$-similar to I. As in Sect. 2, we assume that we have a pointer to the cell C' that must store I'. Below we describe updating of the left marked-ancestor and marked-descendant trees only, as all other structures are either symmetric to those, or are updated exactly as in Sect. 2.

The *deletion* of I causes the following modifications. Let C_ℓ be the cell such that $|C_\ell| = |C|$ and C_ℓ contains the right endpoint of I. Cell C_ℓ is marked by I in either L-MAT_1 or in L-MAT_2. In the latter case C_ℓ gets unmarked, and nothing else should be done. In the former case, in addition to unmarking C_ℓ we must check whether such unmarking causes some cell marked in L-MAT_2 to change the marking level and start being marked in L-MAT_1 instead. Observe, that the only cell that could possibly change the marking level is the lowest marked ancestor C_a of C_ℓ in L-MAT_2. To check whether C_a changes the marking level, we perform the leftmost marked-descendant query for C_a. If that query returned a cell C_d, and I_a entirely covers I_d, where I_a and I_d are the intervals that mark C_a and C_d respectively, then C_a stays marked in $L\text{-}MAT_2$. Otherwise (i.e., if I_a does not cover I_d, or if C_a does not have the leftmost marked descendant), we unmark C_a in $L\text{-}MAT_2$ and mark it in $L\text{-}MAT_1$.

The *insertion* of I' causes the following modifications. The quadtree cells of size $|C'|$ intersected by I' get marked. Let C'_ℓ be the quadtree cell such that $|C'_\ell| = |C'|$ and C'_ℓ contains the right endpoint of I'. Cell C'_ℓ should be marked in one of the levels of the left marked-ancestor trees. To decide in which, we perform the leftmost marked-descendant query for C'_ℓ. If the query returns a cell C'_d, we check whether I' covers the interval I'_d that marks C'_d. If this is the case, C'_ℓ should be marked in $L\text{-}MAT_2$. Otherwise, C'_ℓ gets marked in $L\text{-}MAT_1$. In the latter case, we also check for the lowest ancestor C'_a of C'_ℓ in $L\text{-}MAT_1$, whether the marking of C'_ℓ causes a change of marking level of C'_a. Namely, if I' is contained in the interval I'_a that marks C'_a, then C'_a gets unmarked in $L\text{-}MAT_1$ and marked in $L\text{-}MAT_2$ instead.

Whenever a cell C is marked or unmarked in $L\text{-}MAT_1$, it is marked or unmarked in the left marked-descendant tree.

Correctness of the above procedure follows from Lemma 7. The time required by the procedure is dominated by the time for a constant number of lowest marked-ancestor queries and thus is $O(\log n/\log\log n)$, including the overhead due to compression, see Lemma 8. Classic updates are exactly the same as in Sect. 2. We conclude.

Theorem 2. *A set S of n intervals in $\mathbb{R}^1$, such that the ply of S is at most two, can be stored in a data structure of size $O(n)$, subject to stabbing queries, insertion and deletion of intervals in $O(\log n)$ time, and local updates in $O(\log n/\log\log n)$ time. The bound is amortized for the insertion, and worst-case for all other operations.*

3.2 Intervals with Higher Ply

For a set of intervals with ply $k \geq 2$, the above data structure can be generalized, resulting in the following.

Theorem 3. *A set S of n intervals in $\mathbb{R}^1$, that may overlap, can be stored in a data structure of size $O(n)$, subject to stabbing queries, insertion and deletion of intervals in $O(\log n + k\log n/\log\log n)$ time, and local updates in $O(k\log n/\log\log n)$ time, where k is the ply of S at the time of the operation. The bound is amortized for the insertion, and worst-case for all other operations.*

Proof. (sketch). We maintain k levels of marked-ancestor trees. For a quadtree cell C, if there is an interval I such that Condition 2(i) is satisfied for C and I, then C is marked in $L\text{-}MAT_j$, for some $1 \leq j \leq k$. Cell C is marked in $L\text{-}MAT_i$, $1 < i \leq k$, if there is a sequence $C_1, \ldots, C_{i-1}$ of descendants of C and a sequence $I_1, \ldots, I_{i-1}$ of intervals in S such that $I_1 \subset I_2 \subset \ldots \subset I_{i-1} \subset I$ and Condition 2(i) is satisfied for each pair C_j, I_j, $1 \leq j \leq i - 1$. Otherwise, C is marked in $L\text{-}MAT_1$.

We also maintain $k - 1$ marked-descendant trees, where the left marked-descendant tree of level i, $1 \leq i < k$ is defined as in Sect. 3.1 for the cells marked in $L\text{-}MAT_i$.

The operations and queries are simple generalizations of the ones from Sect. 3.1. For example, while doing the stabbing queries for a point q, we search in each marked-ancestor tree for at most k lowest marked ancestors of C, where C is the quadtree leaf containing q; the total number of marked-ancestor queries performed is $O(k)$ due to a generalization of Lemma 6. $\square$

4   Compressing the Quadtree

The compressed quadtree contains a-compressed cells for some large constant a. An a-compressed cell C has only one child C', such that $|C'| \leq |C|/a$, and for any interval I in S whose center point is contained in $C \backslash C'$, I entirely covers C. The compressed nodes cut the quadtree into several *regular components* that are smaller (uncompressed) quadtrees. We need to modify our data structure from Sects. 2 and 3 to handle the presence of compressed cells in the quadtree. Lemma 8 discusses the complexity of such modifications. In particular, each update operation needs additional time to restore the properties of a compressed quadtree. We refer to such additional time as *overhead*.[5] Notice that stabbing queries do not modify the quadtree, therefore they do not cause any overhead.

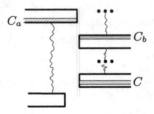

Fig. 7. Regular components of the quadtree (bounded by bold lines), and paths that connect the compressed nodes with their unique children (zigzag lines). Cell C_b is marked in $L'\text{-}MAT$ by the compressed cell C_a. For the cell C, its lowest marked ancestor in $L'\text{-}MAT$ is C_b.

Lemma 8. *Compression in the quadtree for intervals that possibly overlap, can be maintained in $O(n)$ space. The time overhead is $O(\log n / \log \log n)$ per local update operation, and $O(\log \log n)$ per classic update operation.*

Proof. Consider a local update that turns an interval I into an interval I', $O(1)$-similar to I. We need to quickly find the cell where I' should be stored. That cell should be a constant number of cells away from C, and in a non-compressed (or in a balanced) quadtree it (or the leaf where it should be inserted) could be

---

[5] Note that the analysis in Sects. 2 and 3 already takes Lemma 8 into account.

found by following the level links from C to the cells of size $|C|$ adjacent to C. In our setting, it may happen that such cell C'' does not exist, and the smallest cell C_a above C'' is a compressed cell. In that case, once C_a is found, the necessary decompression can be done in $O(1)$ time.

To perform the search for C_a efficiently, we maintain two additional marked-ancestor trees on top of the quadtree. These trees are denoted L'-MAT and R'-MAT, and correspond to the following marking rule. A cell C_a is marking a cell C_b in L'-MAT if C_a is the rightmost leaf in its regular quadtree component, and C_b is the largest cell adjacent to the right border of C_a, such that $|C_b| \le |C_a|$, and C_b is to the right of C_a. See Fig. 7. Marking in R'-MAT is symmetric.

For a cell C (see Fig. 7), going to the lowest marked ancestor C_b of C in L'-MAT, and then to the cell that marks C_b, results exactly in the sought compressed cell C_a. Thus the time needed to restore the properties of a compressed quadtree after a local update is dominated by the time to find the lowest marked ancestor in L'-MAT or R'-MAT. By Lemma 3, this is $O(\log n / \log \log n)$.

A classic update causes $O(1)$ compression/decompression operations like in the case of non-overlapping intervals [8], which requires $O(1)$ time, and $O(1)$ marking/unmarking operations in L'-MAT and R'-MAT, which by Lemma 3 can be performed in $O(\log \log n)$ time in total.   ⊓

Acknowledgements. We wish to thank Irina Kostitsyna for helpful discussions.

References

1. Agarwal, P.K., Arge, L., Kaplan, H., Molad, E., Tarjan, R.E., Yi, K.: An optimal dynamic data structure for stabbing-semigroup queries. SIAM J. Comput. **41**(1), 104–127 (2012)
2. Alstrup, S., Husfeldt, T., Rauhe, T.: Marked ancestor problems. In: 39th Annual Symposium on Foundations of Computer Science, pp. 534–543 (1998)
3. Arge, L., Vitter, J.S.: Optimal dynamic interval management in external memory. In: 37th Conference on Foundations of Computer Science, pp. 560–569 (1996)
4. Berg, M., Cheong, O., Kreveld, M., Overmars, M.: Computational Geometry - Algorithms and Applications, 3rd edn. Springer, Heidelberg (2008)
5. Cormen, T.H., Leiserson, C.E., Rivest, R.L., Stein, C.: Introduction to Algorithms, 3rd edn. MIT Press, Cambridge (2009)
6. Edelsbrunner, H.: Dynamic data structures for orthogonal intersection queries. Report F59. Technische Universität Graz (1980)
7. Kaplan, H., Molad, E., Tarjan, R.: Dynamic rectangular intersection with priorities. In: 35th ACM Symposium on Theory of Computing (STOC), pp. 639–648 (2003)
8. Löffler, M., Simons, J.A., Strash, D.: Dynamic planar point location with sub-logarithmic local updates. In: Dehne, F., Solis-Oba, R., Sack, J.-R. (eds.) WADS 2013. LNCS, vol. 8037, pp. 499–511. Springer, Heidelberg (2013). doi:10.1007/978-3-642-40104-6_43
9. McCreight, E.M.: Efficient algorithms for enumerating intersecting intervals and rectangles. report csl-80-9. Technical report, Xerox Palo Alto Res. Center (1980)
10. McCreight, E.M.: Priority search trees. SIAM J. Comput. **14**(2), 257–276 (1985)

11. Nekrich, Y.: Data structures with local update operations. In: Gudmundsson, J. (ed.) SWAT 2008. LNCS, vol. 5124, pp. 138–147. Springer, Heidelberg (2008). doi:10.1007/978-3-540-69903-3_14
12. Nekrich, Y.: A dynamic stabbing-max data structure with sub-logarithmic query time. In: Asano, T., Nakano, S., Okamoto, Y., Watanabe, O. (eds.) ISAAC 2011. LNCS, vol. 7074, pp. 170–179. Springer, Heidelberg (2011). doi:10.1007/978-3-642-25591-5_19
13. Thorup, M.: Space efficient dynamic stabbing with fast queries. In: 35th ACM Symposium on Theory of Computing (STOC), pp. 649–658. ACM Press (2003)

The Transformation Monoid
of a Partially Lossy Queue

Chris Köcher[✉] and Dietrich Kuske

Institut für Theoretische Informatik,
Technische Universität Ilmenau, Ilmenau, Germany
{chris.koecher,dietrich.kuske}@tu-ilmenau.de

Abstract. We model the behavior of a lossy fifo-queue as a monoid of transformations that are induced by sequences of writing and reading. To have a common model for reliable and lossy queues, we split the alphabet of the queue into two parts: the forgettable letters and the letters that are transmitted reliably.

We describe this monoid by means of a confluent and terminating semi-Thue system and then study some of the monoids algebraic properties. In particular, we characterize completely when one such monoid can be embedded into another as well as which trace monoids occur as submonoids. The resulting picture is far more diverse than in the case of reliable queues studied before.

1 Introduction

Queues (alternatively: fifo queues or channels) form a basic storage mechanism that allows to append items at the end and to read the left-most item from the queue. Providing a finite state automaton with access to a queue results in a Turing complete computation model [2] such that virtually all decision problems on such devices become undecidable.

Situation changes to the better if one replaces the reliable queue by some forgetful version. The most studied version are lossy queues that can nondeterministically lose any item at any moment [1,3,7,13]: in that case reachability, safety properties over traces, inevitability properties over states, and fair termination are decidable (although of prohibitive complexity, see, e.g., [4]). A practically more realistic version are priority queues where items of high priority can erase any previous item of low priority. If items of priority i can be erased by subsequent items of priority *at least* i, then safety and inevitability properties are decidable, if items of priority i can be erased by subsequent items of priority *strictly larger than* i, only, then these problems become undecidable [8].

In this paper, we study partially lossy queues that can be understood as a model between lossy and priority queues. Seen as a version of lossy queues, their alphabet is divided into two sets of reliable and forgettable letters where

C. Köcher and D. Kuske—Supported by the DFG-Project "Speichermechanismen als Monoide", KU 1107/9-1.

P. Weil (Ed.): CSR 2017, LNCS 10304, pp. 191–205, 2017.
DOI: 10.1007/978-3-319-58747-9_18

only items from the second set can be lost. Seen as a version of priority queues, partially lossy queues use only two priorities (0 and 1) where items of priority 0 can be erased by any item of priority at least 0 (i.e., by all items) and items of priority 1 can only be erased by items of strictly larger priority (which do not exist).

We describe the behavior of such a partially lossy queue by a monoid as was done, e.g., for pushdowns in [10] and for reliable queues in [9,12]: A partially lossy queue is given by its alphabet A as well as the subset $X \subseteq A$ of letters that the queue will transmit reliably. Note that writing a symbol into a queue is always possible (resulting in a longer queue), but reading a symbol is possible only if the symbol is at the beginning of the queue (or is preceded by forgettable symbols, only). Thus, basic actions define partial functions on the possible queue contents. The generated transformation monoid is called *partially lossy queue monoid* or *plq monoid* $\mathcal{Q}(A, X)$. Then $\mathcal{Q}(A, A)$ models the behaviour of a reliable queue with alphabet A [9,12] and $\mathcal{Q}(A, \emptyset)$ the fully lossy queue that can forget any symbol [11].

The first part of this paper presents a complete infinite semi-Thue system for the monoid $\mathcal{Q}(A, X)$. The resulting normal forms imply that two sequences of actions are equivalent if their subsequences of write and of read actions, respectively, coincide and if the induced transformations agree on the shortest queue that they are defined on.

This result is rather similar, although technically more involved, than the corresponding result on the monoid $\mathcal{Q}(A, A)$ of the reliable queue from [9]. In that paper, it is also shown that $\mathcal{Q}(A, A)$ embeds into $\mathcal{Q}(B, B)$ provided B is not a singleton. This is an algebraic formulation of the wellknown fact that the reliable queue with two symbols can simulate any other reliable queue. The second part of the current paper is concerned with the embeddability relation between the monoids $\mathcal{Q}(A, X)$. Clearly, the monoid $\mathcal{Q}(A, \emptyset)$ of the fully lossy queue embeds into $\mathcal{Q}(B, \emptyset)$ whenever $|A| \leq |B|$ by looking at A as a subset of B. Joining this almost trivial idea with the (nontrivial) idea from [9], one obtains an embedding of $\mathcal{Q}(A, X)$ into $\mathcal{Q}(B, Y)$ provided the second queue has at least as many forgettable letters as the first and its number of unforgettable letters is at least the number of unforgettable letters of the first queue or at least two (i.e., $|A \backslash X| \leq |B \backslash Y|$ and $\min\{|X|, 2\} \leq |Y|$). We prove that, besides these cases, an embedding exists only in case the second queue has precisely one non-forgettable letter and properly more forgettable letters than the first queue (i.e., $|Y| = 1$ and $|A \backslash X| < |B \backslash Y|$). As for the reliable queue, this algebraically mirrors the intuition that a partially lossy queue can simulate another partially lossy queue in these cases, only. In particular, a reliable queue does not simulate a fully lossy queue and vice versa and a fully lossy queue cannot simulate another fully lossy queue with more (forgettable) letters.

These results show that the class of submonoids of a plq monoid $\mathcal{Q}(A, X)$ depends heavily on the number of forgettable and non-forgettable letters. In [9], it is shown that the direct product of two free monoids embeds into the monoid of the reliable queue $\mathcal{Q}(A, A)$ (with $|A| \geq 2$). The paper [12] elaborates on this and characterizes the class of trace monoids $\mathbb{M}(\Gamma, I)$ [6] that embed into $\mathcal{Q}(A, A)$. In particular, it shows that $\mathbb{N}^3$ is not a submonoid of $\mathcal{Q}(A, A)$. The final section of this paper studies this question for plq monoids. The – at least for the authors – surprising answer is that, provided the queue has at least one non-forgettable or at least three forgettable letters, a trace monoid embeds into $\mathcal{Q}(A, X)$ if and only if it embeds into $\mathcal{Q}(A, A)$. By [12], this is the case if all letters in the independence alphabet (Γ, I) have degree at most 1 or the independence alphabet is a complete bipartite graph with some additional isolated vertices. We provide a similar characterization for trace monoids embedding into $\mathcal{Q}(\{a, b\}, \emptyset)$: here, the complete bipartite component is replaced by a star graph. In any case, the direct product of $(\mathbb{N}, +)$ and $\{a, b\}^*$ embeds into $\mathcal{Q}(A, X)$. Since in this direct product, the inclusion problem for rational sets is undecidable (cf. [15]), the same applies to $\mathcal{Q}(A, X)$.

In summary, we study properties of the transformation monoid of a partially lossy queue that were studied for the reliable queue in [9,12]. We find expected similarities (semi-Thue system), differences (embeddability relation) and surprising similarities (trace submonoids).

2 Preliminaries

At first we need some basic definitions. So let A be an alphabet. A word $u \in A^*$ is a *prefix* of $v \in A^*$ iff $v \in uA^*$. Similarly, u is a *suffix* of v iff $v \in A^*u$. Furthermore u is a *subword* of v iff there are $k \in \mathbb{N}$, $a_1, \ldots, a_k \in A$ and $w_1, \ldots, w_{k+1} \in A^*$ such that $u = a_1 \ldots a_k$ and $v = w_1 a_1 w_2 a_2 \ldots w_k a_k w_{k+1}$, i.e., we obtain u if we drop some letters from v. In this case we write $u \preceq v$. Note that $\preceq$ is a partial ordering on A^*. Let $X \subseteq A$. Then we define the *projection* $\pi_X : A^* \to X^*$ on X by

$$\pi_X(\varepsilon) = \varepsilon \text{ and } \pi_X(au) = \begin{cases} a\pi_X(u) & \text{if } a \in X \\ \pi_X(u) & \text{otherwise} \end{cases}$$

for each $a \in A$ and $u \in A^*$. Moreover, u is an *X-subword* of v (denoted $u \preceq_X v$) if $\pi_X(v) \preceq u \preceq v$, i.e., if we obtain u from v by dropping some letters not in X. Note that $\preceq_\emptyset$ is the subword relation $\preceq$ and $\preceq_A$ is the equality relation.

2.1 Definition of the Monoid

We want to model the behaviour of an unreliable queue that stores entries from the alphabet A. The unreliability of the queue stems from the fact that it can forget certain letters that we collect in the set $A \backslash X$. In other words, letters from $X \subseteq A$ are *non-forgettable* and those from $A \backslash X$ are *forgettable*. Note that this unreliability extends the approach from [9] where we considered reliable queues (i.e., $A = X$).

So let A be an alphabet of possible queue entries and let $X \subseteq A$ be the set of non-forgettable letters. The states of the queue are the words from A^*.

Furthermore we have some basic controllable actions on these queues: writing of a symbol $a \in A$ (denoted by a) and reading of $a \in A$ (denoted by $\overline{a}$). Thereby we assume that the set $\overline{A}$ of all these reading operations $\overline{a}$ is a disjoint copy of A. So $\Sigma := A \cup \overline{A}$ is the set of all controllable operations on the partially lossy queue. For a word $u = a_1 \ldots a_n \in A^*$ we write $\overline{u}$ for the word $\overline{a_1} \, \overline{a_2} \ldots \overline{a_n}$.

Formally, the action $a \in A$ appends the letter a to the state of the queue. The action $\overline{a} \in \overline{A}$ tries to cancel the letter a from the beginning of the current state of the queue. If this state does not start with a then the operation $\overline{a}$ is not defined. The lossiness of the queue is modeled by allowing it to forget arbitrary letters from $A \backslash X$ of its content at any moment.

This semantics is similar to the "standard semantics" from [4, Appendix A] where a lossy queue can lose any message at any time. The main part of that paper considers the "write-lossy semantics" where lossiness is modeled by the effect-less writing of messages into the queue. The authors show that these two semantics are equivalent [4, Appendix A] and similar remarks can be made about priority queues [8]. A third possible semantics could be termed "read-lossy semantics" where lossiness is modeled by the loss of any messages that reside in the queue before the one that shall be read. In that case, the queue forgets letters only when necessary and this necessity occurs when one wants to read a letter that is, in the queue, preceded by some forgettable letters.

In the complete version of this paper, we define both, the "standard semantics" and the "read-lossy semantics" and prove that the resulting transformation monoids are isomorphic; here, we only define the "read-lossy semantics" as this semantics is more convenient for our further considerations.

Definition 2.1. *Let $X \subseteq A$ be two finite sets and $\bot \notin A$. Then the map $\circ_X : (A^* \cup \{\bot\}) \times \Sigma^* \to (A^* \cup \{\bot\})$ is defined for each $q \in A^*$, $a \in A$ and $u \in \Sigma^*$ as follows:*

(i) $q \circ_X \varepsilon = q$

(ii) $q \circ_X au = qa \circ_X u$

(iii) $q \circ_X \overline{a}u = \begin{cases} q' \circ_X u & \text{if } q \in (A \backslash (X \cup \{a\}))^* \, a \, q' \\ \bot & \text{otherwise} \end{cases}$

(iv) $\bot \circ_X u = \bot$

Consider the definition of $q \circ_X \overline{a}u$. There, the word aq' is the smallest suffix of q that contains all the occurrences of the letter a (it follows that the operation $\circ_X$ is welldefined) and the complementary prefix consists of forgettable entries, only. Hence, to apply $\overline{a}$, the queue first "forgets" the prefix and then "delivers" the letter a that is now at the first position.

Lemma 2.2. *Let $q, u \in A^*$ such that $q \circ_X \overline{u} \neq \bot$. Then $q \circ_X \overline{u}$ is the longest suffix of q with $\pi_X(p) \preceq u \preceq p$ where p is the complementary prefix.*

Example 2.3. Let $a \in A \backslash X$, $b \in X$, $q = aabaabba$ and $u = aba$. Then we have $q \circ_X \bar{u} = abaabba \circ_X \overline{ba} = aabba \circ_X \bar{a} = abba$.

On the other hand, the words $aaba$ and $aabaa$ are the only prefixes p' of q with $\pi_X(p') \preceq u \preceq p'$. Their complementary suffixes are $abba$ and bba, the longer one equals $q \circ_X \bar{u}$ as claimed by the lemma.

Two sequences of actions that behave the same on each and every queue will be identified:

Definition 2.4. *Let $X \subseteq A$ be two finite sets and $u, v \in \Sigma^*$. Then u and v act equally (denoted by $u \equiv_X v$) if $q \circ_X u = q \circ_X v$ holds for each $q \in A^*$.*

The resulting relation $\equiv_X$ is a congruence on the free monoid Σ^*. Hence, the quotient $\mathcal{Q}(A, X) := \Sigma^* / {\equiv_X}$ is a monoid which we call partially lossy queue monoid or plq monoid induced by (A, X).

Example 2.5. Let $a, b \in A$ be distinct. Then we have $\varepsilon \circ_\emptyset ba\bar{a} = ba \circ_\emptyset \bar{a} = \varepsilon$ and $\varepsilon \circ_\emptyset b\bar{a}a = \bot$ implying $ba\bar{a} \not\equiv_\emptyset b\bar{a}a$.

On the other hand, $\varepsilon \circ_A ba\bar{a} = ba \circ_A \bar{a} = \bot = \varepsilon \circ_A b\bar{a}a$. It can be verified that, even more, $q \circ_A ba\bar{a} = q \circ_A b\bar{a}a$ holds for all $q \in A^*$ (since $a \neq b$) implying $ba\bar{a} \equiv_A b\bar{a}a$.

General Assumption. Suppose $A = \{a\}$ is a singleton. Then $a^{n+1} \circ_X \bar{a} = a^n$ for any $n \geq 0$ (independent of whether $X = A$ or $X = \emptyset$). Hence $\mathcal{Q}(A, A) = \mathcal{Q}(A, \emptyset)$ is the bicyclic semigroup. From now on, we exclude this case and assume $|A| \geq 2$.

2.2 A Semi-Thue System for $\mathcal{Q}(A, X)$

Lemma 2.6. *Let $a, b \in A$, $x \in X$ and $w \in A^*$. Then the following hold:*

(i) $b\bar{a} \equiv_X \bar{a}b$ *if $a \neq b$*   *(iii)* $xw a\bar{a} \equiv_X xw\bar{a}a$
(ii) $a\overline{ab} \equiv_X \bar{a}a\bar{b}$ *(iv)* $aw a\bar{a} \equiv_X aw\bar{a}a$

At first we take a look at equations (i)–(iii) (with $|w|_a = 0$ for simplicity). In order for a queue $q \in A^*$ to be defined after execution of the actions, the letter a must already be contained in q preceded by forgettable letters only. Since, in all cases, $\bar{a}$ is the first read operation, $\bar{a}$ reads this occurrence of a from q. Hence it does not matter whether we write b (a, resp.) before or after this reading of a. In equation (iv) we are in the same situation after execution of the leading write operation a. Therefore we can commute the read and write operations in all these situations.

In case of $X = A$, (iv) is a special case of (iii). Furthermore (i), (ii), and (iii) with $w = \varepsilon$ are exactly the equations that hold in $\mathcal{Q}(A, A)$ by [9, Lemma 3.5].

Ordering the equations from Lemma 2.6, the semi-Thue system $\mathcal{R}_X$ consists of the following rules for $a, b \in A$, $x \in X$ and $w \in A^*$:

(a) $b\bar{a} \rightarrow \bar{a}b$ if $a \neq b$ (c) $xw a\bar{a} \rightarrow xw\bar{a}a$
(b) $a\overline{ab} \rightarrow \bar{a}a\bar{b}$ (d) $aw a\bar{a} \rightarrow aw\bar{a}a$

Since all the rules are length-preserving and move letters from $\overline{A}$ to the left, this semi-Thue system is terminating. Since it is also locally confluent, it is confluent. Hence for any word $u \in \Sigma^*$, there is a unique irreducible word $\mathsf{nf}_X(u)$ with $u \to^* \mathsf{nf}_X(u)$, the *normal form* of u. Let NF_X denote the set of words in normal form.

Proposition 2.7. *Let* $u, v \in \Sigma^*$. *Then* $u \equiv_X v$ *if, and only if,* $\mathsf{nf}_X(u) = \mathsf{nf}_X(v)$.

Recall that $a\overline{a}b \equiv_X \overline{a}ab$ and $a\overline{a} \not\equiv_X \overline{a}a$, i.e., in general, we cannot cancel in the monoid $\mathcal{Q}(A, X)$. Since rules from $\mathcal{R}_X$ move letters from $\overline{A}$ to the left, we obtain the following restricted cancellation property.

Corollary 2.8. *Let* $u, v \in \Sigma^*$ *and* $x, y \in A^*$ *with* $\overline{x}uy \equiv_X \overline{x}vy$. *Then* $u \equiv_X v$.

To describe the shape of words from NF_X we use a special shuffle operation on two words $u, v \in A^*$: Each symbol $\overline{a}$ of $\overline{v}$ is placed directly behind the first occurrence of a such that we preserve the relative order of symbols in $\overline{v}$ and such that there is no symbol from X between the preceding reading symbol and $a\overline{a}$.

Example 2.9. Let $a, b \in A$ with $a \neq b$ and $q = aabb\overline{ab}$. If $a \notin X$ then we have

$$aabb\overline{ab} \to aab\overline{a}b\overline{b} \to aa\overline{a}bb\overline{b} \to a\overline{a}abb\overline{b} \to a\overline{a}abb\overline{b}$$

and therefore $a\overline{a}abb\overline{b} = \mathsf{nf}_X(aabb\overline{ab})$. Otherwise, i.e., if $a \in X$, we can apply rule (c) to $a\overline{a}abb\overline{b}$ and hence obtain $\mathsf{nf}_X(aabb\overline{ab}) = \overline{ab}aabb$.

The "special shuffle" alluded to above in these cases is $\langle\!\langle aabb, \overline{ab} \rangle\!\rangle = a\overline{a}abb\overline{b}$ if $a \notin X$ and $\langle\!\langle aabb, \overline{ab} \rangle\!\rangle = \overline{ab}aabb$ otherwise.

The inductive definition of the special shuffle looks as follows:

Definition 2.10. *Let* $u, v \in A^*$ *and* $a \in A$. *Then we set*

$$\langle\!\langle u, \overline{\varepsilon} \rangle\!\rangle := u$$

$$\langle\!\langle u, \overline{av} \rangle\!\rangle := \begin{cases} u_1 a\overline{a}\langle\!\langle u_2, \overline{v} \rangle\!\rangle & \text{if } u = u_1 a u_2 \text{ where } u_1 \in (A \backslash (X \cup \{a\}))^*, u_2 \in A^* \\ \text{undefined} & \text{otherwise.} \end{cases}$$

By induction on the length of the word v, one obtains that $\langle\!\langle u, \overline{v} \rangle\!\rangle$ is defined if, and only if, u has a prefix u' with $v \preceq_X u'$. We denote this property by $v \leq_X u$ and call v an *X-prefix* of u. Clearly, the binary relation $\leq_X$ is a partial order. Note that $\leq_\emptyset$ is the subword relation $\preceq$ and $\leq_A$ is the prefix relation on A^*.

Definition 2.11. *The projections* $\pi, \overline{\pi} : \Sigma^* \to A^*$ *on write and read operations are defined for any* $u \in \Sigma^*$ *by* $\pi(u) = \pi_A(u)$ *and* $\overline{\pi}(u) = \pi_{\overline{A}}(u)$.

In a nutshell, the projection π deletes all letters from $\overline{A}$ from a word. Dually, the projection $\overline{\pi}$ deletes all letters from A from a word *and then suppresses the overlines*. For instance $\pi(a\overline{a}b) = ab$ and $\overline{\pi}(a\overline{a}b) = a$.

Remark 2.12. Since a word is in normal form if no rule from the semi-Thue system $\mathcal{R}_X$ can be applied to it, we get

$$\mathsf{NF}_X = \{\overline{u}\langle\!\langle v, \overline{w}\rangle\!\rangle \mid u, v, w \in A^*, v \leq_X w\} = \overline{A}^* \left(\bigcup_{a \in A} (A\backslash(X \cup \{a\}))^* a\overline{a}\right)^*.$$

Thus, for $u \in \Sigma^*$, there are unique words $u_1, u_2, u_3 \in A^*$ with $\mathsf{nf}_X(u) = \overline{u_1}\langle\!\langle u_2, \overline{u_3}\rangle\!\rangle$; we set $\overline{\pi}_1(u) = u_1$ and $\overline{\pi}_2(u) = u_3$. As a consequence, we get $\mathsf{nf}_X(u) = \overline{\overline{\pi}_1(u)}\langle\!\langle \pi(u), \overline{\overline{\pi}_2(u)}\rangle\!\rangle$.

While $\overline{\pi}_1(u)$ is defined using the semi-Thue system $\mathcal{R}_X$, it also has a natural meaning in terms of the function $\circ_X$: $\overline{\pi}_1(u) \circ_X u$ is defined and, if $q \circ_X u$ is defined, then $|\overline{\pi}_1(u)| \leq |q|$. Hence $\overline{\pi}_1(u)$ is the shortest queue such that execution of u does not end up in the error state.

Example 2.13. Recall Example 2.9. In case of $a \notin X$ we have $\overline{\pi}_1(q) = \varepsilon$ and $\overline{\pi}_2(q) = ab$. Otherwise we have $\overline{\pi}_1(q) = ab$ and $\overline{\pi}_2(q) = \varepsilon$.

For words $u, v \in A^*$ with $\mathsf{nf}_X(u\overline{v}) = \overline{w_1}\langle\!\langle w_2, \overline{w_3}\rangle\!\rangle$, we have $w_2 = \pi(u\overline{v}) = u$ and $w_1 w_3 = \pi(u\overline{v}) = v$. Hence, to describe the normal form of $u\overline{v}$, we have to determine $w_3 - \overline{\pi}_2(u\overline{v})$ which is accomplished by the following lemma.

Lemma 2.14. *Let $u, v \in A^*$. Then $\overline{\pi}_2(uv)$ is the longest suffix v' of v that satisfies $v' \leq_X u$, i.e., such that $\langle\!\langle u, \overline{v'}\rangle\!\rangle$ is defined.*

3 Fully Lossy Queues

The main result of this section is Theorem 3.4 that provides a necessary condition on a homomorphism into $\mathcal{Q}(A, X)$ to be injective. We derive this condition by considering first queue monoids where all letters are forgettable, i.e., monoids of the form $\mathcal{Q}(A, \emptyset)$. Note that the relations $\preceq_\emptyset$ and $\leq_\emptyset$ are both equal to the subword relation $\preceq$. Hence we will use this in the following statements.

The first result of this section (Theorem 3.2) describes the normal form of the product of two elements from $\mathcal{Q}(A, \emptyset)$ in terms of their normal forms (Lemma 2.14 solves this problem in case the first factor belongs to $[A^*]$ and the second to $[\overline{A}^*]$ for arbitrary sets $X \subseteq A$.)

Definition 3.1. *Let $u, v \in A^*$. The* overlap *of u and v is the longest suffix $\mathsf{ol}(u, v)$ of v that is a subword of u.*

Assuming $X = \emptyset$ the relation $\leq_X$ equals the subword relation $\preceq$. Hence, in this situation, Lemma 2.14 implies $\overline{\pi}_2(u\overline{v}) = \mathsf{ol}(u, v)$ for any words $u, v \in A^*$.

Recall that Lemma 2.14 describes the shape of $\mathsf{nf}_X(uv)$ for arbitrary X, $u \in A^*$ and $v \in \overline{A}^*$. The following Theorem describes this normal form for $X = \emptyset$, but arbitrary $u, v \in \Sigma^*$.

Theorem 3.2. *Let $X = \emptyset$, $u, v \in \Sigma^*$, and $w = \mathsf{ol}(\pi(u), \overline{\pi}_2(u)\overline{\pi}_1(v))$. Then*

$$\overline{\pi}_2(uv) = \qquad w\,\overline{\pi}_2(v) \ and$$
$$\overline{\pi}(u)\overline{\pi}_1(v) = \overline{\pi}_1(uv)\,w.$$

We next infer that if u and v agree in their subsequences of read and write operations, respectively, then they can be equated by multiplication with a large power of one of them.

Proposition 3.3. *Let $u, v \in \Sigma^*$ with $\pi(u) = \pi(v)$, $\overline{\pi}(u) = \overline{\pi}(v)$, and $\overline{\pi}_1(u) \in \overline{\pi}_1(v)A^*$. Then there is a number $i \in \mathbb{N}$ with $\mathsf{nf}_\emptyset(u^i v u^i) = \mathsf{nf}_\emptyset(u^i u u^i)$.*

Proof. If there is $i \geq 1$ with $|\pi(v)| \leq |\overline{\pi}_1(u^i)|$, then $\overline{\pi}_2(vu^i) = \overline{\pi}_2(u\,u^i)$ can be derived from Theorem 3.2 by inductively proving a similar statement for powers of u. Otherwise, let $i \geq 1$ such that $|\overline{\pi}_1(u^i)|$ is maximal (this maximum exists since $|\overline{\pi}_1(u^j)| < |\pi(v)|$ for any $j \in \mathbb{N}$). Again by Theorem 3.2, one obtains $\overline{\pi}_2(u^i v) = \overline{\pi}_2(u^i u)$. Hence, in any case, $\overline{\pi}_2(u^i v u^i) = \overline{\pi}_2(u^i u\, u^i)$. Note that $\pi(u^i v u^i) = \pi(u^i u u^i)$ follows from $\pi(u) = \pi(v)$ and similarly for $\overline{\pi}(u^i v u^i) = \overline{\pi}(u^i u u^i)$. Consequently $\mathsf{nf}_\emptyset(u^i v u^i) = \mathsf{nf}_\emptyset(u^i u u^i)$. □

From this proposition, we can infer the announced necessary condition for a homomorphism into $\mathcal{Q}(A, X)$ to be injective. This condition will prove immensely useful in our investigation of submonoids of $\mathcal{Q}(A, X)$ in the following two sections. It states that if the images of x and y under an embedding ϕ perform the same sequences of read and write operations, respectively, then x and y can be equated by putting them into a certain context.

Theorem 3.4. *Let $\mathcal{M}$ be a monoid, $\phi : \mathcal{M} \hookrightarrow \mathcal{Q}(A, X)$ an embedding, and $x, y \in \mathcal{M}$ such that $\pi(\phi(x)) = \pi(\phi(y))$ and $\overline{\pi}(\phi(x)) = \overline{\pi}(\phi(y))$.*
Then there is $z \in \mathcal{M}$ with $zxz = zyz$.

Proof. For notational simplicity, let $\phi(x) = [u]$ and $\phi(y) = [v]$.

We can, without loss of generality, assume that $|\overline{\pi}_1(u)| \leq |\overline{\pi}_1(v)|$. Since $\overline{\pi}(u) = \overline{\pi}(v)$, the word $\overline{\pi}_1(u)$ is a prefix of the word $\overline{\pi}_1(v)$. By Proposition 3.3, there is $i \in \mathbb{N}$ such that $\mathsf{nf}_\emptyset(u^i v u^i) = \mathsf{nf}_\emptyset(u^i u u^i)$. As the semi-Thue system $\mathcal{R}_X$ contains all the rules from $\mathcal{R}_\emptyset$ we get $\mathsf{nf}_X(u^i v u^i) = \mathsf{nf}_X(u^i u u^i)$ and therefore $u^i v u^i \equiv_X u^i u u^i$. In other words, $\phi(x^i y x^i) = \phi(x^i x x^i)$. The injectivity of ϕ now implies $x^i y x^i = x^i x x^i$. Setting $z = x^i$ yields $zxz = zyz$ as claimed. □

4   Embeddings Between PLQ Monoids

We now characterize when the plq monoid $\mathcal{Q}(A, X)$ embeds into $\mathcal{Q}(B, Y)$.

Theorem 4.1. *Let A, B be alphabets with $|A|, |B| \geq 2$, $X \subseteq A$ and $Y \subseteq B$. Then $\mathcal{Q}(A, X) \hookrightarrow \mathcal{Q}(B, Y)$ holds iff all of the following properties hold:*

(A) $|A \backslash X| \leq |B \backslash Y|$, i.e., (B, Y) has at least as many forgettable letters as (A, X).

(B) If $Y = \emptyset$, then also $X = \emptyset$, i.e., if (B, Y) consists of forgettable letters only, then so does (A, X).

(C) If $|Y| = 1$, then $|A \backslash X| < |B \backslash Y|$ or $|X| \leq 1$, i.e., if (B, Y) has exactly one non-forgettable letter and exactly as many forgettable letters as (A, X), then A contains at most one non-forgettable letter.

In particular, $\mathcal{Q}(A, A)$ embeds into $\mathcal{Q}(B, B)$ whenever $|B| \geq 2$, i.e., this theorem generalizes [9, Corollary 5.4]. We prove it in Propositions 4.2 and 4.5.

4.1   Preorder of Embeddability

The embeddability of monoids is reflexive and transitive, i.e., a preorder. Before diving into the proof of Theorem 4.1, we derive from it an order-theoretic description of this preorder on the class of all plq monoids (see the reflexive and transitive closure of the graph on the right). The plq monoid $\mathcal{Q}(A, X)$ is, up to isomorphism, completely given by the numbers $m = |X|$ and $n = |A \backslash X|$ of unforgettable and of forgettable letters, respectively. Therefore, we describe this preorder in terms of pairs of natural numbers. We write $(m, n) \rightarrow (m', n')$ if

$$\mathcal{Q}([m + n], [m]) \hookrightarrow \mathcal{Q}([m' + n'], [m'])$$

where, as usual, $[n] = \{1, 2, \ldots, n\}$. Then Theorem 4.1 reads as follows: If $m, n, m', n' \in \mathbb{N}$ with $m + n, m' + n' \geq 2$, then $(m, n) \rightarrow (m', n')$ iff all of the following properties hold:

(A) $n \leq n'$
(B) If $m' = 0$, then $m = 0$
(C) If $m' = 1$, then $m \leq 1$ or $n < n'$

Then we get immediately for all appropriate natural numbers $m, n, n' \in \mathbb{N}$:

- if $k \geq 2$, then $(2, n) \rightarrow (k, n) \rightarrow (2, n)$
- $(2, n) \rightarrow (2, n')$ iff $n \leq n'$
- $(1, n) \rightarrow (2, n')$ iff $n \leq n'$
- $(0, n) \rightarrow (2, n')$ iff $n \leq n'$
- $(2, n) \rightarrow (1, n')$ iff $n < n'$
- $(1, n) \rightarrow (1, n')$ iff $n \leq n'$
- $(0, n) \rightarrow (1, n')$ iff $n \leq n'$
- $(2, n) \nrightarrow (0, n')$
- $(1, n) \nrightarrow (0, n')$ iff $n \leq n'$
- $(0, n) \rightarrow (0, n')$ iff $n \leq n'$

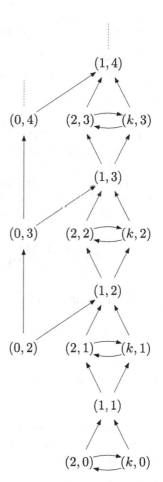

These facts allow to derive the above graph (where k stands for an arbitrary number at least 3).

First look at the nodes not of the form $(0, n)$. They form an alternating chain of infinite equivalence classes $\{(k, n) \mid k \geq 2\}$ and single nodes $(1, n)$. The infinite equivalence class at the bottom corresponds to the monoids of fully reliable queues considered in [9].

The nodes of the form $(0, n)$ also form a chain of single nodes (these nodes depict the fully lossy queue monoids from [11]). The single node number n (i.e., $(0, 2 + n)$) from this chain is directly below the single node number $2 + n$ (i.e., $(1, 2 + n)$) of the alternating chain.

4.2   Sufficiency in Theorem 4.1

Proposition 4.2. *Let A, B be non-singleton alphabets, $X \subseteq A$, $Y \subseteq B$ satisfying Conditions (A)–(C) from Theorem 4.1. Then $\mathcal{Q}(A, X)$ embeds into $\mathcal{Q}(B, Y)$.*

Proof. First suppose $|X| \leq |Y|$. By Condition (A), we can assume $A \backslash X \subseteq B \backslash Y$ and $X \subseteq Y$. Then Proposition 2.7 implies that $\mathcal{Q}(A, X)$ is a submonoid of $\mathcal{Q}(B, Y)$ since the rules of the semi-Thue system only permute letters in words.

Now assume $|X| > |Y|$. By Condition (A), there exists an injective mapping $\phi_1 : A \backslash X \hookrightarrow B \backslash Y$. Since $|X| > |Y|$, Condition (B) implies $Y \neq \emptyset$. Let $b_1 \in Y$ be arbitrary. If $|Y| > 1$, then choose $b_2 \in Y \backslash \{b_1\}$. Otherwise, we have $1 = |Y| < |X|$. Hence, by Condition (C), the mapping ϕ_1 is not surjective. So we can choose $b_2 \in B \backslash (Y \cup \{\phi_1(a) \mid a \in A \backslash X\})$. With $X = \{x_1, x_2, \ldots, x_n\}$, we set (for $a \in A$)

$$\phi'(a) = \begin{cases} \phi_1(a) & \text{if } a \in A \backslash X \\ b_1^{|A|+i} b_2 b_1^{|A|-i} b_2 & \text{if } a = x_i \end{cases} \quad \text{and} \quad \phi'(\bar{a}) = \overline{\phi'(a)}.$$

Then ϕ' maps $(A \cup \overline{A})^*$, A^*, and $\overline{A}^*$ injectively into $(B \cup \overline{B})^*$, B^*, and $\overline{B}^*$, respectively.

We prove that ϕ' induces an embedding $\phi : \mathcal{Q}(A, X) \hookrightarrow \mathcal{Q}(B, Y)$ by $\phi([u]) = [\phi'(u)]$.

First let $u \equiv_X v$ with $u, v \in (A \cup \overline{A})^*$ be any of the equations in Lemma 2.6. In each of the four cases, one obtains $\phi'(u) \equiv_Y \phi'(v)$. Consequently, $u \equiv_X v$ implies $\phi'(u) \equiv_Y \phi'(v)$ for any $u, v \in (A \cup \overline{A})^*$ by Proposition 2.7. Hence ϕ is welldefined.

To prove its injectivity, let $u, v \in (A \cup \overline{A})^*$ with $\phi'(u) \equiv_Y \phi'(v)$.

Set $\mathsf{nf}_X(u) = \overline{u_1} \langle\!\langle u_2, \overline{u_3} \rangle\!\rangle$ and similarly $\mathsf{nf}_X(v) = \overline{v_1} \langle\!\langle v_2, \overline{v_3} \rangle\!\rangle$. The crucial part of the proof demonstrates that ϕ' commutes with the shuffle operation, more precisely, $\phi'(\langle\!\langle u_2, \overline{u_3} \rangle\!\rangle) \equiv_Y \langle\!\langle \phi'(u_2), \overline{\phi'(u_3)} \rangle\!\rangle$ and similarly for v.

Since ϕ' is a homomorphism, we then get

$$\phi'(\overline{u_1}) \langle\!\langle \phi'(u_2), \overline{\phi'(u_3)} \rangle\!\rangle \equiv_Y \phi'(\overline{u_1} \langle\!\langle u_2, \overline{u_3} \rangle\!\rangle) \equiv_Y \phi'(u)$$

and similarly $\phi'(v) \equiv_Y \phi'(\overline{v_1}) \langle\!\langle \phi'(v_2), \overline{\phi'(v_3)} \rangle\!\rangle$.

Thus the words $\phi'(\overline{u_1})\langle\!\langle\phi'(u_2), \overline{\phi'(u_3)}\rangle\!\rangle$ and $\phi'(\overline{v_1})\langle\!\langle\phi'(v_2), \overline{\phi'(v_3)}\rangle\!\rangle$ in normal form are equivalent and therefore equal. Hence we get

$$\phi'(\overline{u_1}) = \phi'(\overline{v_1}), \quad \phi'(u_2) = \phi'(v_2) \text{ and } \phi'(\overline{u_3}) = \phi'(\overline{v_3}).$$

Since $\phi' : (A \cup \overline{A})^* \to (B \cup \overline{B})^*$ is injective, this implies $u_1 = v_1$, $u_2 = v_2$, and $u_3 = v_3$ and therefore $u \equiv_X \mathsf{nf}(u) = \mathsf{nf}(v) \equiv_X v$. Thus, indeed, ϕ is an embedding of $\mathcal{Q}(A, X)$ into $\mathcal{Q}(B, Y)$.   □

4.3   Necessity in Theorem 4.1

Now we have to prove the other implication of the equivalence in Theorem 4.1. Recall the embedding ϕ we constructed in the proof of Proposition 4.2. In particular, it has the following properties:

(1) If $a \in A$, then $\phi(a) \in [B^+]$ and $\phi(\overline{a}) = \overline{\phi(a)}$. In particular, the image of every write operation a performs write operations, only, and the image of every read operation $\overline{a}$ is the "overlined version of the image of the corresponding read operation" and therefore performs read operations, only.
(2) If $a \in A\backslash X$, then $\phi(a) \in [B\backslash Y]$. In particular, the image of every write operation of a forgettable letter writes only forgettable letters.
(3) If $x \in X$, then $\phi(x) \in [B^*YB^*]$. In particular, the image of every write operation of a non-forgettable letter writes at least one non-forgettable letter.

The proof of the necessity in Theorem 4.1 first shows that any embedding satisfies slightly weaker properties. We start with our weakenings of properties (1) and (2). The first statement of the following lemma is a weakening of (1) since it only says something about the letters in $\phi(a)$ and $\phi(\overline{a})$ but not that these two elements are dual. Similarly the second statement is a weakening of (2) since it does not say anything about the length of $\phi(a)$ but only something about the letters occurring in $\phi(a)$.

Lemma 4.3. *Let A, B be non-singleton alphabets, $X \subseteq A$, $Y \subseteq B$, and ϕ an embedding of $\mathcal{Q}(A, X)$ into $\mathcal{Q}(B, Y)$. Then the following holds:*

(i) $\phi(a) \in [B^+]$ and $\phi(\overline{a}) \in [\overline{B}^+]$ for each $a \in A$.
(ii) $\phi(a) \in [(B\backslash Y)^*]$ for each $a \in A\backslash X$.

Proof. To prove (i), let $a \in A$ and suppose $\phi(a) \notin [B^*]$. One first shows that $\phi(a)$ performs at least one write operation, i.e., $\phi(a) \notin [\overline{B}^*]$. Let $p, q \in B^+$ be the primitive roots of the nonempty words $\pi(\phi(a))$ and $\overline{\pi}(\phi(a))$, respectively.

Since $|A| \geq 2$, there exist distinct letters $a_1, a_2 \in A$. A crucial property of ϕ is that then $\pi(\phi(\overline{a_i})) \in p^*$ and $\overline{\pi}(\phi(\overline{a_i})) \in q^*$. Consequently, $\pi(\phi(\overline{a_1 a_2})) = \pi(\phi(\overline{a_1}))\,\pi(\phi(\overline{a_2})) = \pi(\phi(\overline{a_2}))\,\pi(\phi(\overline{a_1})) = \pi(\phi(\overline{a_2 a_1}))$ (the equality $\overline{\pi}(\phi(\overline{a_1 a_2})) = \overline{\pi}(\phi(\overline{a_2 a_1}))$ follows similarly). Since ϕ is an embedding, Theorem 3.4 implies the existence of $u \in (A \cup \overline{A})^*$ with $u\overline{a_1}\,\overline{a_2}u \equiv_X u\overline{a_2}\,\overline{a_1}u$. It follows that these two words have the same sequence of read operations and therefore in particular

$a_1 a_2 = a_2 a_1$. But this implies $a_1 = a_2$ which contradicts our choice of these two letters. Hence, indeed, $\phi(a) \in [B^*]$ which proves the first claim from (i), the second follows similarly.

Statement (ii) is shown by contradiction. Let $a \in A\backslash X$ with $\phi(a) \notin [(B\backslash Y)^*]$. Since $|A| \geq 2$, there exists a distinct letter $b \in A\backslash\{a\}$. Using (i) and the assumption on $\phi(a)$, one obtains $\phi(a^n b\bar{b}) = \phi(a^n \bar{b}b)$ with n the length of $\phi(\bar{b})$. Injectivity of ϕ and Proposition 2.7 lead to a contradiction. $\qquad \square$

We next come to property (3) that we prove for every embedding.

Lemma 4.4. *Let A, B be non-singleton alphabets, $X \subseteq A$, $Y \subseteq B$, and ϕ an embedding of $\mathcal{Q}(A, X)$ into $\mathcal{Q}(B, Y)$. Then we have $\phi(x) \in [B^* Y B^*]$ for each $x \in X$.*

Proof. Let $x \in X$. Since $|A| \geq 2$, there is a distinct letter $a \in A\backslash\{x\}$. By Lemma 4.3, there are words $u, v, w \in B^+$ with $\phi(a) = [u]$, $\phi(\bar{a}) = [\bar{v}]$ and $\phi(x) = [w]$. One then shows $\overline{\pi}_2(wu\bar{v}) = \varepsilon \neq \overline{\pi}_2(u\bar{v})$.

By Lemma 2.14, $v' = \overline{\pi}_2(u\bar{v}) \neq \varepsilon$ is a suffix of v with $v' \leq_X u'$ for some prefix u' of u implying $v' \preceq wu$. Since $\overline{\pi}_2(wu\bar{v}) = \varepsilon$, Lemma 2.14 implies $\pi_Y(wu') \neq \pi_Y(v') = \pi_Y(u')$, i.e., w contains some letter from Y. $\qquad \square$

Finally we obtain the remaining implication in Theorem 4.1.

Proposition 4.5. *Let A and B be non-singleton alphabets, $X \subseteq A$ and $Y \subseteq B$ such that $\mathcal{Q}(A, X) \hookrightarrow \mathcal{Q}(B, Y)$. Then the Conditions (A)–(C) from Theorem 4.1 hold.*

Proof. First suppose $X \neq \emptyset$. Then, $Y \neq \emptyset$ by Lemma 4.4, i.e., we have (B).

Condition (A) is trivial if $A\backslash X = \emptyset$. If $A\backslash X$ is a singleton, then Lemma 4.3(ii) implies $B\backslash Y \neq \emptyset$ and therefore $|A\backslash X| \leq |B\backslash Y|$. So it remains to consider the case that $A\backslash X$ contains at least two elements. One then shows that the last letters of the words $\overline{\pi}(\phi(\bar{a}))$ for $a \in A\backslash X$ are mutually distinct.

To prove Condition (C), suppose $Y = \{y\}$ and $|A\backslash X| = |B\backslash Y|$. One then proves $|X| \leq 1$ by considering the last letters of $\overline{\pi}(\phi(\bar{x}))$ for $x \in X$. $\qquad \square$

5   Embeddings of Trace Monoids

Corollary 5.4 from [9] implies that all reliable queue monoids $\mathcal{Q}(A, A)$ for $|A| \geq 2$ have the same class of submonoids. Our Theorem 4.1 shows that this is not the case for all plq monoids $\mathcal{Q}(A, X)$ (e.g., $\mathcal{Q}(A, A)$ does not embed into $\mathcal{Q}(A, \emptyset)$ and vice versa). This final section demonstrates a surprising similarity among all these monoids, namely the trace monoids contained in them.

These trace (or free partially commutative) monoids are used for modeling concurrent systems where the concurrency is governed by the use of joint resources (cf. [14]). Formally such a system is a so called *independence alphabet*, i.e., a tuple (Γ, I) of a non-empty finite set Γ and a symmetric, irreflexive relation $I \subseteq \Gamma^2$, i.e., (Γ, I) can be thought of as an undirected graph. Given an independence alphabet (Γ, I), we define the relation $\equiv_I \subseteq (\Gamma^*)^2$ as the least congruence satisfying $ab \equiv_I ba$ for each $(a, b) \in I$. The induced *trace monoid* is $\mathbb{M}(\Gamma, I) := \Gamma^*/_{\equiv_I}$. See [5,6,14] for further information on trace monoids.

5.1   Large Alphabets

Theorem 2.7 from [12] describes when the trace monoid $\mathrm{M}(\Gamma, I)$ embeds into the queue monoid $\mathcal{Q}(A, A)$ for $|A| \geq 2$. The following theorem shows that this is the case if, and only if, it embeds into $\mathcal{Q}(A, X)$ provided $|A| + |X| \geq 3$.

Theorem 5.1. *Let A be an alphabet and $X \subseteq A$ with $|A|+|X| \geq 3$. Furthermore let (Γ, I) be an independence alphabet. Then the following are equivalent:*

(A) $\mathrm{M}(\Gamma, I)$ embeds into $\mathcal{Q}(A, X)$.
(B) $\mathrm{M}(\Gamma, I)$ embeds into $\mathcal{Q}(A, A)$.
(C) $\mathrm{M}(\Gamma, I)$ embeds into $\{a, b\}^* \times \{c, d\}^*$.
(D) One of the following conditions holds:
  (D.a) All nodes in (Γ, I) have degree ≤ 1.
  (D.b) The only non-trivial connected component of (Γ, I) is complete bipartite.

Since $X \subseteq A$, the condition $|A| + |X| \geq 3$ implies in particular $|A| \geq 2$. Hence the equivalence between (B), (C), and (D) follows from [12, Theorem 2.7].
    For the implication "(C) $\Rightarrow$ (A)", one considers the two cases $|A| \geq 3$ and $|A| = 2$, $X \neq \emptyset$ separately. In the first case, one chooses pairwise distinct $a, b, c \in A$ and sets $\phi(a, \varepsilon) = a$, $\phi(b, \varepsilon) = b$, $\phi(\varepsilon, c) = \overline{a}\overline{c}$, and $\phi(\varepsilon, d) = \overline{b}\overline{c}$. In the second case, the embedding is similar to the one from [9, Proposition 8.3] (proving the implication "(C) $\Rightarrow$ (B)").
    The implication "(A) $\Rightarrow$ (D)" is proved under the slightly more general assumption $|A| \geq 2$. It is, by far, more involved. We nevertheless only give an overview here since it follows the proof of the implication "(B) $\Rightarrow$ (D)" from [12] rather closely:
    Suppose ϕ embeds the trace monoid $\mathrm{M}(\Gamma, I)$ into the plq monoid $\mathcal{Q}(A, X)$ with $|A| \geq 2$. This defines a partition of the independence alphabet into the sets $\Gamma_+ := \{\alpha \in \Gamma \mid \phi(\alpha) \in [A^+]\}$, $\Gamma_- := \{\alpha \in \Gamma \mid \phi(\alpha) \in [\overline{A}^+]\}$, and $\Gamma_\pm := \Gamma \setminus (\Gamma_+ \cup \Gamma_-)$. The crucial steps are then to verify the following properties:

  (i) $(\Gamma_+ \cup \Gamma_-, I)$ is complete bipartite with the partitions Γ_+ and Γ_-.
  (ii) Let $a \in \Gamma_+ \cup \Gamma_-$ and $b, c \in \Gamma$ with $(b, c) \in I$. Then $(a, b) \in I$ or $(a, c) \in I$.
  (iii) Let $a \in \Gamma_\pm$. Then a has degree ≤ 1 in the undirected graph (Γ, I).
  (iv) (Γ, I) is P_4-free, i.e., the path on four vertices is no induced subgraph.

    The proof of [12, Theorem 4.14] shows that any graph $(\Gamma_+ \uplus \Gamma_- \uplus \Gamma_\pm, I)$ satisfying these graph theoretic properties also satisfies (D.a) or (D.b).

5.2   The Binary Alphabet

In Theorem 5.1 we have only considered partially lossy queues with $|A| > 2$ or $|X| \neq 0$. For a complete picture, it remains to consider the case $|A| = 2$ and $|X| = 0$. The following theorem implies in particular that $\mathcal{Q}(\{\alpha, \beta\}, \emptyset)$ does not contain the direct product of two free monoids, i.e., it contains properly less trace monoids than $\mathcal{Q}(A, X)$ with $|A| + |X| \geq 3$.

Theorem 5.2. *Let A be an alphabet with $|A| = 2$ and (Γ, I) be an independence alphabet. Then the following are equivalent:*

(A) $\mathbb{M}(\Gamma, I)$ *embeds into* $\mathbb{Q}(A, \emptyset)$.
(B) *One of the following conditions holds:*
    (B.1) *All nodes in (Γ, I) have degree ≤ 1.*
    (B.2) *The only non-trivial connected component of (Γ, I) is a star graph.*

For the proof of the implication "(B) $\Rightarrow$ (A)", one provides the embeddings as follows (with $A = \{\alpha, \beta\}$):

(B.1) It suffices to consider the case that (Γ, I) is the disjoint union of the edges (a_i, b_i) for $1 \leq i \leq n$. Then we define $w_i = \alpha^i \beta$ for $1 \leq i \leq n$ and the embedding ϕ is given by $\phi(a_i) = [\overline{w_i} w_i]$ and $\phi(b_i) = [\overline{w_i}\, \overline{w_i} w_i]$.

(B.2) Let c be the center of the star graph, s_i for $1 \leq i \leq m$ its neighbors, and r_i for $1 \leq i \leq n$ the isolated nodes of (Γ, I). Then the embedding ϕ is given by $\phi(c) = [\alpha]$, $\phi(s_i) = [\overline{w_i}]$ and $\phi(r_j) = [w_j \beta \beta]$.

Note that these embeddings map letters to sequences containing both, read and write operations.

For the more involved implication "(A) $\Rightarrow$ (B)", suppose Γ has a node of degree ≥ 2 and, towards a contradiction, (B) does not hold. Since we proved the implication "(A) $\Rightarrow$ (D)" in Theorem 5.1 under the assumption $|A| \geq 2$, we obtain that (Γ, I) has a single nontrivial connected component $C \subseteq \Gamma_+ \cup \Gamma_-$. Furthermore, there are $a, b \in \Gamma_+$ distinct and $c \in \Gamma_-$ such that $(a, c), (c, b) \in I$. Using Lemma 2.14, one arrives at $ab \equiv_I ba$ which contradicts $a \neq b$.

References

1. Abdulla, P.A., Jonsson, B.: Verifying programs with unreliable channels. Inf. Comput. **127**(2), 91–101 (1996)
2. Brand, D., Zafiropulo, P.: On communicating finite-state machines. J. ACM **30**(2), 323–342 (1983)
3. Cécé, G., Finkel, A., Iyer, S.P.: Unreliable channels are easier to verify than perfect channels. Inf. Comput. **124**(1), 20–31 (1996)
4. Chambart, P., Schnoebelen, P.: The ordinal recursive complexity of lossy channel systems. In: LICS 2008, pp. 205–216. IEEE Computer Society Press (2008)
5. Diekert, V.: Combinatorics on Traces, vol. 454. Springer, Heidelberg (1990)
6. Diekert, V., Rozenberg, G.: The Book of Traces. World Scientific, Singapore (1995)
7. Finkel, A.: Decidability of the termination problem for completely specified protocols. Distrib. Comput. **7**(3), 129–135 (1994)
8. Haase, C., Schmitz, S., Schnoebelen, P.: The power of priority channel systems. Log. Methods Comput. Sci. **10**(4), 4 (2014)
9. Huschenbett, M., Kuske, D., Zetzsche, G.: The monoid of queue actions. In: Semigroup Forum (2017, to appear)
10. Kambites, M.: Formal languages and groups as memory. Commun. Algebra **37**(1), 193–208 (2009)

11. Köcher, C.: Einbettungen in das Transformationsmonoid einer vergesslichen Warteschlange. Master's thesis, TU Ilmenau (2016)
12. Kuske, D., Prianychnykova, O., The trace monoids in the queue monoid, in the direct product of two free monoids. arXiv preprint arXiv:1603.07217 (2016)
13. Masson, B., Schnoebelen, P.: On verifying fair lossy channel systems. In: Diks, K., Rytter, W. (eds.) MFCS 2002. LNCS, vol. 2420, pp. 543–555. Springer, Heidelberg (2002). doi:10.1007/3-540-45687-2_45
14. Mazurkiewicz, A.: Concurrent program schemes and their interpretations. DAIMI Rep. Ser. **6**(78), 1–51 (1977)
15. Muscholl, A., Petersen, H.: A note on the commutative closure of star-free languages. Inf. Process. Lett. **57**(2), 71–74 (1996)

Approximation Algorithms for the Maximum Carpool Matching Problem

Gilad Kutiel[(⊠)]

Department of Computer Science, Technion, Haifa, Israel
gkutiel@cs.technion.ac.il

Abstract. The MAXIMUM CARPOOL MATCHING problem is a star packing problem in directed graphs. Formally, given a directed graph $G = (V, A)$, a capacity function $c : V \to \mathbb{N}$, and a weight function $w : A \to \mathbb{R}$, a feasible *carpool matching* is a triple (P, D, M), where P (passengers) and D (drivers) form a partition of V, and M is a subset of $A \cap (P \times D)$, under the constraints that for every vertex $d \in D$, $deg_{in}^M(d) \le c(d)$, and for every vertex $p \in P$, $deg_{out}^M(p) \le 1$. In the MAXIMUM CARPOOL MATCHING problem we seek for a matching (P, D, M) that maximizes the total weight of M.

The problem arises when designing an online carpool service, such as Zimride [1], that tries to connect between passengers and drivers based on (arbitrary) similarity function. The problem is known to be NP-hard, even for uniform weights and without capacity constraints.

We present a 3-approximation algorithm for the problem and 2-approximation algorithm for the unweighted variant of the problem.

1 Introduction

Carpooling, is the sharing of car journeys so that more than one person travels in a car. Knapen et al. [7] describe an automatic service to match commuting trips. Users of the service register their personal profile and a set of periodically recurring trips, and the service advises registered candidates on how to combine their commuting trips by carpooling. The service acts in two phases.

In the first phase, the service estimates the probability that a person a traveling in person's b car will be satisfied by the trip. This is done based on personal information and feedback from users on past rides. The second phase is about finding a carpool matching that maximizes the global (total expected) satisfaction.

The second phase can be modeled in terms of graph theory. Given a directed graph $G = (V, A)$. Each vertex $v \in V$ corresponds to a user of the service and an arc (u, v) exists if the user corresponding to vertex u is willing to commute with the user corresponding to vertex v. A capacity function $c : V \to \mathbb{N}$ is defined according to the number of passengers each user can drive if she was selected as a driver. A weight function $w : A \to \mathbb{R}$ defines the amount of satisfaction $w(u, v)$, that user u gains when riding with user v.

© Springer International Publishing AG 2017
P. Weil (Ed.): CSR 2017, LNCS 10304, pp. 206–216, 2017.
DOI: 10.1007/978-3-319-58747-9_19

A feasible *carpool matching* (matching) is a triple (P, D, M), where P and D form a partition of V, and M is a subset of $A \cap (P \times D)$, under the constraints that for every driver $d \in D$, $deg_{in}^M(d) \leq c(d)$, and for every passenger $p \in P$, $deg_{out}^M(p) \leq 1$. In the MAXIMUM CARPOOL MATCHING problem we seek for a matching (P, D, M) that maximizes the total weight of M. In other words, the MAXIMUM CARPOOL MATCHING problem is about finding a set of (directed toward the center) vertex disjoint stars that maximizes the total weights on the arcs. Figure 1 is an example of the MAXIMUM CARPOOL MATCHING problem.

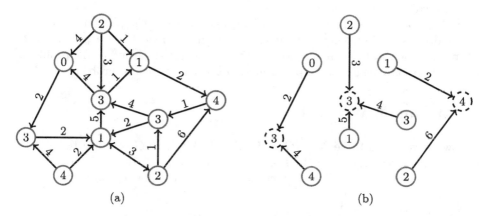

(a)   (b)

Fig. 1. A carpool matching example: (a) a directed graph with capacities on the vertices and weights on the arcs. (b) a feasible matching with total weight of 26. P is the set of blue vertices, and D is the set of red, dashed vertices. (Color figure online)

Hartman [5] introduced the *Maximum Carpool Matching* problem and proved it to be NP-hard. She also proved that the problem remains NP-hard even for a binary weight function when the capacity function $c(v) \leq 2$ for every vertex in V. It is also worth mentioning, that in the undirected, uncapacitated, unweighted variant of the problem, the set of drivers in an optimal solution form a minimum dominating set. When the set of drivers is known in advance, however, the problem becomes tractable and can be solved using a reduction to a flow network problem.

Agatz et al. [2] outlined the optimization challenges that arise when developing technology to support ride-sharing and survey the related operations research models in the academic literature. Hartman et al. [6] designed several heuristic algorithms for the Maximum Carpool Matching problem and compared their performance on real data. Other heuristic algorithms were developed as well [8]. Arkin et al. [3], considered other variants of capacitated star packing where a capacity vector is given as part of the input and capacities need to be assigned to vertices.

Nguyen et al. [9] considered the SPANNING STAR FOREST problem (the undirected, uncapacitated, unweighted variant of the problem). They proved the following results: 1. there is a polynomial-time approximation scheme for planar graphs; 2. there is a polynomial-time $\frac{3}{5}$-approximation algorithm for graphs; 3. there is a polynomial-time $\frac{1}{2}$-approximation algorithm for weighted graphs. They also showed how to apply the spanning star forest model to aligning multiple genomic sequences over a tandem duplication region. Chen et al. [4] improved the approximation ratio to 0.71, and also showed that the problem can not be approximated to within a factor of $\frac{31}{32} + \epsilon$ for any $\epsilon > 0$ under the assumption that P $\neq$ NP. It is not clear, however, if any of the technique used to address the SPANNING STAR FOREST problem can be generalized to approximate the directed capacitated variant.

In Sect. 3 we present an exact, efficient algorithm for the problem when the set of drivers and passengers is given in advance. In Sect. 4 we present a 2-approximation local search algorithm for the unweighted variant of the problem. Finally in Sect. 5 we give a 3-approximation algorithm for the problem.

2   Maximum Weight Flow

A flow network is a tuple $N = (G = (V, A), s, t, c)$, where G is a directed graph, $s \in V$ is a source vertex, $t \in V$ is a target vertex, and $c : A \to \mathbb{R}$ is a capacity function. A flow $f : A \to \mathbb{R}$ is a function that has the following properties:

- $f(e) \leq c(e), \quad \forall e \in A$
- $\sum_{(u,v) \in A} f(u, v) = \sum_{(v,w) \in A} f(v, w), \quad \forall v \in V \setminus \{s, t\}$

Given a flow function f, and a weight function $w : A \to \mathbb{R}$, the *flow weight* is defined to be: $\sum_{e \in A} w(e)f(e)$. A flow with a maximum weight (*maximum weight flow*) can be efficiently found by adding the arc (t, s), with $c(t, s) = \infty$, and $w(t, s) = 0$ and reducing the problem (by switching the sign of the weights) to the minimum cost circulation problem [10]. When the capacity function c is integral, a maximum weight integral flow can be efficiently found.

3   Fixed Maximum Carpool Matching

In the FIXED MAXIMUM CARPOOL MATCHING problem, P and D are given, and the goal is to find M that maximizes the total weight. This variant of the problem can be solved efficiently[1], by reducing it to a maximum weight flow (flow) problem as follow: Let $(G = (V, A), c, w)$ be a Maximum Carpool Matching instance, let (P, D) be a partition of V, let $N = (G' = (V', A'), s, t, c')$ be a flow network, and let $w' : A \to \mathbb{N}$ be a weight function, where

---

[1] A solution to this variant of the problem was already proposed in [6]. For the sake of completeness, however, we describe a detailed solution for this variant. More importantly, the described solution helps us develop the intuition and understand the basic idea behind the approximation algorithm described in Sect. 5.

$$V' = P \cup D \cup \{s, t\}$$
$$A' = A_{sp} \cup A_{pd} \cup A_{dt}$$
$$A_{sp} = \{(s, p) : p \in P\}$$
$$A_{pd} = A \cap (P \times D)$$
$$A_{dt} = \{(d, t) : d \in D\}$$
$$c'(u, v) = \begin{cases} c(u) & \text{if } (u, v) \in A_{dt} \\ 1 & \text{otherwise} \end{cases}$$
$$w'(e) = \begin{cases} w(e) & \text{if } e \in A_{pd} \\ 0 & \text{otherwise} \end{cases}$$

The flow network is described in Fig. 2.

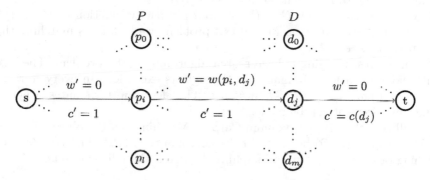

Fig. 2. Illustration of a flow network corresponding to a Fixed Maximum Carpool Matching instance.

Observation 1. *For every integral flow f in N, there is a carpool matching M on G with the same weight.*

Proof. Consider the carpool matching (P, D, M^f), where

$$M^f = \{(p, d) \in A_{pd} : f(p, d) = 1\}$$

one can verify that this is indeed a matching with the same weight as f. □

Observation 2. *For every carpool matching (P, D, M) on G, there exists a flow f on N with the same weight.*

Proof. Consider the flow function

$$f(s, p_i) = deg_{out}^M(p_i)$$
$$f(p_i, d_j) = \begin{cases} 1 & \text{if } (p_i, d_j) \in M \\ 0 & \text{otherwise} \end{cases}$$
$$f(d_j, t) = deg_{in}^M(d_j)$$

It is easy to verify, that f is indeed a flow function. Also, observe, that by construction, the weight of f equals to the weight of the matching.   □

As we mentioned, the maximum weight flow problem can be solved efficiently, and so is the Fixed Maximum Carpool Matching problem. It is worth mentioning, that it is possible that in a maximum weight flow, some of the arcs will have no flow at all, that is, it is possible that in a Fixed Maximum Carpool Matching some of the passengers and drivers will be unmatched.

4   Unweighted Carpool Matching

In this section we present a local search algorithm for the unweighted variant of the problem. We show that the approximation ratio of this algorithm is 2 and give an example to show that our analysis is tight.

Given a directed graph $G = (V, A)$, and a capacity function $c : V \to \mathbb{N}$, In the UNWEIGHTED CARPOOL MATCHING problem, we seek for a matching that maximizes the size of M.

We now present a simple local search algorithm for the problem. The algorithm maintains a feasible matching through its execution. In every iteration of the algorithm, the size of M increases. The algorithm terminates, when no further improvement can be made.

Recall that the Fixed Maximum Carpool Matching can be solved efficiently. Let $M = \text{opt}_{fixed}(P, D)$ be an optimal solution of the Fixed Maximum Carpool Matching problem. For a given matching M, define the following sets:

- $P^M = \{v : deg_{out}^M(v) = 1\}$
- $D^M = \{v : deg_{in}^M(v) > 0\}$
- $D_c^M = \{v : deg_{in}^M(v) = c(v)\}$
- $F^M = \{v : deg_{in}^M(v) = deg_{out}^M(v) = 0\}$

We refer to the vertices in these sets as, *passenger, driver, saturated driver,* and *free vertex* respectively. The local search algorithm, in every iteration, tries to improve the current matching, by switching a passenger or a free vertex into a driver and compute an optimal fixed matching. The local search algorithm is described in Algorithm 1.

First, observe that the outer loop on line 2 of the local search algorithm can be executed at most n times, where n is the total number of vertices, this is because the loop is executed only when there was an improvement, and this can happen at most n times. Also, observe that the body of this loop can be computed in polynomial time, and we can conclude that Algorithm 1 runs in polynomial time.

We now prove that the local search algorithm achieves an approximation ratio of 2. Let M be a matching found by the local search algorithm, and let M^* be an arbitrary but fixed optimal matching. Observe that the optimal solution cannot match two free vertices to each other, formally:

Algorithm 1. Local Search

Input: $G = (V, A)$, $c : V \to \mathbb{N}$
Output: M

```
1  M ← ∅
2  repeat
3  │   done ← true
4  │   for v ∈ (V \ D^M) do
5  │   │   D ← D^M ∪ {v}
6  │   │   P ← V \ D
7  │   │   M' = opt_{fixed}(P, D)
8  │   │   if |M'| > |M| then
9  │   │   │   M ← M'
10 │   │   │   done ← false
11 │   │   │   break
12 │   │   end
13 │   end
14 until done;
15 return M
```

Observation 3. *If $(u, v) \in M^*$, then $\{u, v\} \cap (P^M \sqcup D^M) \neq \emptyset$.*

Proof. If this is not the case, Algorithm 1 can improve M by adding the arc (u, v).

Now, with respect to M, the optimal solution can not match two free vertices to the same passenger, formally:

Observation 4. *If $(p, d) \in M$, $f_1, f_2 \in F^M$, and $(f_1, p), (f_2, p) \in M^*$, then $f_1 = f_2$.*

Proof. If this is not the case, Algorithm 1 can improve M by removing the arc (p, d) and adding the arcs $(f_1, p), (f_2, p)$.

Finally, with respect to M, the optimal solution can not match a free vertex to a driver that is not saturated, formally:

Observation 5. *If $(f, d) \in M^*$, $f \in F^M$, and $d \in D^M$, then $d \in D_c^M$.*

Proof. If this is not the case, once again, Algorithm 1 can improve M by adding the arc (f, d).

To show that Algorithm 1 is 2-approximation, consider the charging scheme that is illustrated in Fig. 3: Load every arc $(p, d) \in M$ with 2 coins, place one coin on p and one coin on d. Observe that every vertex $p \in P^M$ is loaded with one coin, and every vertex $d \in D^M$ is loaded with $deg_{in}^M(d)$ coins. Now, pay one coin for every $(u, v) \in M^*$, charge u if $u \in P^M \cup D^M$, otherwise ($v \in P^M \cup D^M$) charge v. Clearly, every arc in M^* is paid. We claim that no vertex is overcharged.

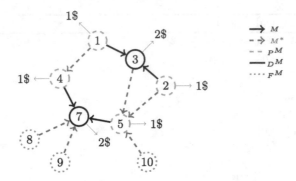

Fig. 3. Charging Scheme: 1. vertices $1, 2, 4, 5$ are loaded with 1\$ each and vertices $3, 7$ with 2\$ each. 2. vertex 1 pays for the arc $(1, 4)$. 3. vertex 5 pays for the arc $(10, 5)$. 4. vertex 7 is saturated. It pays for arcs $(8, 7)$ and $(9, 7)$.

Observation 6. *If $u \in P^M$, then u is not overcharged.*

Proof. If $u \in P^{M^*}$, then it is only charged once, otherwise, if $u \in D^{M^*}$, then it is only charged for arcs (w, u) where $w \in F^M$, and by Observation 4, there is at most one such arc. □

Observation 7. *If $u \in D^M$, then u is not overcharged.*

Proof. If $u \in P^{M^*}$, then it is only charged once, if $u \in D^{M^*}$, then it is only charged for arcs (w, u) where $w \in F^M$, if such arcs exists, then by observation 5, u is saturated, and can not be overcharged. □

Theorem 1. *Algorithm 1 is 2-approximation.*

Proof. We use a charging scheme where we manage to pay 1 coin for each arc in M^* by using at most $2|M|$ coins. □

To conclude this section, we show that our analysis is tight. Consider the example given in Fig. 4. Assume, in this example, that there are no capacity constraints, if the local search algorithm starts by choosing vertex 3 to be a driver, then the returned matching is the single arc $(2, 3)$. At this point, no further improvement can be done. The optimal matching, on the other hand, is $\{(1, 2), (3, 2)\}$. The path in the example can be duplicated to form an arbitrary large graph (forest).

Fig. 4. Local search - worst case example

5   Maximum Carpool Matching

5.1   Super Matching

A super-matching is a relaxed variant of the Maximum Carpool Matching problem where every node can act both as a driver and as a passenger. Formally, given a directed graph $G = (V, A)$, a capacity function $c : V \to \mathbb{N}$, and a weight function $w : A \to \mathbb{R}$, a *super-matching* is a set $M \subseteq A$, under the constraint that $\forall v \in V$, $deg_{in}^M(v) \leq c(v)$, and $deg_{out}^M(v) \leq 1$. Clearly, the following observation holds:

Observation 8. *Every matching* (P, V, M) *is a super-matching* M.

A maximum super matching can be found efficiently by the following reduction to a maximum weight flow problem: Let $N = (G', s, t, c', w')$ be a flow network, where

$$G' = (P \cup D \cup \{s, t\}, A_{sp} \cup A_{pd} \cup A_{dt})$$
$$P = \{p_v : v \in V\}$$
$$D = \{d_v : v \in V\}$$
$$A_{sp} = \{(s, p_v) : p_v \in P\}$$
$$A_{pd} = \{((p_u, d_v)) : (u, v) \in A\}$$
$$A_{dt} = \{(d_v, t) : d_v \in D\}$$
$$c'(s, p_v) = c'(p_u, d_v) = 1$$
$$c'(d_v, t) = c(v)$$
$$w'(p_u, d_v) = \begin{cases} w(u, v) & \text{if } (p_u, d_v) \in A_{pd} \\ 0 & \text{otherwise} \end{cases}$$

That is, we construct a bipartite graph where the left side represents each vertex in V being a passenger, and the right side represents each vertex in V being a driver. Figure 5 illustrates this flow network. One can verify that this is indeed a (integral) flow network and that there is a straight forward translation between a flow and a super matching with the same weight.

5.2   3-Approximation

We now present a 3-approximation algorithm for the MAXIMUM CARPOOL MATCHING problem. This algorithm acts in two phases. In the first phase it computes a maximum super-matching of G, in the second phase it decomposes the super-matching into 3 feasible carpool matchings and outputs the best of them.

We now describe how a super-matching can be decomposed into 3 feasible carpool matching. First, consider the graph obtained by an optimal super-matching. Recall that in a super matching the out degree of every vertex is at most 1, that is, the graph obtained by an optimal super matching is a pseudoforest - every connected component has at most one cycle. We now eliminate cycles from the

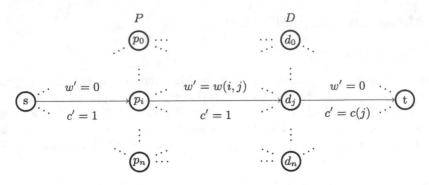

Fig. 5. Illustration of the flow network that is used to find a super-matching.

Algorithm 2. SuperMatching

Input: $G = (V, A), c : V \to \mathbb{N}, w : A \to \mathbb{R}$
Output: $(M \subseteq A)$
1 $M \leftarrow \emptyset$
2 $G' = (V, A') \leftarrow \text{superMatching}(G)$
3 **for** *every connected component* $C_i = (V_i, A_i) \in G'$ **do**
4   Eliminate the cycle in C_i by removing an arc a_i
5   Decompose the remaining in-tree into two solutions, M_1^i, M_2^i
6     $M \leftarrow M \cup \arg\max_{F \in \{\{e\}, M_1, M_2\}} w(F)$
7 **end**
8 **return** M

---

super-matching by removing one edge from every connected component. It is easy to see that the resulting graph is a forest of in-trees. Each of these in-trees can be, in turn, decomposed into two disjoint feasible carpool matchings. This can be done, for example, by coloring each such in-tree with two colors, say red and blue, and then consider the two solutions: one where the green nodes are the drivers, and the other where the red nodes are the drivers. We describe the algorithm in Algorithm 2, and illustrate it in Fig. 6.

Theorem 2. *Algorithm 2 achieves a 3-approximation ratio.*

Proof. Let $M_a = \bigcup_i \{a_i\}$ be the set of all removed arcs in the cycle elimination phase. Let $M_1 = \bigcup_i M_1^i$, and $M_2 = \bigcup_i M_2^i$. Clearly, $M_a \cup M_1 \cup M_2 = A'$, and that $\max(w(M_a), w(M_1), w(M_2)) \geq \frac{w(A')}{3}$. The observation that the weight of a maximum super-matching is an upper bound on the weight of a maximum carpool matching finishes the proof.   □

To see that our analysis is tight, consider the example in Fig. 7. Assume, for the given graph in the figure, that all weights are 1 and that there is no capacity constraint. The maximum matching, then, is 3 ($\{(1,4), (2,4), (3,4)\}$), but the algorithm can return the super matching $\{(1,2), (2,3), (3,1)\}$ from which only one arc can survive.

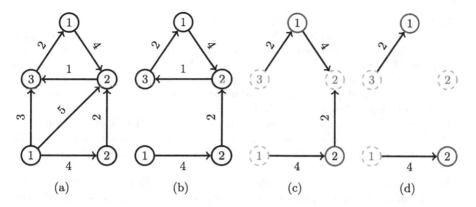

Fig. 6. Illustration of the SuperMatching algorithm: (a) a directed graph. (b) a maximum super-matching. (c) an in-tree: M_1 is the set of arcs exiting red, dashed vertices, and M_2 is the set of arcs exiting blue vertices. (d) a feasible carpool matching with total value of 6. (Color figure online)

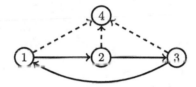

Fig. 7. Super Matching algorithm, worst case example

6 Conclusion

We study the weighted and unweighted variants of the Maximum Carpool Matching problem and present a min-cost flow based 3-approximation, and a local search based 2-approximation algorithms for the two variants of the problem respectively. To the best of our knowledge, these are the first approximation algorithms to the problem.

References

1. Zimride by enterprise. https://zimride.com/
2. Agatz, N., Erera, A., Savelsbergh, M., Wang, X.: Optimization for dynamic ridesharing: A review. Eur. J. Oper. Res. **223**(2), 295–303 (2012)
3. Arkin, E.M., Hassin, R., Rubinstein, S., Sviridenko, M.: Approximations for maximum transportation with permutable supply vector and other capacitated star packing problems. Algorithmica **39**(2), 175–187 (2004)
4. Chen, N., Engelberg, R., Nguyen, C.T., Raghavendra, P., Rudra, A., Singh, G.: Improved approximation algorithms for the spanning star forest problem. In: Charikar, M., Jansen, K., Reingold, O., Rolim, J.D.P. (eds.) APPROX/RANDOM -2007. LNCS, vol. 4627, pp. 44–58. Springer, Heidelberg (2007). doi:10.1007/978-3-540-74208-1_4

5. Hartman, I.B.-A.: Optimal assignment for carpooling-draft. Draft (2013)
6. Hartman, I.B.-A., Keren, D., Dbai, A.A., Cohen, E., Knapen, L., Janssens, D., et al.: Theory and practice in large carpooling problems. Procedia Comput. Sci. **32**, 339–347 (2014)
7. Knapen, L., Keren, D., Cho, S., Bellemans, T., Janssens, D., Wets, G., et al.: Estimating scalability issues while finding an optimal assignment for carpooling. Procedia Comput. Sci. **19**, 372–379 (2013)
8. Knapen, L., Yasar, A., Cho, S., Keren, D., Dbai, A.A., Bellemans, T., Janssens, D., Wets, G., Schuster, A., Sharfman, I., et al.: Exploiting graph-theoretic tools for matching in carpooling applications. J. Ambient Intell. Humaniz. Comput. **5**(3), 393–407 (2014)
9. Nguyen, C.T., Shen, J., Hou, M., Sheng, L., Miller, W., Zhang, L.: Approximating the spanning star forest problem and its application to genomic sequence alignment. SIAM J. Comput. **38**(3), 946–962 (2008)
10. Tardos, É.: A strongly polynomial minimum cost circulation algorithm. Combinatorica **5**(3), 247–255 (1985)

The Conjugacy Problem in Free Solvable Groups and Wreath Products of Abelian Groups is in TC⁰

Alexei Miasnikov[1], Svetla Vassileva[2], and Armin Weiß[1][(✉)]

[1] Stevens Institute of Technology, Hoboken, NJ, USA
weiss@fmi.uni-stuttgart.de
[2] Champlain College, St-lambert, QC, Canada

Abstract. We show that the conjugacy problem in a wreath product $A \wr B$ is uniform-TC⁰-Turing-reducible to the conjugacy problem in the factors A and B and the power problem in B. Moreover, if B is torsion free, the power problem for B can be replaced by the slightly weaker cyclic submonoid membership problem for B, which itself turns out to be uniform-TC⁰-Turing-reducible to the conjugacy problem in $A \wr B$ if A is non-abelian.

Furthermore, under certain natural conditions, we give a uniform TC⁰ Turing reduction from the power problem in $A \wr B$ to the power problems of A and B. Together with our first result, this yields a uniform TC⁰ solution to the conjugacy problem in iterated wreath products of abelian groups – and, by the Magnus embedding, also for free solvable groups.

Keywords: Wreath products · Conjugacy problem · Word problem · TC⁰ · Free solvable group

1 Introduction

The conjugacy problem is one of Dehn's fundamental algorithmic problems in group theory [2]. It asks on input of two group elements (given as words over a fixed set of generators) whether the two group elements are conjugate. The conjugacy problem can be seen as a generalization of the word problem, which on input of one word asks whether the word represents the identity element of the group. In recent years the conjugacy problem gained an increasingly important role in non-commutative cryptography; see for example [4,9,22,26]. These applications use the fact that it is easy to create elements which are conjugated, but to check whether two given elements are conjugated might be difficult even if the word problem is easy. In fact, there are groups where the word problem is in polynomial time but the conjugacy problem is undecidable [16]. Moreover, there are natural classes like polycyclic groups which have a word problem in uniform TC⁰ [21], but conjugacy problem not even known to be in NP. Another example for such a huge contrast is the Baumslag group, whose word problem

© Springer International Publishing AG 2017
P. Weil (Ed.): CSR 2017, LNCS 10304, pp. 217–231, 2017.
DOI: 10.1007/978-3-319-58747-9_20

is decidable in polynomial time, but the conjugacy problem is conjectured to be non-elementary [3].

The class TC^0 is a very low complexity class consisting of those problems which can be recognized by a family of constant depth and polynomial size Boolean circuits which also may use majority gates. We only consider uniform TC^0 (and subsequently simply write TC^0 for uniform TC^0). The word problem of abelian groups as well as integer arithmetic (iterated addition, multiplication, division) are problems in TC^0. However, there are not many groups known to have conjugacy problem in TC^0. Indeed, without the results of this paper, the Baumslag-Solitar groups $\mathbf{BS}_{1,q}$ [3] are the only natural examples we are aware of besides abelian and nilpotent groups. On the other hand, there is a wide range of groups having word problem in TC^0: all polycyclic groups [21] and, more generally, by a recent result all solvable linear groups [10]. Also iterated wreath products of abelian groups are known to have word problem in TC^0 [11].

The study of the conjugacy problem in wreath products has quite a long history: in [15] Matthews proved that a wreath product $A \wr B$ has decidable conjugacy problem if, and only if, both A and B have decidable conjugacy problem and B has decidable *cyclic subgroup membership problem* (note that in [15] this is called *power problem*). As a consequence, she obtained a solution to the conjugacy problem in free metabelian groups. Kargapolov and Remeslennikov generalized the result by establishing decidability of the conjugacy problem in free solvable groups of arbitrary solvability degree [8].

A few years later Remeslennikov and Sokolov [20] also generalized Matthews results to iterated wreath products by solving the cyclic subgroup membership problem in these groups. They also showed that the Magnus embedding [14] of free solvable groups into iterated wreath products of abelian groups preserves conjugacy – thus, giving a new proof for decidability of the conjugacy problem in free solvable groups.

Later, in [23] it is shown that the criterion of [15] can be actually checked in polynomial time for iterated wreath products of abelian groups – this also yields a polynomial time algorithm for the conjugacy problem in free solvable groups. In [18] this has been further improved to LOGSPACE. Recently, in [5], Matthews result has been generalized to a wider class of groups without giving precise complexity bounds – see the discussion in last section.

In this work we use the same fundamental theoretical result as in [15,18] in order to give a precise complexity version of Matthews result. Even though we follow the same scheme as [18], we need to overcome additional technical difficulties related to TC^0 complexity. Moreover, we extend the result of [18] also in several directions. At some points we need a stronger hypothesis than in [15] though: it is not sufficient to assume that the cyclic subgroup membership problem is decidable in TC^0 in order to reduce the conjugacy problem in a wreath product to the factors. Instead, we need the stronger power problem to be in TC^0: on input of two group elements b and c compute an integer k such that $b^k = c$. More precisely, we establish the following results:

- There is a uniform TC^0 Turing reduction from the conjugacy problem in $A \wr B$ to the conjugacy problems in A and B together with the power problem in B. If B is torsion-free, the power problem can be replaced by the cyclic submonoid membership problem.
- Suppose the orders of torsion elements of B are β-smooth for some $\beta \in \mathbb{N}$. Then, the power problem in $A \wr B$ is uniform-TC^0-Turing-reducible to the power problems in A and B. As a corollary we obtain that iterated wreath products of abelian groups have conjugacy problem in uniform TC^0. Using the Magnus embedding [14,20], also the conjugacy problem in free solvable groups is in uniform TC^0.

Notice that images of group elements under the Magnus embedding can be computed in TC^0 (since any image under homomorphisms of finitely generated monoids can be computed in TC^0 [12]). Thus, for free solvable groups as well as for iterated wreath products of abelian groups, our results nail down the complexity of conjugacy precisely. This is because already the word problem in $\mathbb{Z}$ is hard for TC^0 (and so the conjugacy problem in free solvable groups is TC^0-complete). Also for wreath products $A \wr B$ with A non-abelian and B torsion-free, we have a tight complexity bound because in this case there is a reduction from the cyclic submonoid membership problem in B to the conjugacy problem in $A \wr B$.

To solve the conjugacy problem, we first deal with the word problem. For a free solvable group of solvability degree d, we obtain a circuit of majority depth d. It is not clear how a circuit of smaller majority depth could be constructed. On the other hand, the algorithm for the word problem in [17] runs in cubic time independently of the solvability degree. This gives rise to the question whether the depth (or the size) of circuits for the word and conjugacy problem of free solvable groups could be bounded uniformly independent of the solvability degree. Note that a negative answer to this question would imply that $\mathsf{TC}^0 \neq \mathsf{NC}^1$.

We want to emphasize that throughout we assume that the groups are finitely generated. As wreath products we consider only *restricted* wreath products, that is the underlying functions are required to have finite support.

Outline. Section 2 introduces some notation and recalls some basic facts on complexity. Then in Sect. 3, we define wreath products and discuss the solution to the word problem. Sections 4 and 5, the main parts, examine the conjugacy problem in wreath products resp. iterated wreath products. In order to do so, we deal with the power problem in iterated wreath products in Sect. 5. Finally, in Sect. 6, we discuss some open problems. Most proofs are omitted due to space constraints – they can be found in the full version on arXiv [19].

2 Preliminaries

Words. An *alphabet* is a (finite or infinite) set Σ; an element $a \in \Sigma$ is called a *letter*. The free monoid over Σ is denoted by Σ^*; its elements are finite sequences of letters and they are called *words*. The multiplication of the monoid is concatenation of words. The identity element is the empty word ε.

Groups. We consider a finitely generated (f. g.) group G together with a surjective homomorphism $\eta : \Sigma^* \to G$ (a *monoid presentation*) for some finite alphabet Σ. Throughout, all groups we consider are finitely generated even if not mentioned explicitly. In order to keep notation simple, we suppress the homomorphism η and consider words also as group elements. We write $w =_G w'$ as a shorthand for $\eta(w) = \eta(w')$ and $w \in_G A$ instead of $\eta(w) \in \eta(A)$ for $A \subseteq \Sigma^*$ and $w \in \Sigma^*$. Whenever it is clear that we deal with group elements $g, h \in G$, we simply write $g = h$ for equality in G.

We say two group elements $g, h \in G$ are conjugate, and we write $g \sim h$, if there exists an element $x \in G$ such that $g^x = x^{-1}gx = h$. Similarly, we say two words u and v in generators of G are conjugate, and we write $u \sim_G v$, if the elements of G represented by u and v are conjugate as elements of G. We denote by $\mathrm{ord}(g)$ the order of a group element g (i. e., the smallest positive integer d such that $g^d = 1$, or ∞ if no such integer exists). For $g \in G$, the cyclic subgroup generated by g is denoted by $\langle g \rangle$. A d-fold commutator is a group element of the form $x^{-1}y^{-1}xy$ for $(d-1)$-fold commutators x and y; a 0-fold commutator is any group element. The *free solvable group* of *solvability degree* d is the group subject only to the relations that all d-fold commutators are trivial.

2.1   Complexity

Computation or decision problems are given by functions $f : \Delta^* \to \Sigma^*$ for some finite alphabets Δ and Σ. A decision problem (or formal language) L is identified with its characteristic function $\chi_L : \Delta^* \to \{0, 1\}$ with $\chi_L(x) = 1$ if, and only if, $x \in L$.

Computational Problems in Group Theory. Let G be a group with finite generating set Σ. We define the following algorithmic problems in group theory.

- The *word problem* WP(G) of G, is the set of all words representing the identity in G.
- The *conjugacy problem* CP(G) is the set of all pairs (v, w) such that $v \sim_G w$.
- The *cyclic subgroup membership problem* CSMP(G): the set of all pairs (v, w) such that $w \in \langle v \rangle$ (i. e., there is some $k \in \mathbb{Z}$ with $v^k =_G w$).
- The *cyclic submonoid membership problem* CSMMP(G): the set of all pairs (v, w) such that $w \in_G \{v\}^*$ (i. e., there is some $k \in \mathbb{N}$ with $v^k =_G w$).
- The *power problem* PP(G): on input of some $(v, w) \in \Sigma^* \times \Sigma^*$ decide whether w is a power of v that is whether there is some $k \in \mathbb{Z}$ such that $v^k =_G w$. In the "yes" case compute this k in binary representation. If v has finite order in G, the computed k has to be the smallest non-negative such k.

Whereas the first four of these problems are decision problems, the last one is an actual computation problem. Be aware that sometimes in literature the power problem is defined as what we refer to as cyclic subgroup membership problem.

Circuit Classes. The class AC^0 is defined as the class of functions computed by families of circuits of constant depth and polynomial size with unbounded fan-in Boolean gates (and, or, not). TC^0 additionally allows majority gates. A majority gate (denoted by Maj) returns 1 if the number of 1s in its input is greater or equal to the number of 0s. In the following, we always assume that the alphabets Δ and Σ are encoded over the binary alphabet $\{0,1\}$ such that each letter uses the same number of bits. Moreover, we assume that also the empty word ε has such a encoding over $\{0,1\}$, which is denoted by ε as well (be aware of the slight ambiguity). The empty word letter is used to pad outputs of circuits to fit the full number of output bits; still we do not forbid to use it in the middle. We say a function f is AC^0-*computable* (resp. TC^0-*computable*) if $f \in AC^0$ (resp. $f \in TC^0$).

In the following, we only consider Dlogtime-uniform circuit families. Dlogtime-uniform means that there is a deterministic Turing machine which decides in time $\mathcal{O}(\log n)$ on input of two gate numbers (given in binary) and the string 1^n whether there is a wire between the two gates in the n-input circuit and also decides of which type some gates is. Note that the binary encoding of the gate numbers requires only $\mathcal{O}(\log n)$ bits – thus, the Turing machine is allowed to use time linear in the length of the encodings of the gates. For more details on these definitions we refer to [24]. In order to keep notation simple we write AC^0 (resp. TC^0) for Dlogtime-uniform AC^0 (resp. Dlogtime-uniform TC^0) throughout. We have the following inclusions (note that even $TC^0 \subseteq P$ is not known to be strict):

$$AC^0 \subsetneq TC^0 \subseteq LOGSPACE \subseteq P.$$

The following facts are well-known and will be used in the following without further reference:

- Barrington, Immerman, and Straubing [1] showed that $TC^0 = FO(+, *, Maj)$ i.e., TC^0 comprises exactly those languages which are defined by some first order formula with majority quantifiers where positions may be compared using $+$, $*$ and $<$. In particular, if we can give a formula with majority quantifiers using only addition and multiplication predicates, we do not need to worry about uniformity.
- Homomorphisms can be computed in TC^0 [12]: on input of two alphabets Σ and Δ (coded over the binary alphabet), a list of pairs (a, v_a) with $a \in \Sigma$ and $v_a \in \Delta^*$ such that each $a \in \Sigma$ occurs in precisely one pair, and a word $w \in \Sigma^*$, the image $\varphi(w)$ under the homomorphism φ defined by $\varphi(a) = v_a$ can be computed in TC^0. Moreover, if φ is length-multiplying (that is $\varphi(a)$ and $\varphi(b)$ have the same length for all $a, b \in \Sigma$), the computation is in AC^0.
- Iterated addition is the following problem: given n numbers $a_1, \ldots, a_n$ (in binary), compute $\sum_{i=1}^{n} a_i$ (as binary number). This is known to be in TC^0.

Example 1. Finitely generated abelian groups have word problem in TC^0: the word problem of $\mathbb{Z}$ is in TC^0 using iterated addition (since it is essentially summing up numbers 1 and -1), the word problem of finite cyclic groups is in TC^0 by then calculating modulo; and, finally, a word in a direct product is the identity if, and only if, it is the identity in all components.

Reductions. Let $K \subseteq \Delta^*$ and $L \subseteq \Sigma^*$ be languages and $\mathcal{C}$ a complexity class. Then K is called $\mathcal{C}$-*many-one-reducible* to L if there is a $\mathcal{C}$-computable function $f : \Delta^* \to \Sigma^*$ such that $w \in K$ if, and only if, $f(w) \in L$. In this case, we write $K \leq_{\mathrm{m}}^{\mathcal{C}} L$.

A function f is AC^0 -*(Turing)-reducible* to a function g if there is a Dlogtime-uniform family of AC^0 circuits computing f which, in addition to the Boolean gates, also may use oracle gates for g (i. e., gates which on input x output $g(x)$). This is expressed by $f \in \mathsf{AC}^0(g)$ or $f \leq_T^{\mathsf{AC}^0} g$. For a group G, we write $\mathsf{AC}^0(G)$ as shorthand of $\mathsf{AC}^0(\mathrm{WP}(G))$. Likewise TC^0 (Turing) reducibility is defined. Note that if $L_1, \ldots, L_k$ are in TC^0, then $\mathsf{TC}^0(L_1, \ldots, L_k) = \mathsf{TC}^0$ (see e. g. [24]).

Remark 1. The cyclic subgroup membership problem, in particular, allows to solve the word problem: some group element is in the cyclic subgroup generated by the identity if, and only if, it is the identity. Moreover, the cyclic subgroup membership problem for (v, w) can be decided by two calls to the cyclic sub-monoid membership problem (for (v, w) as well as for (v^{-1}, w)). Also, the power problem is a stronger version of the cyclic submonoid membership problem (simply check the sign of the output of the power problem). Thus, we have

$$\mathrm{WP}(G) \leq_{\mathrm{m}}^{\mathsf{AC}^0} \mathrm{CSMP}(G) \leq_T^{\mathsf{AC}^0} \mathrm{CSMMP}(G) \leq_T^{\mathsf{AC}^0} \mathrm{PP}(G).$$

Moreover, the power problem enables to decide whether an element is of finite order (just compute the k such that $g^k = g^{-1}$ – if this is a positive number, then g is of finite order, otherwise not).

3   Wreath Products and the Word Problem

Let A and B be groups. For a function $f : B \to A$ the *support* of f is defined as $\mathrm{supp}(f) = \{b \in B \mid f(b) \neq 1\}$. For two groups A and B, the set of functions from B to A with finite support is denoted by $A^{(B)}$; it forms a group under point-wise multiplication. Mapping $a \in A$ to the function $a(b)$ which has a value of a if $b = 1$ and 1 otherwise, gives an embedding of A into $A^{(B)}$, termed the canonical embedding. In what follows we identify A with its canonical image in $A^{(B)}$. The *wreath product* $A \wr B$ of A and B is defined as the semi-direct product $B \ltimes A^{(B)}$, where the action of $b \in B$ on a function $f \in A^{(B)}$ is defined by $f^b(x) = f(xb^{-1})$ (note that this is also sometimes referred to as restricted wreath product). We identify B and $A^{(B)}$ (and hence also A) with their canonical images in $A \wr B$. Thus, for the multiplication in $A \wr B$ we have the following rules

$$(b, f)(c, g) = (bc, f^c g), \qquad (b, f)^{-1} = (b^{-1}, (f^{-1})^{b^{-1}})$$

for $b, c \in B$ and $f, g \in A^{(B)}$. Here f^{-1} is the pointwise inverse (i. e., $f^{-1}(b) = (f(b))^{-1}$ for all $b \in B$).

Let Σ_A and Σ_B be fixed generating sets of A and B, correspondingly. Then, $A \wr B$ is generated by $\Sigma = \Sigma_A \cup \Sigma_B$ (using the embedding of A into $A \wr B$). Given a word $w \in \Sigma^*$ of length n, we can group it as $w = a_1 b_1 \cdots a_m b_m$ with $a_i \in \Sigma_A^*$, $b_i \in \Sigma_B^*$ and $m \leq n$. Introducing factors $bb^{-1} \in \Sigma_B^*$, we can rewrite this as follows:

$$
\begin{aligned}
w &=_G a_1 b_1 \cdots a_m b_m =_G b_1 b_1^{-1} a_1 b_1 \cdots a_m b_m =_G b_1 a_1^{b_1} a_2 b_2 \cdots a_m b_m \\
&=_G b_1 b_2 (a_1^{b_1} a_2)^{b_2} \cdots a_m b_m =_G b_1 b_2 (a_1^{b_1 b_2} a_2^{b_2}) \cdots a_m b_m \\
&=_G b_1 \cdots b_m \cdot a_1^{b_1 \cdots b_m} \cdots a_m^{b_m}
\end{aligned}
$$

Thus, we have $w =_G (b, f)$ with $b = b_1 \cdots b_m$ and $f = a_1^{b_1 \cdots b_m} \cdots a_m^{b_m}$. Since a^c and $a'^{c'}$ commute for distinct $c, c' \in B$ and for any $a, a' \in A$, we can reorder this product to ensure that the exponents are distinct: whenever we have $b_i \cdots b_m =_B b_j \cdots b_m$ for $i < j$, we combine the corresponding terms into a single term $(a_i a_j)^{b_i \cdots b_m}$. Thus, we can rewrite f as the product $\tilde{a}_1^{\tilde{b}_1} \ldots \tilde{a}_k^{\tilde{b}_k}$, where $\tilde{a}_1, \ldots, \tilde{a}_k \in \Sigma_A^*$, and $\tilde{b}_1, \ldots, \tilde{b}_k \in \Sigma_B^*$ all represent distinct elements of B. Moreover, we can assume that all $\tilde{a}_i$ represent non-trivial elements of A. With this notation, we have $f(\tilde{b}_i) = \tilde{a}_i \neq 1$ and $f(c) = 1$ for $c \notin \{\tilde{b}_1, \ldots, \tilde{b}_k\} = \text{supp}(f)$. Furthermore, f is completely given by the set of pairs $\{(\tilde{b}_1, \tilde{a}_1), \ldots, (\tilde{b}_k, \tilde{a}_k)\}$.

In the following, we always assume that functions $f \in A^{(B)}$ are represented as a list of pairs $f = ((\tilde{b}_1, \tilde{a}_1), \ldots, (\tilde{b}_k, \tilde{a}_k))$. The order of the pairs does not matter – but they are written down in some order. We also assume for an input w of length n, that $k = m = n$ and that every word $\tilde{b}_i, \tilde{a}_i$ has length n. This is achieved by padding with pairs $(\varepsilon, \varepsilon)$ (where ε is the letter representing the empty word).

Lemma 1. *Let A and B be finitely generated groups and let $G = A \wr B$. There is an $\mathsf{AC}^0(A, B)$ circuit family which on input $w \in \Sigma^*$ computes (b, f) with $w =_G (b, f)$.*

Proof. For an input word $w = w_1 \cdots w_n \in \Sigma^*$, we first calculate the image under the projection $\pi_B : a \mapsto \varepsilon$ for $a \in \Sigma_A$. Since ε is a letter in our alphabet, this is a length-preserving homomorphism, and thus, can be computed in AC^0 [12]. We have $b = \pi_B(w)$. Next, define the following equivalence relation $\approx$ on $\{1, \ldots, n\}$:

$$
i \approx j \iff \pi_B(w_{i+1} \cdots w_n) =_B \pi_B(w_{j+1} \cdots w_n)
$$

After the computation of π_B it can be checked for all pairs i, j whether $i \approx j$ using $\binom{n}{2}$ oracle calls in parallel to the word problem of B. Let $[i]$ denote the equivalence class of i. Now, $b^{-1} w$ is in the (finite) direct product $\prod_{[i]} A^{\pi_B(w_{i+1} \cdots w_n)} \leq A^{(B)}$ (this is well-defined by the definition of $\approx$). The projection to component associated to $[i]$ is computed by replacing all w_j by ε whenever $w_j \in \Sigma_B$ or $j \not\approx i$. As before, this computation is in AC^0. As a representative of $[i]$, we choose the smallest $i \in [i]$. Now, the preliminary output is the pair $(b, (f_1, \ldots, f_n))$ with

$$
f_i = \begin{cases} \left(\pi_B(w_{i+1} \cdots w_n), \prod_{j \in [i]} w_j \right) & \text{if } i = \min[i], \\ (\varepsilon, \varepsilon) & \text{otherwise.} \end{cases}
$$

Up to the calculation of $\approx$, everything can be done in AC^0 (checking $i = \min[i]$ amounts to $\bigwedge_{j<i} \neg(i \approx j)$). Finally, pairs $f_i = (b_i, a_i)$ with $a_i =_A 1$ are replaced by $(\varepsilon, \varepsilon)$. This requires an additional layer of calls to the word problem of A.

If we assign appropriate gate numbers corresponding to the description of our circuit (e. g. concatenation of the number of the layer and the indices i, j), it is easy to see that it can be checked in linear time on input of two binary gate numbers if the two gates are connected. This establishes uniformity of the circuit.

Theorem 1. $\mathrm{WP}(A \wr B) \in \mathsf{AC}^0(A, B)$.

Proof. This is an immediate consequence of Lemma 1 since $(b, f) =_G 1$ if, and only if, $b =_B 1$ (can be checked using the word problem of B) and $f = ((\varepsilon, \varepsilon), \ldots, (\varepsilon, \varepsilon))$.

Note that Theorem 1 is a stronger version of [25] where NC^1 reducibility is shown.

Definition 1. *Let $d \in \mathbb{N}$. We define the* left-iterated wreath product, $A \overset{d}{\wr} B$, *and the* right-iterated wreath product $A \wr^d B$ *of two groups A and B inductively:*

- $A \overset{1}{\wr} B = A \wr B$
- $A \overset{d}{\wr} B = A \wr (A \overset{d-1}{\wr} B)$

- $A \wr^1 B = A \wr B$
- $A \wr^d B = (A \wr^{d-1} B) \wr B$

Let $S_{d,r}$ denote the free solvable group of solvability degree d and rank r. The Magnus embedding [14] is an embedding $S_{d,r} \to \mathbb{Z}^r \wr S_{d-1,r}$. By iterating the construction, we obtain an embedding $S_{d,r} \to \mathbb{Z}^r \overset{d}{\wr} 1$. For the purpose of this paper, we do not need to know the homomorphism explicitly – it suffices to know that it is an embedding and that it preserves conjugacy [20]. The following corollary is also a consequence of [11] since a wreath product can be embedded into the corresponding block product.

Corollary 1. *Let A and B be f. g. abelian groups and let $d \geq 1$. The word problems of $A \wr^d B$ and of $A \overset{d}{\wr} B$ are in TC^0. In particular, the word problem of a free solvable group is TC^0-complete.*

Note that here the groups A, B and the number d of wreath products are fixed. Indeed, if there were a single TC^0 circuit which works for free solvable groups of arbitrary solvability degree, this circuit would also solve the word problem of the free group, which is NC^1-hard – thus, showing $\mathsf{TC}^0 = \mathsf{NC}^1$.

Proof. The first statement follows from Theorem 1 because f. g. abelian groups have word problem in TC^0 (see Example 1). The second statement then follows by the Magnus Embedding [14] and the fact that homomorphisms can be computed in TC^0. The hardness-part is due to the fact that the free solvable group has an element of infinite order, i. e., a subgroup $\mathbb{Z}$, whose word problem is hard for TC^0.

Remark 2. For a TC^0 circuit, the *majority depth* is defined as the maximal number of majority gates on any path from an input to an output gate (see e. g. [13]). Assume that $\mathrm{WP}(A), \mathrm{WP}(B) \in \mathsf{TC}^0$. The circuit in the proof of Lemma 1 contains one layer of gates to the word problem of B followed by one gate to the word problem of A. The additional check for $b =_B 1$ in the proof of Theorem 1 can be done in parallel to the computation of Lemma 1; thus, it can be viewed as part or the layer of calls to $\mathrm{WP}(B)$. Since Lemma 1 is an AC^0 reduction, the majority depth of the resulting circuit is at most $m_A + m_B$ where m_A (resp. m_B) is the majority depth of the circuit family for $\mathrm{WP}(A)$ (resp. $\mathrm{WP}(B)$).

Starting with the word problem of free abelian groups, which is in TC^0 with majority depth one, we see inductively that a d-fold iterated wreath product – and thus the free solvable group of solvability degree d – has word problem in TC^0 with majority depth at most d. On the other hand, we do not see a method how to improve this bound any further. In [11] a similar observation was stated for iterated block products (into which wreath products can be embedded). There the question was raised how the depth of the circuit for the word problem (or more general any problem recognized by the block product) is related to the number of block products in an iterated block product (the so-called block-depth).

Question 1. Can the word problem of a free solvable group of solvability degree d be decided in TC^0 with majority depth less than d?

We want to point out that Question 1 is related to an important question in complexity theory: as outlined in [13], a negative answer would imply that $\mathsf{LC}^0 \neq \mathsf{NC}^1$. Nevertheless, the following observations point rather towards a positive answer of Question 1: the word problem of free solvable groups is decidable in time $\mathcal{O}(n^3)$ – regardless of the solvability degree d [17, 23]. Moreover, the circuit for linear solvable groups (*not* for free solvable groups with $d > 2$) from [10] can be arranged with majority depth bounded uniformly for all groups. This is because every matrix entry in a product of upper triangular matrices can be obtained as iterated addition of iterated multiplication of the entries of the original matrices (for the precise formula, see [10]). These operations have circuits of uniformly bounded depth (also for f.g. field extensions). Hence, only the size of the circuits, but not the depth, depends on the solvability degree.

4   The Conjugacy Problem in Wreath Products

In order to decide conjugacy of two elements $(b, f), (c, g)$ in a wreath product $A \wr B$ we will study the behavior of f and g on cosets of $\langle b \rangle \leq B$. A *full system of $\langle b \rangle$ -coset representatives* is a set $T \subseteq B$ of such that $\langle b \rangle t \cap \langle b \rangle t' = \emptyset$ for $t \neq t' \in T$ and $B = \langle b \rangle T$. Let T be a full system of $\langle b \rangle$-coset representatives in B. For $\gamma, b \in B$, $f \in A^{(B)}$, and $t \in T$, we define

$$
\pi_t^{(\gamma)}(f) = \begin{cases} \displaystyle\prod_{j=0}^{N-1} f(tb^j\gamma^{-1}) & \text{if } \mathrm{ord}(b) = N < \infty, \\[2ex] \displaystyle\prod_{j=-\infty}^{\infty} f(tb^j\gamma^{-1}) & \text{if } \mathrm{ord}(b) = \infty, \end{cases}
$$

which is an element of A. We denote $\pi_t^{(1)}(f)$ by $\pi_t(f)$. The definition of the π_t depends on the order of the element b. First, observe that in the case when the order of b is infinite the product is finite, since the function f is of finite support. In fact, it is the product of all possible non-trivial factors of the form $f(tb^j\gamma^{-1})$ multiplied in increasing order of j. The same is true in the case when the order of b is finite. So in order to compute $\pi_t^{(\gamma)}$, we need to find all the elements of the form $tb^j\gamma^{-1}$ for which f is non-trivial, arrange them in increasing order of j and concatenate the respective a_j.

Lemma 2. *The computation of $\pi_t^{(\gamma)}(f)$ is in $\mathsf{TC}^0(\mathrm{PP}(B))$ (more precisely, the input is $b, t, \gamma \in \Sigma_B^*$ and a function $f = ((b_1, a_1), \ldots, (b_n, a_n))$, the output is $\pi_t^{(\gamma)}(f)$ given as a word over Σ_A). Moreover, if B is torsion-free, then it is actually in $\mathsf{TC}^0(\mathrm{CSMMP}(B))$.*

The proof of Lemma 2 is easy: using oracle gates for $\mathrm{PP}(B)$, for all j some $k_j \in \mathbb{Z}$ such that $t^{-1}b_j\gamma = b^{k_j}$ is computed (and checked if there is any); then the tuples (b_j, a_j) are sorted with respect to that power k_j. Notice that it suffices to decide whether $k_i < k_j$ for $i \neq j$, which, in the torsion-free case, can be done using the cyclic submonoid membership problem. For more details, see [19].

In [15], Matthews provides a criterion for testing whether two elements of a wreath product are conjugate. Based on this criterion, in [18] the following criterion for conjugacy is derived, which is more suitable for working in TC^0 or Logspace. Here, $\tilde{\pi}_s$ is defined analogously to π_t, but with respect to a set of $\langle c \rangle$-coset representatives S. In the appendix of our full version [19] we show how to derive Theorem 2 from [15].

Theorem 2 ([15]/[18]). *Let $x = (b, f)$ and $y = (c, g)$ be two elements of $A \wr B$ and let T and S be full systems of $\langle b \rangle$- (resp. $\langle c \rangle$-) coset representatives. If b and c are not conjugate in B, then x and y are not conjugate in $A \wr B$. Otherwise, we distinguish the following cases:*

(i) Suppose $g = 1$. Then $x \sim y$ if, and only if, $\pi_t(f) = 1$ for all $t \in T$.

(ii) Suppose $g \neq 1$, but $\pi_t(f) = 1$ for all $t \in T$. Then $x \sim y$ if, and only if, $\tilde{\pi}_s(g) = 1$ for all $s \in S$.

(iii) Suppose $g \neq 1$ and there exists a $t \in T$ such that $\pi_t(f) \neq 1$. Denote $\mathrm{supp}(g) = \{\beta_1, \ldots, \beta_m\}$. Then $x \sim y$ if, and only if, some $d \in \{\beta_1^{-1}t, \ldots, \beta_m^{-1}t\}$ satisfies $db = cd$ and

(a) $\pi_{t'}(f) = \pi_{t'}^{(d)}(g)$ for all $t' \in T$ if $\mathrm{ord}(b) = \infty$, or

(b) $\pi_{t'}(f) \sim \pi_{t'}^{(d)}(g)$ for all $t \in T'$ if $\mathrm{ord}(b)$ is finite.

Since T and S in Theorem 2 are infinite in general, for algorithmic purposes we need to restrict to a finite subset containing those $t \in T$ where any of the $\pi_t(f)$ or $\pi_t^{(d)}(g)$ is non-trivial (and likewise or S):

Lemma 3. *Let $\mathrm{supp}(f) = \{b_1, \ldots, b_n\}$ and $\mathrm{supp}(g) = \{\beta_1, \ldots, \beta_m\}$. Theorem 2 is still true if we replace the full system of $\langle b \rangle$-coset representatives T by*

$$\widetilde{T} = \{\beta_i\beta_j^{-1}b_k \mid 1 \leq i, j \leq m, 1 \leq k \leq n\}.$$

Theorem 3. *Let A and B be arbitrary finitely generated groups. Then, we have $\mathrm{CP}(A \wr B) \in \mathsf{TC}^0(\mathrm{CP}(A), \mathrm{CP}(B), \mathrm{PP}(B))$. If, moreover, B is torsion-free, then $\mathrm{CP}(A \wr B) \in \mathsf{TC}^0(\mathrm{CP}(A), \mathrm{CP}(B), \mathrm{CSMMP}(B))$.*

The proof of Theorem 3 is straightforward (see [19]) because the condition of Theorem 2 is a simple Boolean combination of several conditions which can be all checked in parallel using the conjugacy problem of A and B after the values $\pi_t(f)$, $\pi_t^{(d)}(g)$, and $\tilde{\pi}_s(g)$ for $t \in \widetilde{T}$ (as defined in Lemma 3) and $s \in \mathrm{supp}(g)$ have been computed, which can be done in TC^0 by Lemma 2. The following quite trivial observation turns out to be very useful:

Lemma 4. *Let G be f. g. by Σ and let the order of its torsion elements be uniformly bounded. Suppose there is a polynomial $p(n)$ such that for every $w \in \Sigma^*$ which is non-torsion, the inequality $k \le p(\|w^k\|)$ is satisfied, where $\|w^k\|$ denotes the geodesic length of the group element w^k. Then $\mathrm{PP}(G) \in \mathsf{AC}^0(\mathrm{WP}(G))$.*

The second condition of Lemma 4 means that there is a uniform polynomial bound on the distortion of infinite cyclic subgroups. This is satisfied by abelian groups (with p being linear). Since the conjugacy problem in abelian groups is in TC^0 (as it is the word problem), we get the following corollary of Theorem 3.

Corollary 2. *Let A and B be f. g. abelian groups. Then $\mathrm{CP}(A \wr B) \subset \mathsf{TC}^0$.*

The following result shows that a weaker form of the power problem in B is also necessary to solve the conjugacy problem in $A \wr B$. It is a complexity analog of [15, Theorem B], which only considers decidability. It establishes that, at least in the torsion-free case, Theorem 3 is the best that we can get. In the torsion case, we need the power problem in Theorem 3 – on the other hand, we do not see how to reduce $\mathrm{PP}(B)$ to $\mathrm{CP}(A \wr B)$ in TC^0 (or even in polynomial time – since the outputs of the power problem might be super-exponential). Note that for pure decidability, it does not matter if we consider $\mathrm{CSMP}(B)$, $\mathrm{CSMMP}(B)$ or $\mathrm{PP}(B)$ since they can all be reduced to each other.

Theorem 4. *Let A be f. g. and non-trivial. Then $\mathrm{CSMP}(B) \le_{\mathrm{m}}^{\mathsf{AC}^0} \mathrm{CP}(A \wr B)$. If, moreover, A is non-abelian, then $\mathrm{CSMMP}(B) \le_{\mathrm{m}}^{\mathsf{AC}^0} \mathrm{CP}(A \wr B)$.*

Notice that Theorem 4 shows that in the case that A is non-abelian and B torsion-free Theorem 3 is the best possible result one could expect.

Proof. The first statement is simply due to the observation that the construction in [15, Thm. B] can be computed in AC^0. Now, let A be non-abelian. In particular, there are elements $a_1, a_2 \in A$ with $a_1 a_2 \ne_A a_2 a_1$. For $b, c \in \Sigma_B^*$, we define two functions $f, g \in A^{(B)}$ by

$$f(1) = a_1 a_2, \qquad\qquad\qquad f(\beta) = 1 \quad \text{for } \beta \in B \setminus \{1\},$$
$$g(1) = a_1, \qquad g(c) = a_2, \qquad g(\beta) = 1 \quad \text{for } \beta \in B \setminus \{1, c\}.$$

Now, we have $\pi_1(f) = a_1 a_2$ and $\pi_t(f) = 1$ for $t \notin \langle b \rangle$. For g, according to Theorem 2 (iii), we have to consider $\pi_1^{(1)}(g)$ and $\pi_1^{(c)}(g)$. If b has finite order, then $\pi_1^1(g)$ and $\pi_1^{(c)}(g)$ are both one of $a_1 a_2$ or $a_2 a_1$ (which are conjugate) if, and only if, $c \in \langle b \rangle =_G \{b\}^*$ (because b has finite order) – otherwise $\pi_1^1(g) = a_1$ and $\pi_1^{(c)}(g) = a_2$. On the other hand if b has infinite order, we have

$$\pi_1^1(g) = \begin{cases} a_1 a_2 & \text{if } c =_B b^k,\ k > 0, \\ a_2 a_1 & \text{if } c =_B b^k,\ k < 0, \\ a_1 & \text{otherwise}, \end{cases} \qquad \pi_1^{(c)}(g) = \begin{cases} a_1 a_2 & \text{if } c =_B b^k,\ k > 0, \\ a_2 a_1 & \text{if } c =_B b^k,\ k < 0, \\ a_2 & \text{otherwise}. \end{cases}$$

Therefore, by Theorem 2, $(b, f) \sim (b, g)$ if, and only if, $c \in_G \{b\}^*$.

5   Conjugacy and Power Problem in Iterated Wreath Products

In order to solve the conjugacy problem in iterated wreath products, we also need to solve the power problem in wreath products. In general, we do not know whether the power problem in a wreath product is in TC^0 given that the power problem of the factors is in TC^0. The problem is that when dealing with torsion it might be necessary to compute GCDs– which is not known to be in TC^0. By restricting torsion elements to have only smooth orders, we circumvent this issue. Recall that a number is called β-smooth for some $\beta \in \mathbb{N}$ if it only contains prime factors less than or equal to β. The proof of the following lemma is elementary.

Lemma 5. *Let $\beta \in \mathbb{N}$. Suppose the orders of all torsion elements in A and B are β-smooth. Then the orders of all torsion elements in $A \wr B$ are β-smooth.*

Note that we are not aware of any finitely generated group with word problem in TC^0 and torsion elements which are not β-smooth for any β. On the other hand, there are recursively presented such groups: for instance, take the infinite direct sum of cyclic groups of arbitrary order. The next step is to show how to reduce the power problem in $A \wr B$ to the power problems of A and B.

Theorem 5. *Let $\beta \in \mathbb{N}$ and suppose torsion elements in A are β-smooth. Then we have $\mathrm{PP}(A \wr B) \in \mathsf{TC}^0(\mathrm{PP}(A), \mathrm{PP}(B))$.*

For a proof of Theorem 5 see our full version [19]. Here we give a short outline: on input (b, f) and (c, g) first apply the power problem in B to b and c. If there is no solution, then there is also no solution for (b, f) and (c, g). Otherwise, the smallest $k \geq 0$ with $b^k =_B c$ can be computed. If b has infinite order, it remains to check whether $(b, f)^k = (c, g)$. Since k might be too large, this cannot be done by simply applying the word problem. Nevertheless, we only need to establish equality of functions in $A^{(B)}$. We show that it suffices to check equality on certain (polynomially many) "test points". In the case that b has finite order K, we know that if there is a solution to the power problem it must be in $k + K\mathbb{Z}$. Now, similar techniques as in the infinite order case can be applied to find the solution.

By repeated application of Theorems 3, 5 and Lemma 5, we obtain the first statement of Corollary 3 below. The second statement follows since the Magnus embedding preserves conjugacy [20] (that means two elements are conjugate in the free solvable group if, and only if, their images under the Magnus embedding are conjugate).

Corollary 3. *Let A and B be f. g. abelian groups and let $d \geq 1$. The conjugacy problems of $A \wr^d B$ and of $A \, ^d\!\wr B$ are in TC^0. Also, the conjugacy problem of free solvable groups is in TC^0.*

6   Conclusion and Open Problem

As already discussed in Question 1, an important open problem is the dependency of the depth of the circuits for the word problem on the solvability degree.

We have seen how to solve the conjugacy problem in a wreath product in TC^0 with oracle calls to the conjugacy problems of both factors and the power problem in the second factor. However, we do not have a reduction from the power problem in the second factor to the conjugacy problem in the wreath product: we only know that the cyclic submonoid membership problem is necessary to solve the conjugacy problem in the wreath product.

Question 2. Is $\mathrm{CP}(A \wr B) \in \mathsf{TC}^0(\mathrm{CP}(A), \mathrm{CP}(B), \mathrm{CSMMP}(B))$ in general? Moreover, in which cases is $\mathrm{CP}(A \wr B) \in \mathsf{TC}^0(\mathrm{CP}(A), \mathrm{CP}(B), \mathrm{CSMP}(B))$?

For iterated wreath products we needed the power problem to be in TC^0 in order to show that the conjugacy problem is in TC^0. One reason was that we only could reduce the power problem in the wreath product to the power problems of the factors. However, we have seen that in torsion-free groups, we do not need the power problem to solve conjugacy, as the cyclic submonoid membership problem is sufficient. Therefore, it would be interesting to reduce the cyclic submonoid membership problem in a wreath product to the same problem in its factors.

Question 3. Is $\mathrm{CSMMP}(A \wr B) \in \mathsf{TC}^0(\mathrm{CSMMP}(A), \mathrm{CSMMP}(B))$ or similarly is $\mathrm{CSMP}(A \wr B) \in \mathsf{TC}^0(\mathrm{CSMP}(A), \mathrm{CSMP}(B))$?

In [5], Gul, Sohrabi, and Ushakov generalized Matthews result by considering the relation between the conjugacy problem in F/N and the power problem in F/N', where F is a free group with a normal subgroup N and N' is its derived subgroup. They show that $\mathrm{CP}(F/N')$ is polynomial-time-Turing-reducible to $\mathrm{CSMP}(F/N)$ and $\mathrm{CSMP}(F/N)$ is Turing-reducible to $\mathrm{CP}(F/N')$. Moreover, they establish that $\mathrm{WP}(F/N')$ is polynomial-time-Turing-reducible to $\mathrm{WP}(F/N)$.

Question 4. What are the precise relations in terms of complexity between $\mathrm{CP}(F/N')$ and $\mathrm{CSMP}(F/N)$ resp. $\mathrm{WP}(F/N')$ and $\mathrm{WP}(F/N)$?

References

1. Barrington, D.A.M., Immerman, N., Straubing, H.: On uniformity within NC^1. J. Comput. Syst. Sci. **41**(3), 274–306 (1990)
2. Dehn, M.: Über unendliche diskontinuierliche Gruppen. Math. Ann. **71**(1), 116–144 (1911)
3. Diekert, V., Myasnikov, A.G., Weiß, A.: Conjugacy in Baumslag's Group, generic case complexity, and division in power circuits. In: LATIN Symposium, pp. 1–12 (2014)
4. Grigoriev, D., Shpilrain, V.: Authentication from matrix conjugation. Groups Complex. Cryptology **1**, 199–205 (2009)
5. Gul, F., Sohrabi, M., Ushakov, A.: Magnus embedding and algorithmic properties of groups $F/N^{(d)}$. ArXiv e-prints, abs/1501.01001, January 2015
6. Hesse, W.: Division is in uniform TC^0. In: Orejas, F., Spirakis, P.G., Leeuwen, J. (eds.) ICALP 2001. LNCS, vol. 2076, pp. 104–114. Springer, Heidelberg (2001). doi:10.1007/3-540-48224-5_9
7. Hesse, W., Allender, E., Barrington, D.A.M.: Uniform constant-depth threshold circuits for division and iterated multiplication. JCSS **65**, 695–716 (2002)
8. Kargapolov, M.I., Remeslennikov, V.N.: The conjugacy problem for free solvable groups. Algebra i Logika Sem. **5**(6), 15–25 (1966)
9. Ko, K.H., Lee, S.J., Cheon, J.H., Han, J.W., Kang, J., Park, C.: New public-key cryptosystem using braid groups. In: Bellare, M. (ed.) CRYPTO 2000. LNCS, vol. 1880, pp. 166–183. Springer, Heidelberg (2000). doi:10.1007/3-540-44598-6_10
10. König, D., Lohrey, M.: Evaluating matrix circuits. CoRR, abs/1502.03540 (2015)
11. Krebs, A., Lange, K., Reifferscheid, S.: Characterizing TC^0 in terms of infinite groups. Theory Comput. Syst. **40**(4), 303–325 (2007)
12. Lange, K.-J., McKenzie, P.: On the complexity of free monoid morphisms. In: Chwa, K.-Y., Ibarra, O.H. (eds.) ISAAC 1998. LNCS, vol. 1533, pp. 247–256. Springer, Heidelberg (1998). doi:10.1007/3-540-49381-6_27
13. Maciel, A., Thérien, D.: Threshold circuits of small majority-depth. Inf. Comput. **146**(1), 55–83 (1998)
14. Magnus, W.: On a theorem of Marshall Hall. Ann. Math. **40**, 764–768 (1939)
15. Matthews, J.: The conjugacy problem in wreath products and free metabelian groups. Trans. Am. Math Soc. **121**, 329–339 (1966)
16. Miller III, C.F.: On group-theoretic decision problems and their classification, vol. 68. Annals of Mathematics Studies. Princeton University Press (1971)
17. Myasnikov, A., Roman'kov, V., Ushakov, A., Vershik, A.: The word, geodesic problems in free solvable groups. Trans. Amer. Math. Soc. **362**(9), 4655–4682 (2010)
18. Myasnikov, A.G., Vassileva, S., Weiß, A.: Log-space complexity of the conjugacy problem in wreath products. Groups Complex. Cryptol. (2017, to appear)
19. Miasnikov, A., Vassileva, S., Weiß, A.: The conjugacy problem in free solvable groups and wreath product of abelian groups is in TC^0. ArXiv e-prints, abs/1612.05954 (2016)
20. Remeslennikov, V., Sokolov, V.G.: Certain properties of the Magnus embedding. Algebra i logika **9**(5), 566–578 (1970)
21. Robinson,D.: Parallel Algorithms for Group Word Problems. PhD thesis, University of California, San Diego (1993)
22. Shpilrain, V., Zapata, G.: Combinatorial group theory and public key cryptography. Appl. Algebra Engrg. Comm. Comput. **17**, 291–302 (2006)

23. Vassileva, S.: Polynomial time conjugacy in wreath products and free solvable groups. Groups Complex. Cryptol. **3**(1), 105–120 (2011)
24. Vollmer, H.: Introduction to Circuit Complexity. Springer, Berlin (1999)
25. Waack, S.: The parallel complexity of some constructions in combinatorial group theory (abstract). In: Rovan, B. (ed.) MFCS 1990. LNCS, vol. 452, pp. 492–498. Springer, Heidelberg (1990). doi:10.1007/BFb0029647
26. Wang, L., Wang, L., Cao, Z., Okamoto, E., Shao, J.: New constructions of public-key encryption schemes from conjugacy search problems. In: Lai, X., Yung, M., Lin, D. (eds.) Inscrypt 2010. LNCS, vol. 6584, pp. 1–17. Springer, Heidelberg (2011). doi:10.1007/978-3-642-21518-6_1
27. Weiß, A.: A logspace solution to the word and conjugacy problem of generalized Baumslag-Solitar groups. In: Algebra and computer science, vol. 677. Contemporary Mathematics, pp. 185–212. American Mathematical Society, Providence, RI (2016)

On Algorithmic Statistics for Space-Bounded Algorithms

Alexey Milovanov[1,2,3]($\boxtimes$)

[1] National Research University Higher School of Economics, Moscow, Russia
almas239@gmail.com
[2] Moscow Institute of Physics and Technology, Dolgoprudny, Russia
[3] Moscow State University, Moscow, Russia

Abstract. Algorithmic statistics studies explanations of observed data that are good in the algorithmic sense: an explanation should be simple i.e. should have small Kolmogorov complexity and capture all the algorithmically discoverable regularities in the data. However this idea can not be used in practice because Kolmogorov complexity is not computable.

In this paper we develop algorithmic statistics using space-bounded Kolmogorov complexity. We prove an analogue of one of the main result of 'classic' algorithmic statistics (about the connection between optimality and randomness deficiences). The main tool of our proof is the Nisan-Wigderson generator.

Keywords: Algorithmic statistics · Kolmogorov complexity · Nisan-Wigderson generator · Computational complexity theory · Derandomization

1 Introduction

In this section we give an introduction to algorithmic statistics and present our results.

We consider strings over the binary alphabet $\{0, 1\}$. We use $|x|$ to denote the length of a string x. All of the logarithms are base 2. Denote the conditional Kolmogorov complexity[1] of x given y by $C(x|y)$.

1.1 Introduction to Algorithmic Statistics

Let x be some observation data encoded as a binary string, we need to find a suitable explanation for it. An explanation (=model) is a finite set containing x. More specifically we want to find a simple model A such that x is a typical element in A. How to formalize that A is 'simple' and x is a 'typical element' in A? In classical algorithmic statistics a set A is called simple if it has small

---

[1] The definition and basic properties of Kolmogorov complexity can be found in the textbooks [5,13], for a short survey see [11].

© Springer International Publishing AG 2017
P. Weil (Ed.): CSR 2017, LNCS 10304, pp. 232–244, 2017.
DOI: 10.1007/978-3-319-58747-9_21

Kolmogorov complexity $C(A)$[2]. To measure typicality of x in A one can use the *randomness deficiency* of x as an element of A:

$$d(x|A) := \log|A| - C(x|A).$$

The randomness deficiency is always non-negative with $O(\log|x|)$ accuracy, as we can find x from A and the index of x in A. For most elements x in any set A the randomness deficiency of x in A is negligible. More specifically, the fraction of x in A with randomness deficiency greater than β is less than $2^{-\beta}$.

There is another quantity measuring the quality of A as an explanation of x: the *optimality deficiency*:

$$\delta(x, A) := C(A) + \log|A| - C(x).$$

It is also non-negative with logarithmic accuracy (by the same reason). This value represents the following idea: a good explanation (a set) should not only be simple but also should be small.

One can ask: why as explanations we consider only sets—not general probability distributions? This is because for every string x and for every distribution P there exists a set $A \ni x$ explaining x that is not worse than P in the sense of deficiencies defined above[3].

Theorem 1 ([15]). *For every string x and for every distribution P there exists a set $A \ni x$ such that $C(A|P) \leq O(\log|x|)$ and $\frac{1}{|A|} \geq \frac{1}{2}P(x)$.*

Kolmogorov called a string x *stochastic* if there exists a set $A \ni x$ such that $C(A) \approx 0$ and $d(x|A) \approx 0$. The last equality means that $\log|A| \approx C(x|A)$ hence $\log|A| \approx C(x)$ because $C(A) \approx 0$. So, $\delta(x, A)$ is also small.

For example, an incompressible string of length n (i.e. a string whose complexity is close to n) is stochastic—the corresponding set is $\{0, 1\}^n$. Non-stochastic objects also exist, however this fact is more complicated—see [12, 15].

1.2   Space-Bounded Algorithmic Statistics

As mentioned by Kolmogorov in [4], the notion of Kolmogorov complexity $C(x)$ has the following minor point. It ignores time and space needed to produce x from its short description. This minor point can be fixed by introducing space or time bounded Kolmogorov complexity (see, for example, [2] or [14]). In this paper we consider algorithms whose space (not time) is bounded by a polynomial of the length of a string.

The distinguishing complexity of a string x with space bound m is defined as the minimal length of a program p such that

---

[2] Kolmogorov complexity of A is defined as follows. We fix any computable bijection $A \mapsto [A]$ from the family of finite sets to the set of binary strings, called *encoding*. Then we define $C(A)$ as the complexity $C([A])$ of the code $[A]$ of A.

[3] The randomness deficiency of a string x with respect to a distribution P is defined as $d(x|P) := -\log P(x) - C(x|P)$, the optimality deficiency is defined as $\delta(x, P) := C(P) - \log P(x) - C(x)$.

- $p(y) = 1$ if $y = x$;
- $p(y) = 0$ if $y \neq x$;
- p uses at most m bits of memory on every input.

We denote this value by $\mathrm{CD}^m(x)$. If for some x and m such a program p does not exist then $\mathrm{CD}^m(x) := \infty$. We say that p *distinguishes* x (from other strings) if p satisfies the first and the second requirements of the definition.

In this definition $p(y)$ denotes $V(p, y)$ for a universal Turing machine V. A Turing machine is called universal if for every machine U and for every q there exists p such that $V(p, y) = U(q, y)$ for every y, $|p| < |q| + O(1)$ and V uses space at most $O(m)$ if U uses space m on input (q, y). Here the constant in $O(m)$ depends on V and U but does not depend on q[4].

Now we extend this notion to arbitrary finite sets. The distinguishing complexity of a set A with space bound m is defined as the minimal length of a program p such that

- $p(y) = 1$ if $y \in A$;
- $p(y) = 0$ if $y \notin A$;
- p uses space m on every input.

Denote this value as $\mathrm{CD}^m(A)$.

The value $\mathrm{CD}^a(x|A)$ is defined as the minimal length of a program that distinguishes x by using space at most m and uses A as an oracle. The value $\mathrm{CD}^a(B|A)$ for an arbitrary finite set B is defined the same way.

How to define typicality of a string x in a set A? Consider the following resource-bounded versions of randomness and optimal deficiencies:

$$d^a(x|A) := \log|A| - \mathrm{CD}^a(x|A),$$

$$\delta^{b,d}(x, A) := \mathrm{CD}^b(A) + \log|A| - \mathrm{CD}^d(x).$$

One can show that these values are non-negative (with logarithmic accuracy) provided $a \geq p(|x|)$ and $d \geq p(|x| + b)$ for a large enough polynomial p.

We say that a set A is a good explanation for a string x (that belongs to A) if $\mathrm{CD}^r(A) \approx 0$ (with $O(\log|x|)$ accuracy) and $\log|A| \approx \mathrm{CD}^m(x)$. Here r and m are some small numbers. For such A the values $d^m(x|A)$ and $\delta^{r,m}(x, A)$ are small.

It turns out that every string has a good explanation. Indeed, let x be a string such that $\mathrm{CD}^m(x) = k$. Define a set $A \ni x$ as $\{y \mid \mathrm{CD}^m(y) \leq k\}$. The log-size of this set is equal to k up to a non-negative constant and hence $\log|A| = \mathrm{CD}^m(x)$. Note that A can be distinguished by a program of length $O(\log(k + m))$ that uses $\mathrm{poly}(m)$ space.

So, for space-bounded algorithms all strings have good explanations (in other words, they are stochastic).

---

[4] Such an universal machine does exist – see [5].

1.3   Distributions and Sets

Recall that in the classical algorithmic statistics for every distribution P and every x there is a finite set $A \ni x$ that is not worse than P as an explanation for x. It turns out that this the case also for space-bounded algorithmic statistics (otherwise we could not restrict ourselves to finite sets).

   Before we formulate this result we give a definition of the complexity of a probability distribution P with space bound m that is denoted by $C^m(P)$. This value is defined as the minimal length of a program p without input and with the following two properties. First, for every x the probability of the event $[x$ output by $p]$ is equal to $P(x)$. Second, p uses space at most m (always). If such a program does not exist then $C^m(P) := \infty$.

Theorem 2. *There exist a polynomial r and a constant c such that for every string x, for every distribution P and for every m there exists a set $A \ni x$ such that $CD^{r(m+n)}(A) \leq C^m(P) + c\log(n+m)$ and $\frac{1}{|A|} \geq P(x)2^{-c\log n}$. Here n is length of x.*

The main tool of the proof of Theorem 2 is the theorem of Nisan "$RL \subseteq SC$", more precisely its generalization—Theorem 1.2 in [8].

1.4   Descriptions of Restricted Type

So far we considered arbitrary finite sets (or more general distributions) as models (statistical hypotheses). We have seen that for such class of hypotheses the theory becomes trivial. However, in practice we usually have some a priori information about the data. We know that the data was obtained by sampling with respect to an unknown probability distribution from a known family of distributions. For simplicity we will consider only uniform distributions i.e. a family of finite sets $\mathcal{A}$.

   For example, we can consider the family of all Hamming balls as $\mathcal{A}$. (That means we know a priory that our string was obtain by flipping certain number of bits in an unknown string.) Or we may consider the family that consists of all 'cylinders': for every n and for every string u of length at most n we consider the set of all n-bit strings that have prefix u. It turns out that for the second family there exists a string that has no good explanations in this family: the concatenation of an incompressible string (i.e. a string whose Kolmogorov complexity is close to its length) and all zero string of the same length. (We omit the rigorous formulation and the proof.)

   Restricting the class of allowed hypotheses was initiated in [16]. It turns out that there exists a direct connection between randomness and optimality deficiencies in the case when a family is enumerable.

Theorem 3 ([16]). *Let $\mathcal{A}$ be an enumerable family of sets. Assume that every set from $\mathcal{A}$ consists of strings of the same length. Let x be a string of length n contained in $A \in \mathcal{A}$. Then:*

(a) $d(x|A) \leq \delta(x, A) + O(\log(C(A) + n))$.

(b) *There exists $B \in A$ containing x such that:*

$$\delta(x, B) \leq d(x|A) + O(\log(C(A) + n)).$$

In our paper we will consider families with the following properties:

- Every set from $\mathcal{A}$ consists of strings of the same length. The family of all subsets of $\{0, 1\}^n$ that belong to $\mathcal{A}$ is denoted by $\mathcal{A}_n$.
- There exists a polynomial p such that $|\mathcal{A}_n| \leq 2^{p(n)}$ for every n.
- There exists an algorithm enumerating all sets from $\mathcal{A}_n$ in space poly(n).

The last requirement means the following. There exists an indexing of $\mathcal{A}_n$ and a Turing machine M that for a pair of integers $(n; i)$ and a string x in the input outputs 1 if x belongs to i-th set of $\mathcal{A}_n$ and 0 otherwise. On every such input M uses at most poly(n) space.

Any family of finite sets of strings that satisfies these three conditions is called *acceptable*. For example, the family of all Hamming balls is acceptable. Our main result is the following analogue of Theorem 3.

Theorem 4. (a) *There exist a polynomial p and a constant c such that for every set $A \ni x$ and for every m the following inequality holds*

$$d^m(x|A) \leq \delta^{m,p}(x, A) + c \log(C^m(A)).$$

Here $p = p(m + n)$ and n is the length of x.

(b) *For every acceptable family of sets $\mathcal{A}$ there exists a polynomial p such that the following property holds. For every $A \in \mathcal{A}$, for every $x \in A$ and for every integer m there exists a set $B \ni x$ from $\mathcal{A}$ such that*

- $\log |B| \leq \log |A| + 1$;
- $CD^s(B) \leq CD^m(A) - CD^s(A|x) + O(\log(n + m))$.

Here $s = p(m + n)$ and n is the length of x.

A skeptical reader would say that an analogue of Theorem 3(b) should has the following form (and we completely agree with him/her).

Hypothesis 1. *There exist a polynomial p and a constant c such that for every set $A \ni x$ from $\mathcal{A}$ and for every m there exists a set $B \in \mathcal{A}$ such that*

$$\delta^{p,m}(x, B) \leq d^p(x|A) + c \log(n + m).$$

Here $p = p(m + n)$, n is the length of x and $\mathcal{A}$ is an acceptable family of sets.

We argue in Subsect. 2.1 why Theorem 4(b) is close to Hypothesis 1.

2   Proof of Theorem 4

Proof (of Theorem 4(a)). The inequality we have to prove means the following

$$\mathrm{CD}^p(x) \leq \mathrm{CD}^m(x|A) + \mathrm{CD}'''(A) + c\log(\mathrm{CD}^m(A) + n)$$

(by the definitions of optimality and randomness deficiencies).

Consider a program p of length $\mathrm{CD}^m(x|A)$ that distinguishes x and uses A as an oracle. We need to construct a program that also distinguishes x but does not use any oracle. For this add to p a procedure distinguishing A. There exists such a procedure of length $\mathrm{CD}^m(A)$. So, we get a program of the length that we want (additional $O(\log(\mathrm{CD}^m(A)))$ bits are used for pair coding) that uses $\mathrm{poly}(m)$ space.

So, for every x and $A \ni x$ the randomness deficiency is not greater than the optimal deficiency. The following example shows that the difference can be large.

Example 1. Consider an incompressible string x of length n, so $C(x) = n$ (this equality as well as further ones holds with logarithmic precision). Let y be n-bit string that is also incompressible and independent of x, i.e. $C(y|x) = n$. By symmetry of information (see [5,13]) we get $C(x|y) = n$.

Define $A := \{0,1\}^n \backslash \{y\}$. The randomness deficiency of x in A (without resource restrictions) is equal to 0. Hence, this is true for any resource restrictions ($C(x|A)$ is not greater than $\mathrm{CD}^m(x|A)$ for every m). Hence, for any m we have $d^m(x|A) = 0$. On the other hand $\delta_m^p(x, A) = n$ for all p and large enough m. Indeed, take $m = \mathrm{poly}(n)$ such that $\mathrm{CD}^m(x) = n$. Since $C(A) = n$ we have $\mathrm{CD}^q(A) = n$ for every q.

So, we can not just let $A = B$ in Hypothesis 1. In some cases we have to 'improve' A (in the example above we can take $\{0,1\}^n$ as an improved set).

2.1   Sketch of Proof of Theorem 3(b)

The proof of Theorem 4(b) is similar to the proof of Theorem 3(b). Therefore we present the sketch of the proof of Theorem 3(b).

Theorem 3 states that there exists a set $B \in \mathcal{A}$ containing x such that $\delta(x|B) \leq d(x, A)$. (Here and later we omit terms of logarithmic order.) First we derive it from the following statement.

(1) There exists a set $B \in \mathcal{A}$ containing x such that $|B| \leq 2 \cdot |A|$ and $C(B) \leq C(A) - C(A|x)$.

For such B the $\delta(x|B) \leq d(x, A)$ easily follows from the inequality $C(A) - C(A|x) - C(x) \leq -C(x|A)$. The latter inequality holds by symmetry of information.

To prove (1) note that

(2) there exist at least $2^{C(A|x)}$ sets in $\mathcal{A}$ containing x whose complexity and size are at most $C(A)$ and $2 \cdot |A|$, respectively.

Indeed, knowing x we can enumerate all sets from $\mathcal{A}$ containing x whose parameters (complexity and size) are not worse than the parameters of A. Since we can describe A by its ordinal number in this enumeration we conclude that the length of this number is at least $C(A|x)$ (with logarithmic precision).

Now (1) follows from the following statement.

(3) Assume that $\mathcal{A}$ contains at least 2^k sets of complexity at most i and size at most 2^j containing x. Then one of them has complexity at most $i - k$.

(We will apply it to $i = C(A)$, $j = \lceil \log |A| \rceil$ and $k = C(A|x)$.)

So, Theorem 4(b) is an analogue of (2). Despite there is an analogue of symmetry of information for space-bounded algorithms (see [6] and Appendix) Hypothesis 1 does not follow Theorem 4(b) directly. (There is some problem with quantifiers.)

Proof of (3) is the main part of the proof of Theorem 3, the same thing holds for Theorem 4.

In the next subsection we derive Theorem 4(b) from Lemma 1 (this is an analogue of the third statement). In the proof of Lemma 1 we use the Nisan-Wigderson generator.

2.2   Main Lemma

We will derive Theorem 4(b) from the following

Lemma 1. *For every acceptable family of sets $\mathcal{A}$ there exist a polynomial p and a constant c such that the following statement holds for every j.*

Assume that a string x of length n belongs to 2^k sets from $\mathcal{A}_n$. Assume also that every of these sets has cardinality at most 2^j and space-bounded by m complexity at most i. Then one of this set is space-bounded by M complexity at most $i - k + c \log(n + m)$. Here $M = m + p(n)$.

Proof (Theorem 4(b) from Lemma 1).

Denote by $\mathcal{A}'$ the family of all sets in $\mathcal{A}_n$ containing x whose parameters are not worse than those of A.

$$\mathcal{A}' := \{A' \in \mathcal{A}_n \mid x \in A', \mathrm{CD}^m(A) \le \mathrm{CD}^m(A'), \log |A'| \le \lfloor \log |A| \rfloor \}.$$

Let $k = \log \mathcal{A}'$.

We will describe A in $k + O(\log(n + m))$ bits when x is known. The sets in $\mathcal{A}'$ (more specifically, their programs) can be enumerated if n, m and $\log |A|$ are known. This enumeration can be done in space $\mathrm{poly}(m + n)$. We can describe A by its ordinal number of this enumeration, so

$$\mathrm{CD}^s(A|x) \le k + O(\log(n + m)).$$

Here $s = \mathrm{poly}(m + n)$.

Theorem 4(b) follows from Lemma 1 for $i = \mathrm{CD}^m(A)$ and $j = \lfloor \log |A| \rfloor$.

2.3   Nisan-Wigderson Generator. Proof of the Main Lemma

Define

$$\mathcal{A}_{n,m}^{i,j} := \{A' \in \mathcal{A}_n \mid \mathrm{CD}^m(A') \leq i, \log|A'| \leq j\}$$

for an acceptable family of sets $\mathcal{A}$.

Define a probability distribution $\mathcal{B}$ as follows. Every set from $\mathcal{A}_{n,m}^{i,j}$ belongs to $\mathcal{B}$ with probability $2^{-k}(n+2)\ln 2$ independently.

We claim that $\mathcal{B}$ satisfies the following two properties with high probability.
(1) The cardinality of $\mathcal{B}$ is at most $2^{i-k+2} \cdot (n+k)^2 \ln 2$.
(2) If a string of length n is contained in at least 2^k sets from $\mathcal{A}_{n,m}^{i,j}$ then one of these sets belongs to $\mathcal{B}$.

Lemma 2. *The family $\mathcal{B}$ satisfies the properties* (1) *and* (2) *with probability at least* $\frac{1}{2}$.

Proof. Show that $\mathcal{B}$ satisfies every of these two properties with probability at least $\frac{3}{4}$.

For (1) it follows from Markov's inequality: the cardinality of $\mathcal{B}$ exceeds the expectation by a factor of 4 with probability less than $\frac{1}{4}$. (Of course we can get a rather more stronger estimation.)

To prove it for (2) consider a string of length n that belongs to at least 2^k sets from $\mathcal{A}_{n,m}^{i,j}$. The probability of the event [every of these 2^k sets does not belong to $\mathcal{B}$] is at most

$$(1 - 2^{-k}(n+2)\ln 2)^{2^k} \leq 2^{-n-2} \text{ (since } 1 - x \leq e^{-x}).$$

The probability of the sum of such events for all strings of length n is at most $2^n 2^{-n-2} = \frac{1}{4}$.

Using Lemma 2 we can prove existence of a required set whose *unbounded* complexity is at most $i - k + O(\log(n+m))$. Indeed, by Lemma 2 *there exists* a subfamily that satisfies the properties (1) and (2). The lexicographically first such family has small complexity—we need only know i, k, n and m to describe it. Note, that k and i are bounded by poly(n): since $\mathcal{A}$ is acceptable $\log|\mathcal{A}_n| =$ poly(n) and hence k is not greater than poly(n). We can enumerate all sets from $\mathcal{A}_n$, so space-bounded complexity of every element of $\mathcal{A}_n$ (in particular, i) is bounded by polynomial in n. Now we can describe a required set as the ordinal number of an enumeration of this subfamily.

However, this method is not suitable for the polynomial space-bounded complexity: the brute-force search for the finding a suitable subfamily uses too much space (exponential). To reduce it we will use the Nisan-Wigderson generator. The same idea was used in [7].

Theorem 5 ([9,10]). *For every constant d and for every positive polynomial $q(m)$ there exists a sequence of functions $G_m : \{0,1\}^f \rightarrow \{0,1\}^m$ where $f = O(\log^{2d+6} m)$ such that:*

– *Function G_m is computable in space poly(f);*
– *For every family of circuits C_n of size $q(v)$ and depth d and for large enough n it holds that:*

$$| \Pr_x[C_m(G_m(x)) = 1] - \Pr_y[C_m(y) = 1]| < \frac{1}{m},$$

where x is distributed uniformly in $\{0,1\}^f$, and y is distributed uniformly in $\{0,1\}^m$.

We will use this theorem for $m = 2^{i+n}$. Then f is a polynomial in $i + n$ (if d is a constant), hence $f = \text{poly}(n)$. Every element whose complexity is at most i corresponds to a string of length i in the natural way. So, we can assign subfamilies of $\mathcal{A}_{n,m}^{i,j}$ to strings of length m.

Assume that there exists a circuit of size $2^{O(n)}$ and constant depth that inputs a subfamily of $\mathcal{A}_n^{i,j}$ and outputs 1 if this subfamily satisfies properties (1) and (2) from Lemma 2, and 0 otherwise. First we prove Lemma 1 using this assumption.

Compute $G_m(y)$ for all strings y of length f until we find a suitable one, i.e. whose image satisfies our two properties. Such a string exists by Lemma 2, Theorem 5 and our assumption. Note that we can find the lexicographically first suitable string by using space $m + \text{poly}(n)$, so *bounded by space $m + \text{poly}(n)$* the complexity of this string is equal to $O(\log(n + m))$.

So, if we can construct a constant depth circuit of the needed size that verifies properties (1) and (2) then we are happy. Unfortunately we do not know how to construct such a circuit verifying the first property (there exist problems with a computation of threshold functions by constant-depth circuits—see [3]). However, we know the following result.

Theorem 6 ([1]). *For every t there exists a circuit of constant depth and poly(t) size that inputs binary strings of length t and outputs 1 if an input has at most $\log^2 t$ ones and 0 otherwise.*

To use this theorem we make a little change of the first property. Divide $\mathcal{A}_n^{i,j}$ into 2^{i-k} parts of size 2^k. The corrected property is the following.

(1)* The family of sets $\mathcal{B}$ contains at most $(n + k)^2$ sets from each of these parts.

Lemma 3. *The family of sets $\mathcal{B}$ satisfies properties (1)* and (2) with probability at least $\frac{1}{3}$.*

The proof of this lemma is not difficult but uses cumbersome formulas. We present the proof of Lemma 3 in Appendix.

Proof (of Lemma 1). It is clear that property (1)* implies property (1). Hence by using Lemma 1 and the discussion above, it is enough to show that properties (1)* and (2) can be verified by constant depth circuits of size $2^{O(i+n)}$.

Such a circuit exists for property (1)* by Theorem 6.

The second property can be verified by the following 2-depth circuit. For every string of length n containing in 2^k sets from $\mathcal{A}_n^{i,j}$ there exists a corresponding disjunct. All of these disjuncts go to a conjunction gate.

3   Proof of Theorem 2

Theorem 2 would have an easy proof if a program that corresponds to a distribution P could use only poly(n) random bits. Indeed, in such case we can run a program with all possible random bits and so calculate $P(x)$ for every x in polynomial space. Hence, we can describe A as the set of all strings whose the probability of output is at least 2^{-k}, where $2^{-k} \geq P(x) > 2^{-k-1}$.

In the general case (when the number of random bits is exponentially large) we will use the following theorem.

Theorem 7 ([8]). *Let f be a probabilistic program, that uses at most $r(n)$ space on inputs of length n for some polynomial r. Assume that f always outputs 0 or 1 (in particular, f never loops). Then there exists a deterministic program $\widehat{f}$ with the following properties:*

(a) $\widehat{f}$ *uses at most $r^2(n)$ space on inputs of length n;*
(b) *if $Pr[f(x) = 1] > \frac{2}{3}$ then $\widehat{f}(x) = 1$. If $Pr[f(x) = 1] < \frac{1}{3}$ then $\widehat{f}(x) = 0$;*
(c) $|\widehat{f}| \leq |f| + O(1)$.[5]

Proof (of Theorem 2). If the complexity of distribution P (bounded by space m) is equal to infinity then we can take $\{x\}$ as A.

Else P can be specified by a program g. Consider the integer k such that: $2^{-k+1} \geq P(x) \geq 2^{-k}$. We can assume that k is not greater than n—the length of x—else we can take $\{0,1\}^n$ as A.

Note, that we can find a good approximation for $P(y)$ running g exponentially times.

More accurately, let us run g for 2^{100k^2} times. For every string y denote by $\omega(y)$ the frequency of output of y. The following inequality holds by Hoeffding's inequality

$$Pr[|w(y) - P(y)| > 2^{-k-10}] < \frac{1}{3}.$$

Hence by using program g we can construct a program f that uses poly(n) space (on inputs of length n) such that

(1) if $P(y) > 2^{-k-1}$ and $|y| = n$ then $Pr[f(y) = 1] > \frac{2}{3}$;
(2) if $P(y) < 2^{-k-2}$ then $Pr[f(y) = 0] > \frac{2}{3}$.

---

[5] Theorem 1.2 in [8] has another formulation: it does not contain any information about $|\widehat{f}|$. However, from the proof of the theorem it follows that a needed program (denote it as $\widehat{f}_1$) is got from f by using an algorithmic transformation. Therefore there exists a program $\widehat{f}$ that works functionally like $\widehat{f}_1$ such that $|\widehat{f}| \leq |f| + O(1)$.

Also, Theorem 1.2 does not assume that $Pr[f(x)]$ can belong to $[\frac{1}{3}; \frac{2}{3}]$. However, this assumption does not used in the proof of Theorem 1.2.

Now using Theorem 7 for f we get a program $\widehat{f}$ such that $|\widehat{f}| \leq |g| + O(\log n)$. By the first property of f we get $\widehat{f}(x) = 1$. From the second property it follows that the cardinality of the set $\{y \mid \widehat{f}(y) = 1\}$ is not greater than 2^{k+2}. So, this set satisfies the requirements of the theorem.

Remark 1. Another proof of Theorem 2 was done by Ricky Demer at Stackexchange – http://cstheory.stackexchange.com/questions/34896/can-every-distribution-producible-by-a-probabilistic-pspace-machine-be-produced.

Open question

Does Hypothesis 1 hold?

Acknowledgments. I would like to thank Nikolay Vereshchagin and Alexander Shen for useful discussions, advice and remarks.

This work is supported by RFBR grant 16-01-00362 and supported in part by Young Russian Mathematics award and RaCAF ANR-15-CE40-0016-01 grant. The study has been funded by the Russian Academic Excellence Project '5-100'.

Appendix

Symmetry of Information

Define $CD^m(A, B)$ as the minimal length of a program that inputs a pair of strings (a, b) and outputs a pair of boolean values $(a \in A, b \in B)$ using space at most m for every input.

Lemma 4 (Symmetry of information). *Assume $A, B \subseteq \{0,1\}^n$. Then*

(a) $\forall m \ \ CD^p(A, B) \leq CD^m(A) + CD^m(B|A) + O(\log(CD^m(A, B) + m + n))$

for $p = m + poly(n + CD^m(A, B))$.

(b) $\forall m \ \ CD^p(A) + CD^p(B|A) \leq CD^m(A, B) + O(\log(CD^m(A, B) + m + n))$

for $p = 2m + poly(n + CD^m(A, B))$.

Proof (of Lemma 4(a)). The proof is similar to the proof of Theorem 4(a).

Proof (of Lemma 4(b)). Let $k := CD^m(A, B)$. Denote by $\mathcal{D}$ the family of sets (U, V) such that $CD^m(U, V) \leq k$ and $U, V \subseteq \{0,1\}^n$. It is clear that $|\mathcal{D}| < 2^{k+1}$. Denote by $\mathcal{D}_A$ the pairs of $\mathcal{D}$ whose the first element is equal to A. Let t satisfy the inequalities $2^t \leq |\mathcal{D}_A| < 2^{t+1}$.

Let us prove that

- $CD^p(B|A)$ does not exceed t significantly;
- $CD^p(A)$ does not exceed $k - t$ significantly.

Here $p = m + O(n)$.

We start with the first statement. There exists a program that enumerates all sets from $\mathcal{D}_A$ using A as an oracle and that works in space $2m + O(n)$. Indeed, such enumeration can be done in the following way: enumerate all programs of length k and verify the following condition for every pair of n-bit strings. First, a program uses at most m space on this input. Second, if a second n-bit string belongs to A then the program outputs 1, and 0 otherwise. Since some program loops we need additional $m + O(n)$ space to take it into account.

Append to this program the ordinal number of a program that distinguishes (A, B). This number is not greater than $t + 1$. Therefore we have $\mathrm{CD}^p(B|A) \leq t + O(\log(\mathrm{CD}^m(A, B) + m + n))$.

Now let us prove the second statement. Note that there exist at most 2^{k-t+1} sets U such that $|\mathcal{D}_U| \geq 2^t$ (including A). Hence, if we construct a program that enumerates all sets with such property (and does not use much space) then we will win—the set A can be described by the ordinal number of this enumeration.

Let us construct such a program. It works as follows:

enumerate all sets U that are the first elements from $\mathcal{D}$, i.e. we enumerate programs that distinguish the corresponding sets (say, lexicographically). We go to the next step if the following properties holds. First, $|\mathcal{D}_U| \geq 2^t$, and second: we did not meet set U earlier (i.e. every program whose the lexicographical number is smaller does not distinguish U or is not the first element from a set from $\mathcal{D}$).

This program works in $2m + \mathrm{poly}(n + \mathrm{CD}^m(A, B))$ space (that we want) and has length $O(\log(\mathrm{CD}^m(A) + n + m))$.

Proof (of Lemma 3). Let us show that B satisfies property $(1)^*$ with probability at most 2^{-n}. Since B satisfies property (2) with probability at most $\frac{1}{4}$ (see the proof of Lemma 2) it would be enough for us.

For this let us show that every part is 'bad' (i.e. has at least $(n+k)^2 + 1$ sets from B) with probability at most 2^{-2n}. The probability of such event is equal to the probability of the following event: a binomial random variable with parameters $(2^k, 2^{-k}(n+2)\ln 2)$ is greater than $(n+k)^2$. To get the needed upper bound for this probability is not difficult however the correspondent formulas are cumbersome. Take $w := 2^k$, $p := 2^{-k}(n + 2)\ln 2$ and $v := (n + k)^2$. We need to estimate

$$\sum_{i=v}^{w} \binom{w}{i} p^i (1-p)^{w-i} < w \cdot \binom{w}{v} p^v (1-p)^{w-v} < w \cdot \binom{w}{v} p^v < w \frac{(wp)^v}{v!}.$$

The first inequality holds since $wp = (n+2)\ln 2 \leq (n+k)^2 = v$. Now note that $wp = (n+2)\ln 2 < 10n$. So

$$w \frac{(wp)^v}{v!} < \frac{2^k (10n)^{(n+k)^2}}{((n+k)^2)!} \ll 2^{-2n}.$$

References

1. Ajtai, M.: Approximate counting with uniform constant-depth circuits. In: Advanced in Computational Complexity Theory, pp. 1–20. American Mathematical Society (1993)
2. Buhrman, H., Fortnow, L., Laplante, S.: Resource-Bounded Kolmogorov complexity revisited. SIAM J. Comput. **31**(3), 887–905 (2002)
3. Furst, M., Saxe, J.B., Sipser, M.: Math. Syst. Theory **17**(1), 13–27 (1984)
4. Kolmogorov, A.N.: Approaches, three approaches to the quantitative definition of information. Problems Inf. Transmission **1**(1), 4–11 (1965). English translation published in Int. J. Comput. Math. **2**, 157–168 (1968)
5. Li, P., Vitányi, P.: An Introduction to Kolmogorov Complexity and Its Applications, 3rd edn, p. 792. Springer, Heidelberg (1993). 1st edn. 1993; 2nd edn. 1997
6. Longpré, L.: Resource bounded kolmogorov complexity, a link between computational complexity and information theory. Ph. D. Thesis, Cornell University, Ithaca, NY (1986)
7. Musatov, D.: Improving the space-bounded version of muchnik's conditional complexity theorem via "naive" derandomization. Theory Comput. Syst. **55**(2), 299–312 (2014)
8. Nisan, N.: $RL \subseteq SC$. J. Comput. Complex. **4**, 1–11 (1994)
9. Nisan, N.: Pseudorandom bits for constant depth circuits. Combinatorica **11**, 63–70 (1991)
10. Nisan, N., Wigderson, A.: Hardness vs randomness. J. Comput. Syst. Sci. **49**(2), 149–167 (1994)
11. Shen, A., Kolmogorov, A.: Around kolmogorov complexity: basic notions and results. In: Vovk, V., Papadoupoulos, H., Gammerman, A. (eds.) Measures of Complexity: Festschrift for Alexey Chervonenkis. Springer, Heidelberg (2015)
12. Shen, A.: The concept of (α, β)-stochasticity in the Kolmogorov sense, and its properties. Sov. Math. Doklady **271**(1), 295–299 (1983)
13. Shen, A., Uspensky, V., Vereshchagin, N.: Kolmogorov complexity and algorithmic randomness. In: MCCME 2013 (Russian). English translation http://www.lirmm.fr/~ashen/kolmbook-eng.pdf
14. Sipser, M.: A complexity theoretic approach to randomness. In: Proceedings of the 15th ACM Symposium on the Theory of Computing, pp. 330–335 (1983)
15. Vereshchagin, N., Vitányi, P.: Kolmogorov's Structure Functions with an Application to the foundations of model selection. IEEE Trans. Inf. Theory **50**(12), 3265–3290 (2004). Preliminary version: Proceedings of 47th IEEE Symposium on the Foundations of Computer Science, pp. 751–760 (2002)
16. Vereshchagin, N.K., Vitányi, P.M.B.: Rate distortion a nd denoising of individual data using kolmogorov complexity. IEEE Trans. Inf. Theory **56**(7), 3438–3454 (2010)

Popularity in the Generalized Hospital Residents Setting

Meghana Nasre and Amit Rawat[⊠]

Indian Institute of Technology Madras, Chennai, India
amit_rawat@fastmail.fm

Abstract. We consider the problem of computing *popular* matchings in a bipartite graph $G = (\mathcal{R} \cup \mathcal{H}, E)$ where $\mathcal{R}$ and $\mathcal{H}$ denote a set of residents and a set of hospitals respectively. Each hospital h has a positive capacity denoting the number of residents that can be matched to h. The residents and the hospitals specify strict preferences over each other. This is the well-studied Hospital Residents (HR) problem which is a generalization of the Stable Marriage (SM) problem. The goal is to assign residents to hospitals *optimally* while respecting the capacities of the hospitals. Stability is a well-accepted notion of optimality in such problems. However, motivated by the need for larger cardinality matchings, alternative notions of optimality like *popularity* have been investigated in the SM setting. In this paper, we consider a generalized HR setting – namely the Laminar Classified Stable Matchings (LCSM$^+$) problem. Here, additionally, hospitals can specify classifications over residents in their preference lists and classes have upper quotas. We show the following new results: We define a notion of popularity and give a structural characterization of popular matchings for the LCSM$^+$ problem. Assume $n = |\mathcal{R}| + |\mathcal{H}|$ and $m = |E|$. We give an $O(mn)$ time algorithm for computing a maximum cardinality popular matching in an LCSM$^+$ instance. We give an $O(mn^2)$ time algorithm for computing a matching that is popular amongst the maximum cardinality matchings in an LCSM$^+$ instance.

1 Introduction

Consider an academic institution where students credit an elective course from a set of available courses. Every student and every course rank a subset of elements from the other set in a strict order of preference. Each course has a quota denoting the maximum number of students it can accommodate. The goal is to allocate to every student at most one course respecting the preferences. This is the well-studied Hospital Residents problem [7]. We consider its generalization where, in addition, a course can *classify* students – for example, the students may be classified as under-graduates, post-graduates, department-wise, and so on. Depending on the classifications, a student may belong to multiple classes. Apart from the total quota, each course now has a quota for every class. An allocation, in this setting, has to additionally respect the class quotas. This is the Classified Stable Matching problem introduced by Huang [10].

© Springer International Publishing AG 2017
P. Weil (Ed.): CSR 2017, LNCS 10304, pp. 245–259, 2017.
DOI: 10.1007/978-3-319-58747-9_22

Stability is a de-facto notion of optimality in settings where both set of participants have preferences. Informally, an allocation of students to courses is stable if no unallocated student-course pair has incentive to deviate from the allocation. Stability is appealing for several reasons – stable allocations are guaranteed to exist, they are efficiently computable and all stable allocations leave the same set of students unallocated [9]. However, it is known [13] that the cardinality of a stable allocation can be half the size of the largest sized allocation possible. Furthermore, in applications like student-course allocation, leaving a large number of students unallocated is undesirable. Thus, it is interesting to consider notions of optimality which respect preferences but possibly compromise stability in the favor of cardinality. Kavitha and Huang [11,13] investigated this in the Stable Marriage (SM) setting where they considered *popularity* as an alternative to stability. At a high level, an allocation of students to courses is *popular* if no *majority* wishes to deviate from the allocation. Here, we consider popularity in the context of two-sided preferences and one-sided capacities with classifications.

We formally define our problem now – we use the familiar hospital residents notation. Let $G = (\mathcal{R} \cup \mathcal{H}, E)$ be a bipartite graph where $|\mathcal{R} \cup \mathcal{H}| = n$ and $|E| = m$. Here $\mathcal{R}$ denotes the set of residents, $\mathcal{H}$ denotes the set of hospitals and every hospital $h \in \mathcal{H}$ has an upper quota $q(h)$ denoting the maximum number of residents h can occupy. A pair $(r, h) \in E$ denotes that r and h are mutually acceptable to each other. Each resident (resp. hospital) has a strict ordering of a subset of the hospitals (resp. residents) that are acceptable to him or her (resp. it). This ordering is called the preference list of a vertex. An assignment (or a matching) M in G is a subset of E such that every resident is assigned to at most one hospital and a hospital h is assigned at most $q(h)$ residents. Let $M(r)$ (resp. $M(h)$) denote the hospital (resp. the set of residents) which are assigned to r (resp. h) in M. A hospital h is under-subscribed if $|M(h)| < q(h)$. A matching M is *stable* if no unassigned pair (r, h) wishes to deviate from M. The goal is to compute a stable matching in G. We denote this problem as HR^+ throughout the paper [1]. The celebrated deferred acceptance algorithm by Gale and Shapley [7] proves that every instance of the HR^+ problem admits a stable matching.

A generalization of the HR^+ problem is the Laminar Classified Stable Matching (LCSM) problem introduced by Huang [10]. An instance of the LCSM^+ problem is an instance of the HR^+ problem where additionally, each hospital h is allowed to specify a classification over the set of residents in its preference list. A class C_k^h of a hospital h is a subset of residents in its preference list and has an associated upper quota $q(C_k^h)$ denoting the maximum number of residents that can be matched to h in C_k^h. (In the LCSM problem [10], classes can have lower quotas as well.) We assume that the classes of a hospital form a *laminar* set. That is, for any two classes C_j^h and C_k^h, either the two classes are disjoint ($C_j^h \cap C_k^h = \emptyset$), or one is contained inside the other ($C_j^h \subset C_k^h$ or $C_k^h \subset C_j^h$). Huang suitably modified the classical definition of stability to account for the presence of these classifications. He showed that every instance of the LCSM^+

---

[1] We use HR^+ instead of HR for consistency with other problems discussed in the paper. The $^+$ signifies that the hospitals do not specify a lower quota.

problem admits a stable matching which can be computed in $O(mn)$ time [10]. A restriction of the LCSM$^+$ problem, denoted by Partition Classified Stable Matching (PCSM$^+$), is where the classes of every hospital partition the residents in its preference list. The classified versions of the HR problem are natural in scenarios like course allocation, academic hiring [10] and in some of these applications stability may be compromised in favour of cardinality. Motivated by this, we consider computing *popular* matchings in the LCSM$^+$ problem. The notion of popularity uses *votes* to compare two matchings. Before we can define voting in the LCSM$^+$ setting, we discuss voting in the context of the SM problem.

Voting in the SM setting: Let $G = (\mathcal{R} \cup \mathcal{H}, E)$ be an instance of the SM problem and let M and M' be any two matchings in G. A vertex $u \in \mathcal{R} \cup \mathcal{H}$ (where each hospital h has $q(h) = 1$) prefers M over M' and therefore votes for M over M' if either (i) u is matched in M and unmatched in M' or (ii) u is matched in both M and M' and prefers $M(u)$ over $M'(u)$. A matching M is more popular than M' if the number of votes that M gets as compared to M' is greater than the number of votes that M' gets as compared to M. A matching M is popular if there does not exist any matching that is more popular than M. In the SM setting it is known that a stable matching is popular, however it was shown to be *minimum* cardinality popular matching [11]. Huang and Kavitha [11,13] gave efficient algorithms for computing a max-cardinality popular matching and a popular matching amongst max-cardinality matchings in an SM instance.

Voting in the capacitated setting: To extend voting in the capacitated setting, we assign a hospital h as many votes as its upper quota $q(h)$. This models the scenario in which hospitals with larger capacity get a larger share of votes. For the HR$^+$ problem, a hospital h compares the most preferred resident in $M(h) \backslash M'(h)$ to the most preferred resident in $M'(h) \backslash M(h)$ (and votes for M or M' as far as those two residents are concerned) and so on.

We show that the straightforward voting scheme as defined in the HR$^+$ does not suffice for the LCSM$^+$ problem. Therefore, we define a voting scheme which takes into consideration the classifications provided by a hospital as well as ensures that every stable matching in the LCSM$^+$ instance is popular. We show the following results:

- We define a notion of popularity for the LCSM$^+$ problem. Since our definition ensures that stable matchings are popular – this guarantees the existence of popular matchings in the LCSM$^+$ problem.
- We give a characterization of popular matchings for the LCSM$^+$ problem, which is an extension of the characterization result in the SM setting [11].
- We obtain the following algorithmic results. An $O(m + n)$ (resp. $O(mn)$) time algorithm for computing a maximum cardinality popular matching in a PCSM$^+$ (resp. LCSM$^+$) instance. An $O(mn)$ (resp. $O(mn^2)$) time algorithm for computing a popular matching amongst maximum cardinality matchings in a PCSM$^+$ (resp. LCSM$^+$) instance.

Very recently, independent of our work, two different groups [4,12] have considered popular matchings in the one-to-many setting. Brandl and Kavitha [4] have considered computing *popular* matchings in the HR$^+$ problem. In their work as well as ours, a hospital h is assigned as many votes as its capacity to compare two matchings M and M'. In contrast, by the definition of popularity in [4], a hospital h chooses the most adversarial ordering of residents in $M(h)\backslash M'(h)$ and $M'(h)\backslash M(h)$ for comparing M and M'. Interestingly, in an HR$^+$ instance the same matching is output by the algorithm in [4] and our algorithm in Sect. 4. On the other hand, we remark that the model considered in our paper is a more general one than the one considered in [4]. Kamiyama [12] has generalized our work and the results in [4] using a matroid based approach.

We finally remark that one can consider voting schemes where a hospital is given a *single* vote instead of capacity many votes. In one such scheme, a hospital compares the set of residents in $M(h)$ and $M'(h)$ in lexicographic order and votes accordingly. However, when such a voting is used, it is possible to construct instances where a stable matching is *not popular*. The techniques in this paper use the fact that stable matchings are popular (for guaranteed existence), therefore it is not clear if the same techniques apply for such voting schemes.

Related Work: The notion of popularity was introduced by Gärdenfors [8] in the context of stable matchings. In [1] Abraham et al. studied popularity in the one-sided preference list model. As mentioned earlier, our work is inspired by a series of papers by Huang, Kavitha and Cseh [5,11,13] where popularity is considered as an alternative to stability in the stable marriage setting. Biró et al. [3] give several practical scenarios where stability may be compromised in the favor of size. The PCSM$^+$ problem is a special case of the Student Project Allocation (SPA) problem studied by Abraham et al. [2]. They gave a linear time algorithm to compute a stable matching in an instance of the SPA problem. In this paper, we use the algorithms of Abraham et al. [2] and Huang [10] for computing stable matchings in the PCSM$^+$ and LCSM$^+$ problems. Both these algorithms follow the standard *deferred acceptance* algorithm of Gale and Shapley with problem specific modifications. We refer the reader to [2,10] for details.

Overview and Organization of paper: In Sect. 2 we define the notion of popularity, and in Sect. 3 we present the structural characterization of popular matchings. Our algorithms to compute popular matchings in the LCSM$^+$ problem are based on the idea of allowing residents to apply to hospitals multiple times, each time with *increased priority*. In Sect. 4 we present these algorithms as a generic reduction (to be invoked with a suitable parameter) to the problem of computing a stable matching in a modified LCSM$^+$ instance.

2   Stability and Popularity in the LCSM$^+$ Problem

Consider an instance $G = (\mathcal{R} \cup \mathcal{H}, E)$ of the LCSM$^+$ problem. As done in [10], assume that for every $h \in \mathcal{H}$ there is a class C_*^h containing all the residents in the preference list of h and $q(C_*^h) = q(h)$. For a hospital h, let $T(h)$ denote the tree of classes corresponding to h where C_*^h is the root of $T(h)$. The leaf classes in $T(h)$ denote the most refined classifications for a resident whereas as we move up in the tree from a leaf node to the root, the classifications gets coarser.

To define stable matchings in the LCSM problem, Huang introduced the notion of a *blocking group* w.r.t. a matching. Later, Fleiner and Kamiyama [6] defined a blocking pair which is equivalent to a blocking group as defined by Huang. We use the definition of stability from [6] which we recall below. Consider the following LCSM$^+$ instance which will serve as an illustrative example throughout the paper. Here $\mathcal{R} = \{r_1, \ldots, r_4\}$ and $\mathcal{H} = \{h_1, \ldots, h_3\}$ and the preference lists of the residents and hospitals are as given in Fig. 1(a) and (b) respectively. The preferences can be read as follows: resident r_1 has h_1 as his top choice hospital. Resident r_2 has h_2 as its top choice hospital followed by h_1 which is his second choice hospital and so on. For $h \in \{h_2, h_3\}$ we have $q(h) = 1$ and both these hospitals have a single class C^h_* containing all the residents in the preference list of h and $q(C^h_*) = q(h)$. For hospital h_1 we have $q(h_1) = 2$ and the classes provided by h_1 are $C^{h_1}_1 = \{r_1, r_2\}, C^{h_1}_2 = \{r_3, r_4\}, C^{h_1}_* = \{r_1, r_2, r_3, r_4\}$ with quotas as follows: $q(C^{h_1}_1) = q(C^{h_1}_2) = 1$ and $q(C^{h_1}_*) = 2$. Note that the example is indeed a PCSM$^+$ instance. Figure 1(c) shows the tree $T(h_1)$.

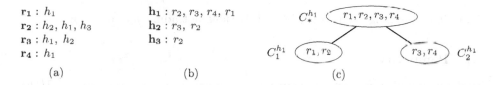

$r_1 : h_1$

$r_2 : h_2, h_1, h_3$

$r_3 : h_1, h_2$

$r_4 : h_1$

(a)

$h_1 : r_2, r_3, r_4, r_1$

$h_2 : r_3, r_2$

$h_3 : r_2$

(b)

(c)

Fig. 1. (a) Resident preferences, (b) Hospital preferences, (c) $T(h_1)$. The matchings $M = \{(r_1, h_1), (r_2, h_2), (r_3, h_1)\}$, $M' = \{(r_2, h_1), (r_3, h_2), (r_4, h_1)\}$, and $M'' = \{(r_1, h_1), (r_2, h_3), (r_3, h_2), (r_4, h_1)\}$ are all feasible in the instance.

Consider the two feasible matchings M and M' defined in Fig. 1. Note that M is stable in the instance whereas the edge (r_3, h_1) blocks M'. While comparing M and M', the vote for every vertex u in the instance except h_1 is clear – u compares $M(u)$ with $M'(u)$ and votes accordingly. In order for h_1 to vote between M and M', the hospital compares between $M(h_1) = \{r_1, r_3\}$ and $M'(h_1) = \{r_2, r_4\}$. A straightforward way is to compare r_3 with r_2 (the most preferred resident in $M(h_1)$ to the most preferred resident in $M'(h_1)$) and then compare r_1 with r_4 (second most preferred resident in $M(h_1)$ to second most preferred resident in $M'(h_1)$). Thus, both the votes of h_1 are in favor of M' when compared with M. Such a comparison has two issues – (i) it ignores the classifications given by h_1, and (ii) the number of votes that M' gets when compared with M is more than the number of votes that M gets as compared to M'. Therefore M' is more popular than M which implies that M (a stable matching) is **not** popular.

We propose a comparison scheme for hospitals which addresses both the issues. In the above example, we note that $r_1 \in M(h)$ has a corresponding resident $r_2 \in M'(h)$ to be compared to in one of the most refined classes $C_1^{h_1}$ (see Fig. 1(c)). Thus, we compare r_1 with r_2. The resident $r_3 \in M(h)$ is compared to $r_4 \in M(h)$ another leaf class $C_2^{h_1}$. According to this comparison, h_1 is indifferent between M and M', and M' is no longer more popular than M. Note that, although in the example, both the comparisons happen in a leaf class, this may not be the case in a general instance. Finally, we note that the matching M'' is a popular matching in the instance and is strictly larger in size than the stable matching M.

A set $S = \{r_1, \ldots, r_l\}$ is *feasible* for a hospital h if $|S| \leq q(h)$ and for every class C_j^h of h (including the root class C_*^h), we have $|C_j^h \cap S| \leq q(C_j^h)$. A matching M in G is feasible if every resident is matched to at most one hospital, and $M(h)$ is feasible for every hospital $h \in \mathcal{H}$. A pair $(r, h) \notin M$ blocks M iff both the conditions below hold:

- r is unmatched in M, or r prefers h over $M(r)$, and
- either the set $M(h) \cup \{r\}$ is feasible for h, or there exists a resident $r' \in M(h)$, such that h prefers r over r', and $(M(h) \setminus \{r'\}) \cup \{r\}$ is feasible for h.

A feasible matching M in G is stable if M does not admit any blocking pair.

2.1   Popularity

To define popularity, we need to specify how a hospital compares two sets $M(h)$ and $M'(h)$ in an LCSM$^+$ setting, where M and M' are two feasible matchings in the instance. Our example shows that if we use a simple voting scheme as in the HR$^+$ problem, a stable matching in the LCSM$^+$ instance is *not popular*. Intuitively, this is because such a voting scheme completely ignores the classifications. To take into account the classifications, for a hospital h and the matchings M and M', we set up a correspondence between residents in $M(h) \setminus M'(h)$ and the residents in $M'(h) \setminus M(h)$. That is, we define:

$$\mathbf{corr} : M(h) \oplus M'(h) \rightarrow M(h) \oplus M'(h) \cup \{\bot\}$$

For a resident $r \in M(h) \oplus M'(h)$ we denote by $\mathbf{corr}(r)$ the corresponding resident to which r gets compared when the hospital h casts its votes. We let $\mathbf{corr}(r) = \bot$ if r does not have a corresponding resident to be compared to from the other matching. The pseudo-code for the algorithm to compute the $\mathbf{corr}$ function is given below.

Algorithm 1. Correspondence between residents of $M(h)$ and $M'(h)$

1: **procedure** FIND-CORRESPONDENCE(h, M, M')
2:     let $T(h)$ be the classification tree associated with h
3:     set $\mathbf{corr}(r) = \perp$ for each $r \in M(h) \oplus M'(h)$
4:     $Y = M(h) \backslash M'(h)$; $Y' = M'(h) \backslash M(h)$
5:     **while** $Y \neq \emptyset$ and $Y' \neq \emptyset$ **do**
6:         **for** each class C_j^h in $T(h)$ **do**
7:             $X_j = C_j^h \cap Y$
8:             $X_j' = C_j^h \cap Y'$
9:         Let C_f^h be one of the most refined classes for which $X_f \neq \emptyset$ and $X_f' \neq \emptyset$.
10:         **for** $k = 1, \ldots, \min(|X_f|, |X_f'|)$ **do**
11:             let r be the k-th most preferred resident in X_f
12:             let r' be the k-th most preferred resident in X_f'
13:             set $\mathbf{corr}(r) = r'$, and $\mathbf{corr}(r') = r$
14:         $Y = Y \backslash \{r\}$; $Y' = Y' \backslash \{r'\}$

The algorithm begins by setting **corr** for every $r \in M(h) \oplus M'(h)$ to $\perp$. The algorithm maintains two sets of residents $Y = M(h) \backslash M'(h)$ and $Y' = M'(h) \backslash M(h)$ for whom **corr** needs to be set. As long as the sets Y and Y' are both non-empty, the algorithm repeatedly computes for every class C_j^h (including the root class C_*^h) the sets $X_j = C_j^h \cap Y$ and $X_j' = C_j^h \cap Y'$. The algorithm then chooses one of the most refined classes, say C_f^h in $T(h)$, for whom X_f and X_f' are both non-empty. Finally, residents in X_f and X_f' are sorted according to the preference ordering of h and the **corr** of the k-th most preferred resident in X_f is set to the k-th most preferred resident in X_f', where $k = 1, \ldots, \min\{|X_f|, |X_f'|\}$.

For $r \in \mathcal{R}$, and any feasible matching M in G, if r is unmatched in M then, $M(r) = \perp$. A vertex prefers any of its neighbours over $\perp$. For a vertex $u \in \mathcal{R} \cup \mathcal{H}$, let $x, y \in N(u) \cup \{\perp\}$, where $N(u)$ denotes the neighbours of u in G.

$$vote_u(x, y) = +1 \quad \text{if } u \text{ prefers } x \text{ over } y$$
$$= -1 \quad \text{if } u \text{ prefers } y \text{ over } x$$
$$= 0 \quad \text{if } x = y$$

Using the above notation, the vote of a resident is easy to define – a resident r prefers M' over M iff the term $\mathcal{V}_r > 0$, where $\mathcal{V}_r = vote_r(M'(r), M(r))$. Recall that a hospital h uses $q(h)$ votes to compare M and M'. Let $q_1(h) = |M(h) \cap M'(h)|$ (number of common residents assigned to h in M and M') and $q_2(h) = q(h) - \max\{|M(h)|, |M'(h)|\}$ (number of unfilled positions of h in both M and M'). Our voting scheme ensures that $q_1(h) + q_2(h)$ votes of h remain unused when comparing M and M'. A hospital h prefers M' over M iff the term $\mathcal{V}_h > 0$, where $\mathcal{V}_h$ is defined as follows:

$$\mathcal{V}_h = (|M'(h)| - |M(h)|) + \sum_{\substack{r \in M'(h) \backslash M(h) \\ \&\& \\ \mathbf{corr}(r) \neq \perp}} vote_h(r, \mathbf{corr}(r))$$

The first term in the definition of $\mathcal{V}_h$ counts the votes of h w.r.t. the residents from either M or M' that did not find correspondence. The second term counts the votes of h w.r.t. the residents each of which has a corresponding resident from the other matching. We note that in the SM setting, $\mathbf{corr}(r)$ will simply be $M(h)$. Thus, our definition of votes in the presence of capacities is a natural generalization of the voting scheme in the SM problem. Let us define the term $\Delta(M', M)$ as the difference between the votes that M' gets over M and the votes that M gets over M'.

$$\Delta(M', M) = \sum_{r \in \mathcal{R}} \mathcal{V}_r + \sum_{h \in \mathcal{H}} \mathcal{V}_h$$

Definition 1. *A matching M is popular in G iff for every feasible matching M', we have $\Delta(M', M) \leq 0$.*

Decomposing $M \oplus M'$: Here, we present a simple algorithm which allows us to decompose edges of components of $M \oplus M'$ in an instance into alternating paths and cycles. This decomposition will be used in the subsequent sections and all of our proofs. Consider the graph $\tilde{G} = (\mathcal{R} \cup \mathcal{H}, M \oplus M')$, for any two feasible matchings M and M' in G. We note that the degree of every resident in $\tilde{G}$ is at most 2 and the degree of every hospital in $\tilde{G}$ is at most $2 \cdot q(h)$. Consider any connected component $\mathcal{C}$ of $\tilde{G}$ and let e be any edge in $\mathcal{C}$. We observe that it is possible to construct a unique maximal M alternating path or cycle ρ containing e using the following simple procedure. Initially ρ contains only the edge e.

1. Let $r \in \mathcal{R}$ be an end point of ρ, and assume that $(r, M(r)) \in \rho$. We grow ρ by adding the edge $(r, M'(r))$ if it exists. Similarly, if an edge from M' is incident on r in ρ, we grow the path by adding the edge $(r, M(r))$ if it exists.
2. Let $h \in \mathcal{H}$ be one of the end points of the path ρ, and assume that $(r, h) \in M \backslash M'$ belongs to ρ. We extend ρ by adding $(\mathbf{corr}(r), h)$ if $\mathbf{corr}(r)$ is not equal to $\perp$. A similar step is performed if the last edge on ρ is $(r, h) \in M' \backslash M$.
3. During this procedure, we may encounter a hospital h which has already occurred on ρ, leading to a possible cycle $\langle h, r, \ldots, r', h \rangle$. We complete the cycle only if r and r' are $\mathbf{corr}$ w.r.t. to h, else we grow the path by creating a copy of the hospital h. We stop the procedure when we complete a cycle or the path can no longer be extended. Otherwise, we go to Step 1 or Step 2 as applicable and repeat.

The above procedure gives us a unique decomposition of a connected component in $\tilde{G}$ into alternating paths and cycles. Note that a hospital may appear multiple times in a single path or a cycle and also can belong to more than one alternating paths and cycles. See [14] for an example. Let $\mathcal{Y}_{M \oplus M'}$ denote the collection of alternating paths and alternating cycles obtained by decomposing every component of $\tilde{G}$. We now state a useful property about any alternating path or cycle in $\mathcal{Y}_{M \oplus M'}$.

Lemma 1. *If ρ is an alternating path or an alternating cycle in $\mathcal{Y}_{M \oplus M'}$, then $M \oplus \rho$ is a feasible matching in G.*

Proof. Let $\langle r', h, r \rangle$ be any sub-path of ρ, where $r' = \mathbf{corr}(r)$, and $(r, h) \in M$. We prove that $(M(h)\backslash\{r\}) \cup \{r'\}$ is feasible for h. Let C_i^h (resp. C_j^h) be the unique leaf class of $T(h)$ containing r (resp. r'). We consider two cases. First, assume that, r and r' belong to the same leaf class in $T(h)$, i.e. $C_i^h = C_j^h$. In this case, it is easy to note that $(M(h)\backslash\{r\}) \cup \{r'\}$ is feasible for h. Secondly assume that, r and r' belong to different leaf classes of $T(h)$, i.e. $C_i^h \neq C_j^h$. Observe that $|(M(h)\backslash\{r\}) \cup \{r'\}|$ can violate the upper quota only for those classes of $T(h)$ which contain r' but do not contain r. Let C_k^h be the least common ancestor of C_i^h and C_j^h in $T(h)$. It suffices to look at any class C_t^h which lies in the path from C_k^h to C_j^h excluding the class C_k^h and show that $|(M(h) \cap C_t^h) \cup \{r'\}| \leq q(C_t^h)$. As $r' = \mathbf{corr}(r)$ and $r \notin C_t^h$, we claim that $|M(h) \cap C_t^h| < |M'(h) \cap C_t^h| \leq q(C_t^h)$. The first inequality is due to the fact that r' did not find a corresponding resident in the set $(M(h)\backslash M'(h)) \cap C_t^h$. The second inequality is because M' is feasible. Thus, $(M(h) \cap C_t^h) \cup \{r'\}$ does not violate the upper quota for C_t^h. Therefore $(M(h)\backslash\{r\}) \cup \{r'\}$ is feasible for h. We note that the hospital h may occur multiple times on ρ. Let $M(h)_\rho$ denote the set of residents matched to h restricted to ρ. To complete the proof of the Lemma, we need to prove that $(M(h)\backslash M(h)_\rho) \cup M'(h)_\rho$ is feasible for h. The arguments for this follow from the arguments given above. ☐

As was done in [13], it is convenient to label the edges of $M'\backslash M$ and use these labels to compute $\Delta(M', M)$. Let $(r, h) \in M'\backslash M$; the label on (r, h) is a tuple:

$$(vote_r(h, M(r)), \quad vote_h(r, \mathbf{corr}(r)))$$

Note that since we are labeling edges of $M'\backslash M$, both entries of the tuple come from the set $\{-1, 1\}$. With these definitions in place, we are ready to give the structural characterization of popular matchings in an LCSM$^+$ instance.

3   Structural Characterization of Popular Matchings

Let $G = (\mathcal{R} \cup \mathcal{H}, E)$ be an LCSM$^+$ instance and let M and M' be two feasible matchings in G. Using the **corr** function, we obtain a correspondence of residents in $M(h) \oplus M'(h)$ for every hospital h in G. Let $\tilde{G} = (\mathcal{R} \cup \mathcal{H}, M \oplus M')$ and let $\mathcal{Y}_{M \oplus M'}$ denote the collection of alternating paths and cycles obtained by decomposing every component of $\tilde{G}$. Finally, we label the edges of $M'\backslash M$ using appropriate votes. The goal of these steps is to rewrite the term $\Delta(M', M)$ as a sum of labels on edges.

We note that the only vertices for whom their vote does not get captured on the edges of $M'\backslash M$ are vertices that are matched in M but not matched in M'. Let $\mathcal{U}$ denote the multi-set of vertices that are end points of paths in $\mathcal{Y}_{M \oplus M'}$ such that there is no M' edge incident on them. Note that the same hospital can belong to multiple alternating paths and cycles in $\mathcal{Y}_{M \oplus M'}$, therefore we need a multi-set. All vertices in $\mathcal{U}$ prefer M over M' and hence we add a -1 while capturing their vote in $\Delta(M', M)$. We can write $\Delta(M', M)$ as:

$$\Delta(M', M) = \sum_{x \in \mathcal{U}} -1 + \sum_{\rho \in \mathcal{Y}_{M \oplus M'}} \left(\sum_{(r,h) \in (M' \cap \rho)} \{vote_r(h, M(r)) + vote_h(r, \mathbf{corr}(r))\} \right)$$

We now delete the edges labeled $(-1, -1)$ from all paths and cycles ρ in $\mathcal{Y}_{M \oplus M'}$. This simply breaks paths and cycles into one or more paths. Let this new collection of paths and cycles be denoted by $\tilde{\mathcal{Y}}_{M \oplus M'}$. Let $\tilde{\mathcal{U}}$ denote the multi-set of vertices that are end points of paths in $\tilde{\mathcal{Y}}_{M \oplus M'}$ such that there is no M' edge incident on them. We rewrite $\Delta(M', M)$ as:

$$\Delta(M', M) = \sum_{x \in \tilde{\mathcal{U}}} -1 + \sum_{\rho \in \tilde{\mathcal{Y}}_{M \oplus M'}} \left(\sum_{(r,h) \in (M' \cap \rho)} \{vote_r(h, M(r)) + vote_h(r, \mathbf{corr}(r))\} \right)$$

Theorem below characterizes a popular matching (see [14] for the full proof).

Theorem 1. *A feasible matching M in G is popular iff for any feasible matching M' in G, the set $\tilde{\mathcal{Y}}_{M \oplus M'}$ does not contain any of the following:*

1. *An alternating cycle with a $(1,1)$ edge,*
2. *An alternating path which has a $(1,1)$ edge and starts with an unmatched resident in M or a hospital which is under-subscribed in M.*
3. *An alternating path which has both its ends matched in M and has two or more $(1,1)$ edges.*

We now prove that every stable matching in an LCSM$^+$ instance is popular.

Theorem 2. *Every stable matching in an LCSM$^+$ instance G is popular.*

Proof. Let M be a stable matching in G. For any feasible matching M' in G consider the set $\mathcal{Y}_{M \oplus M'}$. To prove that M is popular it suffices to show that there does not exist a path or cycle $\rho \in \mathcal{Y}_{M \oplus M'}$ such that an edge of ρ is labeled $(1,1)$. For the sake of contradiction, assume that ρ is such a path or cycle, which has an edge $(r', h) \in M' \backslash M$ labeled $(1,1)$. Let $r = \mathbf{corr}(r')$, where $(r, h) \in M \cap \rho$. From the proof of Lemma 1 we observe that $(M(h) \backslash \{r\}) \cup \{r'\}$ is feasible for h, therefore the edge (r', h) blocks M contradicting the stability of M. □

4   Popular Matchings in LCSM$^+$ Problem

In this section we present efficient algorithms for computing (i) a maximum cardinality popular matching, and (ii) a matching that is popular amongst all the maximum cardinality matchings in a given LCSM$^+$ instance. Our algorithms are inspired by the reductions of Kavitha and Cseh [5] where they work with a stable marriage instance. We describe a general reduction from an LCSM$^+$ instance G to another LCSM$^+$ instance G_s. Here $s = 2, \dots, |\mathcal{R}|$. The algorithms for the two problems are obtained by choosing an appropriate value of s.

The graph G_s: Let $G = (\mathcal{R} \cup \mathcal{H}, E)$ be the input LCSM$^+$ instance. The graph $G_s = (\mathcal{R}_s \cup \mathcal{H}_s, E_s)$ is constructed as follows: Corresponding to every resident $r \in \mathcal{R}$, we have s copies of r, call them $r^0, \ldots, r^{s-1}$ in $\mathcal{R}_s$. The hospitals in $\mathcal{H}$ and their capacities remain unchanged; however we have additional dummy hospitals each of capacity 1. Corresponding to every resident $r \in \mathcal{R}$, we have $(s-1)$ dummy hospitals $d_r^0, \ldots, d_r^{s-2}$ in $\mathcal{H}_s$. Thus,

$$\mathcal{R}_s = \{\ r^0, \ldots, r^{s-1}\ \mid \forall r \in \mathcal{R}\}; \quad \mathcal{H}_s = \mathcal{H} \cup \{\ d_r^0, \ldots, d_r^{s-2}\ \mid \forall r \in \mathcal{R}\}$$

We use the term level-i resident for a resident $r^i \in \mathcal{R}_s$ for $0 \le i \le s-1$. The preference lists corresponding to s different copies of r in G_s are:

- For a level-0 resident r^0, its preference list in G_s is the preference list of r in G, followed by the dummy hospital d_r^0.
- For a level-i resident r^i, where $1 \le i \le s-2$, its preference list in G_s is d_r^{i-1} followed by preference list of r in G, followed by d_r^i.
- For a level-$(s-1)$ resident r^{s-1}, its preference list in G_s is the dummy hospital d_r^{s-2} followed by the preference list of r in G.

The preference lists of hospitals in G_s are as follows.

- The preference list for a dummy hospital d_r^i is r^i followed by r^{i+1}.
- For $h \in \mathcal{H}$, its preference list in G_s, has level-$(s-1)$ residents followed by level-$(s-2)$ residents, so on upto the level 0 residents in the same order as in h's preference list in G.

Finally, we need to specify the classifications of the hospitals in G_s. For every class C_i^h in the instance G, we have a corresponding class $\bar{C}_i^h = \bigcup_{r \in C_i^h} \{r^0, \ldots, r^{s-1}\}$ in G_s, such that $q(\bar{C}_i^h) = q(C_i^h)$. We note that $|\bar{C}_i^h| = s \cdot |C_i^h|$. Then a stable matching M_s in G_s satisfies the following properties:

$(\mathcal{I}_1)$ Each $d_r^i \in \mathcal{H}_s$ for $0 \le i \le s-2$, is matched to one of $\{r^i, r^{i+1}\}$ in M_s.

$(\mathcal{I}_2)$ The above invariant implies that for every $r \in \mathcal{R}$ at most one of $\{r^0, \ldots, r^{s-1}\}$ is assigned to a non-dummy hospital in M_s.

$(\mathcal{I}_3)$ For a resident $r \in \mathcal{R}$, if r^i is matched to a non-dummy hospital in M_s, then for all $0 \le j \le i-1$, $M_s(r^j) = d_r^j$. Furthermore, for all $i+1 \le p \le s-1$, $M_s(r^p) = d_r^{p-1}$. This also implies that in M_s all residents $r^0, \ldots, r^{s-2}$ are matched and only r^{s-1} can be left unmatched in M_s.

These invariants allow us to naturally map the stable matching M_s to a feasible matching M in G. We define a function $map(M_s)$ as follows.

$$M = map(M_s) = \{(r, h) : h \in \mathcal{H} \text{ and } (r^i, h) \in M_s \text{ for exactly one of } 0 \le i \le s-1\}$$

We outline an algorithm that computes a feasible matching in an LCSM$^+$ instance G. Given G and s, construct the graph G_s from G. Compute a stable matching M_s in G_s. If G is an LCSM$^+$ instance we use the algorithm of Huang [10] to compute a stable matching in G. If G is a PCSM$^+$ instance, it is easy to observe that G_s is also a PCSM$^+$ instance. In that case, we use the algorithm of Abraham et al. [2] to compute a stable matching. (There is a simple reduction from the PCSM$^+$ instance to SPA; see [14]). We output $M = map(M_s)$ whose feasibility is guaranteed by the invariants mentioned earlier. The complexity of our algorithm depends on s and the time required to compute a stable matching in the problem instance.

In the rest of the paper, we denote by M the matching obtained as $map(M_s)$ where M_s is a stable matching in G_s. For any resident $r_i \in \mathcal{R}$, we define map^{-1} function which maps a resident $r_i \in G$ to its unique level-j_i copy in G_s.

$$map^{-1}(r_i, M_s) = r_i^{j_i} \quad \text{where } 0 \leq j_i \leq s - 1 \text{ and } M_s(r_i^{j_i}) \text{ is a non-dummy hospital}$$
$$= r_i^{s-1} \quad \text{otherwise.}$$

Recall by Invariant $(\mathcal{I}_3)$, at most one of the level copy of r_i in G_s is matched to a non-dummy hospital in M_s. For any feasible matching M' in G consider the set $\mathcal{Y}_{M \oplus M'}$ – recall that this is a collection of M alternating paths and cycles in G. For any path or cycle ρ in $\mathcal{Y}_{M \oplus M'}$, let us denote by $\rho_s = map^{-1}(\rho, M_s)$ the path or cycle in G_s obtained by replacing every resident r in ρ by $map^{-1}(r, M_s)$. Recall that if a resident r is present in the class C_j^h defined by a hospital h in G, then in the graph G_s, $r^i \in \bar{C}_j^h$ for $i = 0, \ldots, s - 1$. Using Lemma 1 and these observations we get the following corollary.

Corollary 1. *Let ρ be an alternating path or an alternating cycle in $\mathcal{Y}_{M \oplus M'}$, then $M_s \oplus \rho_s$ is a feasible matching in G_s, where $\rho_s = map^{-1}(\rho, M_s)$.*

The following technical lemma is useful in proving the properties of the matchings produced by our algorithms. All omitted proofs can be found in [14].

Lemma 2. *Let ρ be an alternating path or an alternating cycle in $\mathcal{Y}_{M \oplus M'}$, and $\rho_s = map^{-1}(\rho, M_s)$.*

1. *There cannot be any edge labeled $(1, 1)$ in ρ_s.*
2. *Let $\langle r_a^{j_a}, h, r_b^{j_b} \rangle$ be a sub-path of ρ_s, where $h = M_s(r_b^{j_b})$. Then, the edge $(r_a^{j_a'}, h) \notin \rho_s$ cannot be labeled $(1, 1)$, where $j_a' < j_a$.*

4.1 Maximum Cardinality Popular Matching

Let $G = (\mathcal{R} \cup \mathcal{H}, E)$ be an instance of the LCSM$^+$ problem where we are interested in computing a maximum cardinality popular matching. We use our generic reduction with the value of the parameter $s = 2$. Since G_2 is linear in the size of G, and a stable matching in an LCSM$^+$ instance can be computed in $O(mn)$ time [10], we obtain an $O(mn)$ time algorithm to compute a maximum cardinality popular matching in G. In case G is a PCSM$^+$ instance, we use the linear time algorithm in [2] for computing a stable matching to get a linear time algorithm for our problem. We state the main theorem of this section below.

Theorem 3. *Let* $M = map(M_2)$ *where* M_2 *is a stable matching in* G_2. *Then* M *is a maximum cardinality popular matching in* G.

We break down the proof of Theorem 3 in two parts. Lemma 5 shows that the assignment M satisfies all the conditions of Theorem 1 and therefore is popular in G. Lemmas 3 and 4 show that the matching output is indeed the largest size popular matching in the instance. Let M' be any assignment in G. Recall the definition of $\tilde{\mathcal{Y}}_{M \oplus M'}$ from Sect. 3.

Lemma 3. *There is no augmenting path with respect to* M *in* $\tilde{\mathcal{Y}}_{M \oplus M'}$.

Lemma 4. *There exists no popular matching* M^* *in* G *such that* $|M^*| > |M|$.

Lemma 5. *Let* $M = map(M_2)$ *where* M_2 *is a stable matching in* G_2 *and let* M' *be any feasible assignment in* G. *Consider the set of alternating paths and alternating cycles* $\tilde{\mathcal{Y}}_{M \oplus M'}$. *Then, the following hold:*

1. *An alternating cycle* C *in* $\tilde{\mathcal{Y}}_{M \oplus M'}$, *does not contain any edge labeled* $(1, 1)$.
2. *An alternating path* P *in* $\tilde{\mathcal{Y}}_{M \oplus M'}$ *that starts or ends with an edge in* M', *does not contain any edge labeled* $(1, 1)$.
3. *An alternating path* P *in* $\tilde{\mathcal{Y}}_{M \oplus M'}$ *which starts and ends with an edge in* M, *contains at most one edge labeled* $(1, 1)$.

Proof (Sketch). Assume that $\rho = \langle u_0, v_1, u_1, \ldots, v_k, u_0 \rangle$ is an alternating cycle in $\mathcal{Y}_{M \oplus M'}$, where for each $i = 0, \ldots, k$, $v_i = M(u_i)$ (in case u_i is a hospital, $v_i \in M(u_i)$). We assume for contradiction that ρ contains an edge labeled $(1, 1)$. Using such an edge, we show the existence of an edge $e_2 \in E_2 \backslash M_2$ which is labeled $(1, 1)$, contradicting the stability of M_2. □

4.2 Popular Matching Amongst Maximum Cardinality Matchings

In this section we give an efficient algorithm for computing a matching which is *popular* amongst the set of maximum cardinality matchings. The matching M that we output cannot be beaten in terms of votes by any feasible maximum cardinality matching. Our algorithm uses the generic reduction with a value of $s = |\mathcal{R}| = n_1$ (say). Thus, $|\mathcal{R}_{n_1}| = n_1^2$, and $|\mathcal{H}_{n_1}| = |\mathcal{H}| + O(n_1^2)$. Furthermore, $|E_{n_1}| = O(mn_1)$ where $m = |E|$. Thus the running time of the generic algorithm presented earlier with $s = n_1$ for an LCSM$^+$ instance is $O(mn \cdot n_1) = O(mn^2)$ and for a PCSM$^+$ instance is $O(mn_1) = O(mn)$.

To prove correctness, we show that the matching output by our algorithm is (i) maximum cardinality and (ii) popular amongst all maximum cardinality feasible matchings. Let $M = map(M_{n_1})$ and M^* be any maximum cardinality feasible matching in G. Consider the set $\mathcal{Y}_{M \oplus M^*}$, and let ρ be an alternating path or an alternating cycle in $\mathcal{Y}_{M \oplus M^*}$. Let $\rho_{n_1} = map^{-1}(\rho, M_{n_1})$ denote the associated alternating path or cycle in G_{n_1}. We observe that every hospital on the path ρ_{n_1} is a non-dummy hospital since ρ_{n_1} was obtained using the inverse-map of ρ. We observe two useful properties about such a path or cycle ρ_{n_1} in G_{n_1}. We show (using Lemma 6) that if for a hospital $h \in \rho_{n_1}$, the level of the unmatched resident incident on h is greater than the level of the matched resident incident on h, then such a level change is *gradual*, and the associated edge in ρ has the label $(-1, -1)$.

Lemma 6. *Let ρ_{n_1} be an alternating path or an alternating cycle in G_{n_1} and let h be a hospital which has degree two in ρ_{n_1}. Let $\langle r_a^{j_a}, h, r_b^{j_b} \rangle$ be the sub-path containing h where $M(r_b^{j_b}) = h$. If $j_a > j_b$, we claim the following:*

1. $j_a = j_b + 1$.
2. The associated edge $(r_a, h) \in \rho$ is labeled $(-1, -1)$.

We use Lemma 7 to prove that M is a maximum cardinality matching in G.

Lemma 7. *Let M^* be any feasible maximum cardinality matching in G. Then there is no augmenting path with respect to M in $\mathcal{Y}_{M \oplus M^*}$.*

Proof (Sketch). Assume there exists an augmenting path P with respect to M and let $P_{n_1} = map^{-1}(P, M_{n_1})$. We show that the path P_{n_1} starts with two consecutive residents which are level-$(n_1 - 1)$ residents; whereas the path P_{n_1} ends with a level-0 resident. For every hospital on P_{n_1} there can only be a *gradual* increase in the level of the unmatched resident to the level of the matched resident. Using this, and the fact that there are at most $(n_1 - 1)$ residents on the path, we conclude that such a path P_{n_1} cannot exist. This contradicts the existence of the augmenting path P. $\qquad\square$

We can now conclude that the set $\mathcal{Y}_{M \oplus M^*}$ is a set of alternating (and not augmenting) paths and alternating cycles. It remains to show that M is popular amongst all maximum cardinality feasible matchings in G. We show that in an alternating path in $\mathcal{Y}_{M \oplus M^*}$ with exactly one endpoint unmatched in M or an alternating cycle, the number of edges labeled $(1, 1)$ cannot exceed the number of edges labeled $(-1, -1)$ (Lemma 8).

Lemma 8. *Let ρ be an alternating path or an alternating cycle in $\mathcal{Y}_{M \oplus M^*}$. Then the number of edges labeled $(1, 1)$ in ρ is at most the number of edges labeled $(-1, -1)$.*

Thus, we get the following theorem:

Theorem 4. *Let $M = map(M_{n_1})$ where M_{n_1} is a stable matching in G_{n_1}. Then M is a popular matching amongst all maximum cardinality matchings in G.*

Acknowledgement. We thank Prajakta Nimbhorkar for helpful discussions. We thank the anonymous reviewers whose comments have improved the presentation.

References

1. Abraham, D.J., Irving, R.W., Kavitha, T., Mehlhorn, K.: Popular matchings. SIAM J. Comput. **37**(4), 1030–1045 (2007)
2. Abraham, D.J., Irving, R.W., Manlove, D.F.: Two algorithms for the student-project allocation problem. J. Discrete Algorithms **5**(1), 73–90 (2007)
3. Biró, P., Manlove, D., Mittal, S.: Size versus stability in the marriage problem. Theoret. Comput. Sci. **411**(16–18), 1828–1841 (2010)
4. Brandl, F., Kavitha, T.: Popular Matchings with Multiple Partners. CoRR, abs/1609.07531 (2016)
5. Cseh, Á., Kavitha, T.: Popular edges and dominant matchings. In: Proceedings of the Eighteenth Conference on Integer Programming and Combinatorial Optimization, pp. 138–151 (2016)
6. Fleiner, T., Kamiyama, N.: A matroid approach to stable matchings with lower quotas. In: Proceedings of the Twenty-third Annual ACM-SIAM Symposium on Discrete Algorithms, pp. 135–142 (2012)
7. Gale, D., Shapley, L.: College admissions and the stability of marriage. Am. Math. Monthly **69**, 9–14 (1962)
8. Gärdenfors, P.: Match making: assignments based on bilateral preferences. Behav. Sci. **20**, 166–173 (1975)
9. Gusfield, D., Irving, R.W.: The Stable Marriage Problem: Structure and Algorithms. MIT Press, Boston (1989)
10. Huang, C.-C.: Classified stable matching. In: Proceedings of the Twenty-First Annual ACM-SIAM Symposium on Discrete Algorithms, pp. 1235–1253 (2010)
11. Huang, C.-C., Kavitha, T.: Popular matchings in the stable marriage problem. In: Proceedings of 38th International Colloquium on Automata, Languages and Programming, pp. 666–677 (2011)
12. Kamiyama, N.: Popular matchings with two-sided preference lists and matroid constraints. Technical report MI 2016–13 (2016)
13. Kavitha, T.: A size-popularity tradeoff in the stable marriage problem. SIAM J. Comput. **43**(1), 52–71 (2014)
14. Nasre, M., Rawat, A.: Popularity in the Generalized Hospital Residents Setting. CoRR, abs/1609.07650 (2016)

Edit Distance Neighbourhoods of Input-Driven Pushdown Automata

Alexander Okhotin[1(✉)] and Kai Salomaa[2]

[1] St. Petersburg State University, 14th Line V.O., 29B,
Saint Petersburg 199178, Russia
`alexander.okhotin@spbu.ru`
[2] School of Computing, Queen's University,
Kingston, ON K7L 2N8, Canada
`ksalomaa@cs.queensu.ca`

Abstract. Edit distance ℓ-neighbourhood of a formal language is the set of all strings that can be transformed to one of the strings in this language by at most ℓ insertions and deletions. Both the regular and the context-free languages are known to be closed under this operation, whereas the deterministic pushdown automata are not. This paper establishes the closure of the family of input-driven pushdown automata (IDPDA), also known as visibly pushdown automata, under the edit distance neighbourhood operation. A construction of automata representing the result of the operation is given, and close lower bounds on the size of any such automata are presented.

1 Introduction

Edit distance is the standard measure of similarity between two strings: this is the least number of elementary edit operations—such as inserting a symbol, removing a symbol or replacing a symbol with another symbol—necessary to transform one string into another. Algorithms and methods related to the edit distance are useful in numerous applications: whenever DNA sequences are checked for similarity, misspelled words are matched to their most probable spelling, etc.

Many problems involving edit distance are formulated in terms of formal languages. In particular, one can consider the edit distance between a string and a language, which is relevant to assessing the number of syntax errors in an input string, as well as to correcting those errors [13]. There is also a notion of a distance between a pair of languages, studied, in particular, by Chatterjee et al. [6]. The shortest distance between two languages is uncomputable if both languages are given by grammars [11], whereas the distance between a grammar and a regular language is computable [7].

In connection with the distance between a string and a language, there is a convenient notion of *edit distance ℓ-neighbourhood* of a given language: this is a set of all strings at edit distance at most ℓ from some element of that language. The edit distance ℓ-neighbourhood is then an operation on languages.

© Springer International Publishing AG 2017
P. Weil (Ed.): CSR 2017, LNCS 10304, pp. 260–272, 2017.
DOI: 10.1007/978-3-319-58747-9_23

It is known that the regular languages are closed under this operation. In particular, Povarov [20] determined an optimal construction for the 1-neighbourhood of a given automaton; this result was extended to the ℓ-neighbourhood in the papers by Salomaa and Schofield [21] and by Ng, Rappaport and Salomaa [12].

The edit distance operation is no less relevant in formal grammars. For context-free grammars, the work by Aho and Peterson [1] on error recovery in parsers contains a direct construction of a grammar, which is sufficient to prove the closure under edit distance neighbourhood. Also, ℓ-neighbourhood is computable by a nondeterministic finite transducer (NFT), and by the closure of grammars under all such transductions, the closure follows; the same argument applies to all families closed under NFT, such as the linear grammars. On the other hand, for deterministic pushdown automata (DPDA)—or, equivalently, for LR(k) grammars—there is a simple example witnessing their non-closure under the 1-neighbourhood operation: the language $L = \{ ca^nb^n \mid n \geqslant 0 \} \cup \{ da^nb^{2n} \mid n \geqslant 0 \}$ is recognized by a DPDA, whereas its 1-neighbourhood, under intersection with a^*b^*, is the language $\{ a^nb^n \mid n \geqslant 0 \} \cup \{ a^nb^{2n} \mid n \geqslant 0 \}$, which is a classical example of a language not recognized by any DPDA.

This paper investigates ℓ-neighbourhoods for an important subclass of DPDA: the *input-driven pushdown automata* (IDPDA), also known under the name of *visibly pushdown automata*. In these automata, the input symbol determines whether the automaton should push a stack symbol, pop a stack symbol or leave the stack untouched. These symbols are called *left brackets*, *right brackets* and *neutral symbols*, and the symbol pushed at each left bracket is always popped when reading the corresponding right bracket. Input-driven automata are important as a model of hierarchically structured data, such as XML documents or computation traces for recursive procedure calls. They are also notable for their appealing theoretical properties, resembling those of finite automata.

Input-driven automata were first studied by Mehlhorn [10] and by von Braunmühl and Verbeek [4], who determined that the languages they recognize lie in logarithmic space. Von Braunmühl and Verbeek [4] also proved that deterministic and nondeterministic variants of the model are equal in power. Later, Alur and Madhusudan [2,3] reintroduced the model under the names "visibly pushdown automata" and "nested word automata", and carried out its language-theoretic study, in particular, establishing the closure of the corresponding family under the basic operations on languages. Their contribution inspired further work on the closure properties of input-driven automata and on their descriptional complexity [8, 16–18].

The main result of this paper, presented in Sect. 3, is that the family of languages recognized by input-driven automata is closed under the edit distance neighbourhood operation. The main difficulty in the construction is that when the symbol inserted or deleted is a bracket, then adding or removing that symbol changes the bracket structure of the string, so that other brackets may now be matched not to the same brackets as before. It is shown how, given an NIDPDA for the original language, to construct an NIDPDA with one edit operation applied.

The question of whether these constructions are optimal in terms of the number of states is addressed in Sect. 4, where some lower bounds on the worst-case size of an NIDPDA representing the edit distance neighbourhood of an n-state NIDPDA are established. These bounds confirm that the constructions presented in this paper are fairly close to optimal.

In Sect. 5, a similar construction is presented for deterministic input-driven automata. The construction uses exponentially many states, and is accompanied with a fairly close lower bound, showing that a DIDPDA for the edit distance neighbourhood requires $2^{\Omega(n^2)}$ states in the worst case.

2   Input-Driven Automata

An *input-driven pushdown automaton* (IDPDA) [2,3,10] is a special case of a deterministic pushdown automaton, in which the input alphabet Σ is split into three disjoint sets of *left brackets* Σ_{+1}, *right brackets* Σ_{-1} and *neutral symbols* Σ_0. If the input symbol is a left bracket from Σ_{+1}, then the automaton always pushes one symbol onto the stack. For a right bracket from Σ_{-1}, the automaton must pop one symbol. Finally, for a neutral symbol in Σ_0, the automaton may not use the stack. In this paper, symbols from Σ_{+1} and Σ_{-1} shall be denoted by left and right angled brackets, respectively ($<$, $>$), whereas lower-case Latin letters from the beginning of the alphabet ($a, b, c, \ldots$) shall be used for symbols from Σ_0. Input-driven automata may be deterministic (DIDPDA) and nondeterministic (NIDPDA).

Under the simpler definition, input-driven automata operate on input strings, in which the brackets are *well-nested*. When an input-driven automaton reads a left bracket $< \in \Sigma_{+1}$, it pushes a symbol onto the stack. This symbol is popped at the exact moment when the automaton encounters the matching right bracket $> \in \Sigma_{-1}$. Thus, a computation of an input-driven automaton on any well-nested substring leaves the stack contents untouched, as illustrated in Fig. 1.

The more general definition of input-driven automata assumed in this paper also allows ill-nested input strings. For every unmatched left bracket, the symbol pushed to the stack when reading this bracket is never popped, and remains in the stack to the end of the computation. An unmatched right bracket is read with an empty stack: instead of popping a stack symbol, the automaton merely

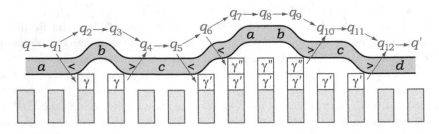

Fig. 1. The computation of an IDPDA on a well-nested string.

detects that the stack is empty and makes a special transition, which leaves the stack empty.

Definition 1 (von Braunmühl and Verbeek [4]; Alur and Madhusudan [2]). *A nondeterministic input-driven pushdown automaton (NIDPDA) over an alphabet $\widetilde{\Sigma} = (\Sigma_{+1}, \Sigma_{-1}, \Sigma_0)$ consists of*

- *a finite set Q of states, with set of initial states $Q_0 \subseteq Q$ and accepting states $F \subseteq Q$;*
- *a finite stack alphabet Γ, and a special symbol $\bot \notin \Gamma$ for the empty stack;*
- *for a neutral symbol $c \in \Sigma_0$, a transition function $\delta_c \colon Q \to 2^Q$ gives the set of possible next states;*
- *for each left bracket symbol $< \in \Sigma_{+1}$, the behaviour of the automaton is described by a function $\delta_< \colon Q \to 2^{Q \times \Gamma}$, which, for a given current state, provides a set of pairs (q, γ), with $q \in Q$ and $\gamma \in \Gamma$, where each pair means that the automaton enters the state q and pushes γ onto the stack;*
- *for every right bracket symbol $> \in \Sigma_{-1}$, there is a function $\delta_> \colon Q \times (\Gamma \cup \{\bot\}) \to 2^Q$ specifying possible next states, assuming that the given stack symbol is popped from the stack (or that the stack is empty).*

A configuration is a triple (q, w, x), with the current state $q \in Q$, remaining input $w \in \Sigma^*$ and stack contents $x \in \Gamma^*$. Possible next configurations are defined as follows.

$$
\begin{aligned}
(q, cw, x) \vdash_A (q', w, x), && c \in \Sigma_0,\ q \in Q,\ q' \in \delta_c(q) \\
(q, {<}w, x) \vdash_A (q', w, \gamma x), && {<} \in \Sigma_{+1},\ q \in Q,\ (q', \gamma) \in \delta_<(q) \\
(q, {>}w, \gamma x) \vdash_A (q', w, x), && {>} \in \Sigma_{-1},\ q \in Q,\ \gamma \in \Gamma,\ q' \in \delta_>(q, \gamma) \\
(q, {>}w, \epsilon) \vdash_A (q', w, \epsilon), && {>} \in \Sigma_{-1},\ q' \in \delta_>(q, \bot)
\end{aligned}
$$

The language recognized by A is the set of all strings $w \in \Sigma^*$, on which the automaton, having begun its computation in the configuration (q_0, w, ϵ), eventually reaches a configuration of the form (q, ϵ, x), with $q \in F$ and with any stack contents $x \in \Gamma^*$.

An NIDPDA is deterministic (DIDPDA), if there is a unique initial state and every transition provides exactly one action.

As shown by von Braunmühl and Verbeek [4], every n-state NIDPDA operating on well-nested strings can be transformed to a 2^{n^2}-state DIDPDA. Alur and Madhusudan [2] extended this construction to allow ill-nested inputs, so that a DIDPDA has 2^{2n^2} states; in the worst case, $2^{\Omega(n^2)}$ states are necessary.

Another basic construction for DIDPDA that will be used in this paper is computing the *behaviour function* of a given DIDPDA by another DIDPDA. When a DIDPDA with a set of states Q processes a well-nested string w and begins in a state q, it finishes reading that string in some state $f(q)$, where $f \colon Q \to Q$ is its *behaviour function on w*, and the stack is left untouched. Thus, f completely characterizes the behaviour of a DIDPDA on w. For any given

DIDPDA A, it is possible to construct an n^n-state DIDPDA, where $n = |Q|$, that reaches the end of an input w in a state representing the behaviour of A on the longest well-nested suffix of w. This construction is necessary for optimal constructions representing operations on DIDPDA [17].

For more details on input-driven automata and their complexity, the readers are directed to a recent survey [15].

3   Edit Distance for Input-Driven Automata

Let Σ be an alphabet, let $a \in \Sigma$ be a symbol. Then, for a string $w \in \Sigma^*$, the set of strings obtained by inserting a at any position is denoted by $\text{insert}_a(w) = \{ uav \mid w = uv \}$. Similarly, the set of strings obtained by erasing a is $\text{delete}_a(w) = \{ uv \mid w = uav \}$. These operations are extended to any language $L \subseteq \Sigma^*$ elementwise, with $\text{insert}_a(L) = \bigcup_{w \in L} \text{insert}_a(w)$ and $\text{delete}_a(L) = \bigcup_{w \in L} \text{delete}_a(w)$.

The set of strings at edit distance at most ℓ from a given string w is called its ℓ-*neighbourhood*, denoted by $E_\ell(w)$ and defined as follows.

$$E_0(w) = \{w\}$$

$$E_{\ell+1}(w) = E_\ell(w) \cup \bigcup_{w' \in E_\ell(w)} \bigcup_{a \in \Sigma} \left(\text{insert}_a(w') \cup \text{delete}_a(w') \right)$$

The ℓ-neighbourhood of a language $L \subseteq \Sigma^*$ is the set of strings at edit distance at most ℓ from any string in L.

$$E_\ell(L) = \bigcup_{w \in L} E_\ell(w)$$

The definition of edit distance often includes the operation of replacing one symbol with another. According to the above definition, replacement can be implemented as a combination of one deletion and one insertion. This difference affects the resulting edit distance. In this paper, the simpler definition is assumed, because it makes the constructions easier; however, the constructions in this paper can be extended to implement replacement as well.

In this paper, the above definitions are applied to languages over an alphabet $\Sigma = \Sigma_{+1} \cup \Sigma_{-1} \cup \Sigma_0$ recognized by an IDPDA, with the intention of constructing another IDPDA that recognizes the edit distance neighbourhood of the given language. A construction shall be obtained by first implementing the elementary operations of inserting or deleting a single symbol. According to the definition of the neighbourhood, all three types of symbols in Σ can be either inserted or deleted. However, since IDPDA handle different types of symbols differently, these six cases generally require separate treatment.

Neutral symbols are the easiest to insert or delete, the construction is the same as for finite automata.

Lemma 1. *Let L be a language recognized by an NIDPDA, let Q be its set of states, and Γ its stack alphabet. Let $c \in \Sigma_0$ be a neutral symbol. Then, both* languages insert$_c(L)$ and delete$_c(L)$ are recognized by NIDPDA with the set of states $Q \cup \widetilde{Q}$, where $\widetilde{Q} = \{\widetilde{q} \mid q \in Q\}$, and with the stack alphabet Γ.

There is also an NIDPDA with the same set of states Q and the same set of stack symbols Γ that recognizes the language $\bigcup_{c \in \Sigma_0} (\text{insert}_c(L) \cup \text{delete}_c(L))$.

The second case is that of inserting a left bracket. The main difficulty is, that once a new left bracket is inserted into a given string, it may match some existing right bracket, which was formerly matched to a different left bracket. This disrupts the operation of the simulated NIDPDA, and requires some efforts to re-create it.

Lemma 2. Let L be a language recognized by an NIDPDA over an alphabet with the set of states Q and with the stack alphabet Γ. Let $\ll \in \Sigma_{+1}$ be a left bracket. Then, the language insert$_\ll(L)$ is recognized by an NIDPDA with the set of states $Q \cup \widetilde{Q} \cup (Q \times \Gamma)$, where $\widetilde{Q} = \{\widetilde{q} \mid q \in Q\}$, and with the stack alphabet $\Gamma \cup \{\Box\} \cup (\Gamma \times \Gamma)$.

There is also an NIDPDA with the same states Q and the same stack symbols Γ that recognizes $\bigcup_{\ll \in \Sigma_{+1}} \text{insert}_\ll(L)$.

Proof. The first two types of states in the new automaton are the states q and $\widetilde{q}$, for any $q \in Q$. In either state, the new automaton simulates the original automaton being in the state q.

In the beginning, the new automaton uses the states from $\widetilde{Q}$ to simulate the operation of the original automaton before it encounters the new left bracket that has been inserted. At some point, the new automaton guesses that the currently observed left bracket is the new one, and executes a special transition: when passing the inserted left bracket ($\ll$) in a state $\widetilde{q}$, the new automaton pushes a special box symbol ($\Box$) into the stack and enters the state q: in these states, the new automaton knows that the inserted symbol has already been encountered, and simulates the original automaton as it is.

Later, when the automaton pops the box ($\Box$) upon reading some right bracket ($>$), it knows that the stack symbol in the original computation corresponding to this bracket lies in its stack one level deeper. Being an input-driven automaton, it cannot pop it yet, but it can guess what that symbol is going to be. If γ is the guessed stack symbol, then the automaton simulates the transition upon popping γ and enters a state of the form (q, γ), where q is the result of the transition, and γ is remembered in the state for later verification.

In states of the form (q, γ), neutral symbols are being read without modifying the remembered stack symbol. Whenever a left bracket ($<$) occurs, and the original automaton would enter a state r and push a stack symbol σ, the new automaton enters the state r and pushes a special stack symbol (σ, γ), which maintains the remembered stack symbol in the second component, and restores it upon reading the well-nested substring.

When, in a state of the form (q, γ), the new automaton reaches a right bracket ($>$), first, it verifies that the symbol being popped is indeed γ. The stack symbol needed to carry out the present transition is again located one level deeper in the stack, and therefore the automaton has to guess another stack symbol γ', and store it in the second component of the pair, etc. This completes the construction.

Since the automaton does not need to know the particular bracket symbol $\ll \in \Sigma_{+1}$ that has been inserted before and after encountering it, the same construction yields an NIDPDA for the language $\bigcup_{\ll \in \Sigma_{+1}} \text{insert}_{\ll}(L)$.   □

The case of erasing a left bracket is carried out slightly differently.

Lemma 3. *Let L be a language recognized by an NIDPDA with the set of states Q and with the stack alphabet Γ. Let $\ll \in \Sigma_{+1}$ be a left bracket. Then, the language $\text{delete}_{\ll}(L)$ is recognized by an NIDPDA with the set of states $Q \cup \widetilde{Q} \cup (Q \times \Gamma)$ and with the stack alphabet $\Gamma \cup (\Gamma \times \Gamma)$.*

Also, the language $\bigcup_{\ll \in \Sigma_{+1}} \text{delete}_{\ll}(L)$, is recognized by an NIDPDA with the same states and stack symbols.

Proof. The plan is that the new automaton is in a state $\widetilde{q}$ before passing the place where a left bracket ($\ll$) was erased. State (q, γ) means the situation after passing the erased left bracket ($\ll$), while remembering the stack symbol that the original automaton would push when reading that erased bracket ($\ll$). This state means that γ is an extra stack symbol simulated on the top of the actual stack. In a state q, the new automaton operates normally, as the erased symbol is no longer expected.

Transitions in the state $\widetilde{q}$ are the same as those in q, except that, upon reading any symbol, the new automaton may decide that directly after that symbol there was a left bracket ($\ll$) that got erased. Then, the new automaton simulates a transition by these two symbols at once, and, assuming that the original automaton's transition upon the left bracket ($\ll$) is to a state r along with pushing a symbol γ, the new automaton enters the state (r, γ).

In a state of the form $(q, \widehat{\gamma})$, upon reading any left bracket ($<$), the automaton pushes a pair of stack symbols $(\gamma, \widehat{\gamma})$, where γ is the symbol that the original automaton would push, and enters the same state that the original automaton would enter. Later, upon reading the matching right bracket ($>$) and popping the pair $(\gamma, \widehat{\gamma})$, the automaton enters the state $(r, \widehat{\gamma})$, assuming that the original automaton would enter the state r.

When the new automaton encounters a right bracket ($>$) in a state of the form $(q, \widehat{\gamma})$, popping a stack symbol γ, it simulates the original automaton's transition in the state q upon popping the stack symbol $\widehat{\gamma}$, and enters the state (r, γ), assuming that r is the state that the original automaton would enter.

In a state of the form $(q, \widehat{\gamma})$, upon reaching the bottom of the stack, the automaton simulates the transition upon popping γ and enters a normal state r, the same that the original automaton would enter.   □

The constructions for insertion and deletion of right brackets are symmetric, the number of states is the same.

Now, an NIDPDA for edit distance 1-neighbourhood can be obtained by using all the six constructions within a single automaton.

Theorem 1. *Let L be a language recognized by an NIDPDA with n states and k stack symbols. Then there exists an NIDPDA recognizing the language $E_1(L)$ that has $10n + 4kn + 1$ states and $k^2 + k + 1$ stack symbols.*

The ℓ-neighbourhood can be obtained by applying this construction ℓ times.

4   Lower Bounds for the Nondeterministic Case

Several constructions of automata have been presented, and the question is, whether those constructions are optimal. This is proved by presenting *witness languages*, that is, families of languages L_n recognized by an NIDPDA of size n, such that every NIDPDA for the corresponding edit distance operation on L_n requires at least $f(n)$ states. The methods for establishing such results were originally developed for finite automata, and later were generalized for NIDPDA.

The *stack height* of a string w is the height of the stack of an NIDPDA after reading w. The height of the stack depends only on w.

Definition 2. *Let $\widetilde{\Sigma} = (\Sigma_{+1}, \Sigma_{-1}, \Sigma_0)$ be an alphabet and lct $L \subseteq \Sigma^*$. A set of pairs $F = \{(x_1, y_1), \ldots, (x_m, y_m)\}$ is said to be a* fooling set of depth k for L, *if each string x_i has stack height k and*

(i) $x_i y_i \in L$ for all $i \in \{1, 2, \ldots, m\}$, and
(ii) for all i, j with $1 \leqslant i < j \leqslant m$, $x_i y_j \notin L$ or $x_j y_i \notin L$.

Lemma 4 ([8,18]).   *Let A be a nondeterministic input-driven pushdown automaton with a set of states Q and a set of stack symbols Γ. If $L(A)$ has a fooling set F of depth k, then $|\Gamma|^k \cdot |Q| \geqslant |F|$.*

First consider the insertion or deletion of a single symbol.
Choose $\Sigma_{+1} = \{<\}$, $\Sigma_{-1} = \{>\}$ and $\Sigma_0 = \{a, b, c, \$\}$. For $n \geqslant 1$ define

$$L_n = \{c^i <c^k a^i b^j \$ b^j >a^i \mid 1 \leqslant i, j \leqslant n, \ k \geqslant 0\}.$$

Lemma 5. *(i) There exists a constant $C \geqslant 1$, such that, for each $n \geqslant 1$, the language L_n is recognized by a DIDPDA A with $C \cdot n$ states and n stack symbols.*
(ii) Any NIDPDA recognizing the language $\text{delete}_<(L_n)$ *needs at least n^2 states.*

Proof. (i) The following discussion assumes that the input string is in $c^+ <c^+ a^+ b^+ \$ b^+ >a^+$. It is easy to see that by increasing the number of states of A by a multiplicative constant, the computation can be made to reject all strings not of this form.

The computation counts the length i of the prefix in c^+ preceding the left bracket $<$ and pushes this value to the stack. If $i > n$, A rejects. Then A skips the following symbols c, checks that the maximal substring in a^+ has length i, and counts the number of b's preceding the marker $\$$. This number is compared with the number of b's after $\$$. At the right bracket $>$ the stack is popped and the computation verifies that the suffix of symbols a has length i.

(ii) Choose

$$S_n = \{(c^{n+1}a^i b^j, \$b^j > a^i) \mid 1 \leqslant i,j \leqslant n\}.$$

For all $i,j \in \{1,\ldots,n\}$, the string $c^{n+1}a^i b^j \cdot \$b^j > a^i$ is obtained from a string from L_n by deleting a left bracket. On the other hand, for $(i,j) \neq (i',j')$, with $i,j,i',j' \in \{1,\ldots,n\}$, the string $c^{n+1}a^i b^j \cdot \$b^{j'} > a^{i'}$ is not in $\text{delete}_<(L_n)$, because $i \neq i'$ or $j \neq j'$. This means that S_n is a fooling set of depth 0 for $\text{delete}_<(L_n)$ and, by Lemma 4, any NIDPDA for $\text{delete}_<(L_n)$ needs $|S_n| = n^2$ states.   □

Lemma 6. *Any NIDPDA recognizing the language* $\text{insert}_>(L_n)$ *needs at least* n^2 *states.*

Proof. Define

$$S'_n = \{(c^i <>ca^i b^j, \$b^j > a^i) \mid 1 \leqslant i,j \leqslant n\}.$$

Again, for all $1 \leqslant i,j \leqslant n$, $c^i <>ca^i b^j \cdot \$b^j > a^i$ is obtained from a string of L_n by inserting a right bracket and, on the other hand, for $(i,j) \neq (i',j')$, $c^i <>ca^i b^j \cdot \$b^{j'} > a^{i'} \notin \text{insert}_>(L_n)$. This means that S'_n is a fooling set for $\text{insert}_>(L_n)$, and the claim follows from Lemma 4.   □

The reversal L^R of the language L recognized by an NIDPDA A can be recognized by an NIDPDA with the same number of states and stack symbols as A, when the left brackets (respectively, right brackets) in the original string are interpreted as right brackets (respectively, left brackets) in the reversed string [3]. Since inserting a left bracket into a language L is the same as inserting a right bracket into L^R, and deleting a right bracket from L is the same as deleting a left bracket from L^R, Lemmas 5 and 6 imply a tight bound on the complexity of inserting left brackets and deleting right brackets in terms of the number of states in NIDPDA.

Corollary 1. *For each* $n \geqslant 1$ *there exists a language* L'_n *recognized by a NIDPDA with* $O(n)$ *states such that any NIDPDA for the neighbourhoods* $\text{delete}_>(L'_n)$ *and* $\text{insert}_<(L'_n)$ *needs* n^2 *states.*

It remains to consider the cases of inserting and deleting a neutral symbol. Povarov [20] has shown that the Hamming neighbourhood of radius r of an n state NFA language can be recognized by an NFA with $n \cdot (r+1)$ states and this number of states is needed in the worst case. Since an input-driven computation on strings consisting of neutral symbols is just an NFA, the lower bound for the number of states applies also for NIDPDAs. Together with Lemma 1 this implies:

Proposition 1. *For an NIDPDA* A *with* n *states and* $\sigma \in \Sigma_0$, *the neighbourhoods* $\text{delete}_\sigma(L(A))$ *and* $\text{insert}_\sigma(L(A))$ *can be recognized by an NIDPDA with* $2 \cdot n$ *states and this number of states is needed in the worst case.*

The construction of Lemma 5 can be extended to yield a lower bound for the cost of deleting multiple symbols. The result is stated in terms of neighbourhoods of a given radius.

Choose $\Sigma_{+1} = \{<\}$, $\Sigma_{-1} = \{>\}$ and $\Sigma_0 = \{a, b, c, \$\}$. For $n \geqslant 1$ define

$$H_n = \{<a^{i_1}c<a^{i_2}c\cdots<a^{i_r}cb^j\$b^j>a^{i_r}>a^{i_{r-1}}\cdots>a^{i_1} \mid$$

$$r \geqslant 1,\ i_1,\ldots,i_r, j \in \{1,\ldots,n\} \}.$$

Lemma 7. *(i) The language H_n can be recognized by an NIDPDA with $C \cdot n$ states and n stack symbols, for some constant C.*

(ii) For $r \geqslant 1$, any NIDPDA for the neighbourhood $E_r(H_n)$ needs at least n^{r+1} states.

5   The Deterministic Case

The construction for the edit distance neighbourhood given in the previous section produces an NIDPDA out of an NIDPDA. If the goal is to obtain a deterministic automaton, then the resulting NIDPDA can of course be determinized, at the cost of a $2^{\Theta(n^2)}$ blow-up in size. This section presents some preliminary results on a direct construction for this operation, which transforms a DIDPDA to a DIDPDA for the language with one left bracket erased.

Lemma 8. *Let L be a language recognized by a DIDPDA with the set of states Q and with the stack alphabet Γ. Let $\ll \in \Sigma_{+1}$ be a left bracket. Then, the language delete$_{\ll}(L)$ is recognized by a DIDPDA with the set of states $Q' = Q \times 2^{Q \times (\Gamma \cup \{\perp\})} \times Q^Q$ and with the stack alphabet $\Gamma' = \Sigma_{+1} \times \Gamma \times 2^{Q \times (\Gamma \cup \{\perp\})} \times Q^Q$, where Q^Q denotes the set of all functions from Q to Q.*

Proof (sketch). At each level of brackets, the new automaton simulates the normal operation of the first automaton (Q), as well as constructs two data structures. The first data structure $(2^{Q \times (\Gamma \cup \{\perp\})})$ is a set of pairs of a state q and a stack symbol γ, each representing a situation when the computation on this level, having processed some erased bracket $(\ll)$ at some position, has pushed γ upon reading that bracket, and finished reading the substring on this level in the state q. The second data structure (Q^Q) is the behaviour function for the well-nested substring at the current level.   □

There is a close lower bound for this construction. Let the alphabet be $\Sigma_{+1} = \{<\}$, $\Sigma_{-1} = \{>\}$ and $\Sigma_0 = \{a, b, c, d\}$. For each $n \geqslant 1$, the language K_n is defined as follows.

$$K_n = \{uc<vd^{|u|_a+|v|_b \bmod n}>a^{|u|_a \bmod n} \mid u, v \in \{a, b, c\}^*\}$$

Lemma 9. *The language K_n is recognized by a DIDPDA with $C \cdot n$ states.*

Proof. First, the DIDPDA counts the number of symbols a in u modulo n. Then, upon encountering the left bracket ($<$) and verifying that it is preceded by c, it pushes the count of symbols a modulo n to the stack, and continues the counting modulo n on the string v, this time counting the symbols b. After reading v, the automaton remembers the sum $|u|_a + |v|_b$ modulo n, and can then test that the number of symbols d is correct. Finally, upon reading the right bracket ($>$), the automaton pops the number $|u|_a$ modulo n from the stack and checks this number against the suffix $a^{|u|_a \bmod n}$. $\square$

Lemma 10. *Every DIDPDA recognizing* $\text{delete}_<(K_n)$ *needs at least* 2^{n^2} *states.*

Proof (Sketch of proof). A DIDPDA is faced with recognizing the following language.

$$K'_n = \{ wd^{i+j}>a^i \mid w \in \{a, b, c\}^*, \text{ and there exists a partition } w = ucv,$$
$$\text{with } i = |u|_a \bmod n \text{ and } j = |v|_b \bmod n \}$$

In the absence of left brackets, the automaton is essentially a DFA. The idea of the lower bound argument is that a DFA should remember all pairs (i, j) corresponding to different partitions of w as $w = ucv$. $\square$

This was just one of the four interesting cases of edit operations on DIDPDA. The other three cases shall be dealt with in the full version of this paper. However, this single case alone already implies a lower bound on the complexity of edit distance 1-neighbourhood of DIDPDA: indeed, any DIDPDA recognizing $E_1(K_n)$ needs at least 2^{n^2} states.

It can be concluded that the edit distance neighbourhood can be efficiently expressed in nondeterministic IDPDA, and incurs a significant blow-up in the deterministic case.

6   Future Work

It would be interesting to consider the edit distance neighbourhood operation for other automaton models related to IDPDA that are relevant to processing hierarchical data. Among such models, there are, in particular, the *transducer-driven automata* (TDPDA), introduced independently by Caucal [5] (as synchronized pushdown automata) and by Kutrib et al. [9].

In addition to the input-driven automaton models, the same question of the expressibility of edit distance neighbourhood would be interesting to investigate for other families of formal grammars besides the ordinary "context-free" grammars. The families proposed for investigation are the *multi-component grammars* [22], which are an established model in computational linguistics and have good closure properties, and the *conjunctive grammars*, which extend the ordinary grammars with a conjunction operation. In particular, it would be interesting to investigate the edit distance for the *linear conjunctive grammars* [14], which are notable for their equivalence with one-way real-time cellular automata, as well as for their non-trivial expressive power [23].

References

1. Aho, A.V., Peterson, T.G.: A minimum distance error-correcting parser for context-free languages. SIAM J. Comput. **1**(4), 305–312 (1972). http://dx.doi.org/doi/10.1137/0201022
2. Alur, R., Madhusudan, P.: Visibly pushdown languages. In: ACM Symposium on Theory of Computing, STOC 2004, Chicago, USA 13–16 June 2004, pp. 202–211 (2004). http://dx.doi.org/10.1145/1007352.1007390
3. Alur, R., Madhusudan, P.: Adding nesting structure to words. J. ACM **56**(3) (2009). http://dx.doi.org/10.1145/1516512.1516518
4. von Braunmühl, B., Verbeek, R.: Input driven languages are recognized in log n space. Ann. Discrete Math. **24**, 1–20 (1985). http://dx.doi.org/10.1016/S0304-0208(08)73072-X
5. Caucal, D.: Synchronization of pushdown automata. In: Ibarra, O.H., Dang, Z. (eds.) DLT 2006. LNCS, vol. 4036, pp. 120–132. Springer, Heidelberg (2006). doi:10.1007/11779148_12
6. Chatterjee, K., Henzinger, T.A., Ibsen-Jensen, R., Otop, J.: Edit distance for pushdown automata. In: Halldórsson, M.M., Iwama, K., Kobayashi, N., Speckmann, B. (eds.) ICALP 2015. LNCS, vol. 9135, pp. 121–133. Springer, Heidelberg (2015). doi:10.1007/978-3-662-47666-6_10
7. Han, Y.-S., Ko, K., Salomaa, K.: Approximate matching between a context-free grammar and a finite-state automaton. Inf. Comput. **247**, 278–289 (2016). http://dx.doi.org/10.1016/j.ic.2016.02.001
8. Han, Y.-S., Salomaa, K.: Nondeterministic state complexity of nested word automata. Theoret. Comput. Sci. **410**, 2961–2971 (2009)
9. Kutrib, M., Malcher, A., Wendlandt, M.: Tinput-driven pushdown automata. In: Durand-Lose, J., Nagy, B. (eds.) MCU 2015. LNCS, vol. 9288, pp. 94–112. Springer, Cham (2015). doi:10.1007/978-3-319-23111-2_7
10. Mehlhorn, K.: Pebbling mountain ranges and its application to DCFL-recognition. In: Bakker, J., Leeuwen, J. (eds.) ICALP 1980. LNCS, vol. 85, pp. 422–435. Springer, Heidelberg (1980). doi:10.1007/3-540-10003-2_89
11. Mohri, M.: Edit-distance of weighted automata: general definitions and algorithms. Int. J. Found. Comput. Sci. **14**(6), 957–982 (2003)
12. Han, Y.-S., Ko, S.-K., Salomaa, K.: Generalizations of code languages with marginal errors. In: Potapov, I. (ed.) DLT 2015. LNCS, vol. 9168, pp. 264–275. Springer, Cham (2015). doi:10.1007/978-3-319-21500-6_21
13. Ng, T., Rappaport, D., Salomaa, K.: Descriptional complexity of error detection. In: Adamatzky, A. (ed.) Emergent Computation. ECC, vol. 24, pp. 101–119. Springer, Cham (2017). doi:10.1007/978-3-319-46376-6_6
14. Okhotin, A.: Input-driven languages are linear conjunctive. Theoret. Comput. Sci. **618**, 52–71 (2016). http://dx.doi.org/10.1016/j.tcs.2016.01.007
15. Okhotin, A., Salomaa, K.: Complexity of input-driven pushdown automata. SIGACT News **45**(2), 47–67 (2014). http://doi.acm.org/10.1145/2636805.2636821
16. Okhotin, A., Salomaa, K.: Descriptional complexity of unambiguous input-driven pushdown automata. Theoret. Comput. Sci. **566**, 1–11 (2015). http://dx.doi.org/10.1016/j.tcs.2014.11.015
17. Okhotin, A., Salomaa, K.: State complexity of operations on input-driven pushdown automata. J. Comput. Syst. Sci. **86**, 207–228 (2017). http://dx.doi.org/10.1016/j.jcss.2017.02.001

18. Piao, X., Salomaa, K.: Operational state complexity of nested word automata. Theoret. Comput. Sci. **410**, 3290–3302 (2009). http://dx.doi.org/10.1016/j.tcs.2009.05.002

19. Pighizzini, G.: How hard is computing the edit distance? Inf. Comput. **165**, 1–13 (2001)

20. Povarov, G.: Descriptive complexity of the Hamming neighborhood of a regular language. In: LATA 2007, pp. 509–520 (2007)

21. Salomaa, K., Schofield, P.N.: State complexity of additive weighted finite automata. Int. J. Found. Comput. Sci. **18**(6), 1407–1416 (2007)

22. Seki, H., Matsumura, T., Fujii, M., Kasami, T.: On multiple context-free grammars. Theoret. Comput. Sci. **88**(2), 191–229 (1991). http://dx.doi.org/10.1016/0304-3975(91)90374-B

23. Terrier, V.: Recognition of linear-slender context-free languages by real time one-way cellular automata. In: Kari, J. (ed.) AUTOMATA 2015. LNCS, vol. 9099, pp. 251–262. Springer, Heidelberg (2015). doi:10.1007/978-3-662-47221-7_19

The (Minimum) Rank of Typical Fooling-Set Matrices

Mozhgan Pourmoradnasseri$^{(\boxtimes)}$ and Dirk Oliver Theis

Institute of Computer Science, University of Tartu,
Ülikooli 17, 51014 Tartu, Estonia
{mozhgan,dotheis}@ut.ee
http://ac.cs.ut.ee/

Abstract. A fooling-set matrix is a square matrix with nonzero diagonal, but at least one in every pair of diagonally opposite entries is 0. Dietzfelbinger et al. '96 proved that the rank of such a matrix is at least $\sqrt{n}$, for a matrix of order n. It is known that the bound is tight (up to a multiplicative constant).

We ask for the *typical* minimum rank of a fooling-set matrix: For a fooling-set zero-nonzero pattern chosen at random, is the minimum rank of a matrix with that zero-nonzero pattern over a field $\mathbb{F}$ closer to its lower bound $\sqrt{n}$ or to its upper bound n? We study random patterns with a given density p, and prove an $\Omega(n)$ bound for the cases when
(a) p tends to 0 quickly enough;
(b) p tends to 0 slowly, and $|\mathbb{F}| = \mathcal{O}(1)$;
(c) $p \in]0,1]$ is a constant.
We have to leave open the case when $p \to 0$ slowly and $\mathbb{F}$ is a large or infinite field (e.g., $\mathbb{F} = GF(2^n)$, $\mathbb{F} = \mathbb{R}$).

1 Introduction

Let $f\colon X \times Y \to \{0,1\}$ be a function. A *fooling set* of size n is a family $(x_1,y_1),\ldots,(x_n,y_n) \in X \times Y$ such that $f(x_i,y_i) = 1$ for all i, and for $i \neq j$, at least one of $f(x_i,y_j)$ or $f(x_j,y_i)$ is 0. Sizes of fooling sets are important lower bounds in Communication Complexity (see, e.g., [12,13]) and the study of extended formulations (e.g., [1,4]).

There is an *a priori* upper bound on the size of fooling sets due to Dietzfelbinger et al. [3], based on the rank of a matrix associated with f. Let $\mathbb{F}$ be an arbitrary field. The following is a slight generalization of the result in [3].

Lemma 1. *No fooling set in f is larger than the square of $\min_A \mathrm{rk}_{\mathbb{F}}(A)$, where the minimum ranges[1] over all $X \times Y$-matrices A over $\mathbb{F}$ with $A_{x,y} = 0$ iff $f(x,y) = 0$.*

Supported by the Estonian Research Council, ETAG (*Eesti Teadusagentuur*), through PUT Exploratory Grant #620, and by the European Regional Development Fund through the Estonian Center of Excellence in Computer Science, EXCS.

[1] This concept of *minimum rank* differs from the definition used in the context of index coding [8,10]. It is closer to the minimum rank of a graph, but there the matrix A has to be symmetric while the diagonal entries are unconstrained.

P. Weil (Ed.): CSR 2017, LNCS 10304, pp. 273–284, 2017.
DOI: 10.1007/978-3-319-58747-9_24

Proof. Let $(x_1, y_1), \ldots, (x_n, y_n) \in X \times Y$ be a fooling set in f, and let A be a matrix over $\mathbb{F}$ with $A_{x,y} = 0$ iff $f(x,y) = 0$. Consider the matrix $B := A \otimes A^\top$. This matrix B contains a permutation matrix of size n as a submatrix: for $i = 1, \ldots, n$, $B_{(x_i,x_i),(y_i,y_i)} = A_{x_i,y_i} A_{y_i,x_i} = 1$ but for $i \neq j$, $B_{(x_i,x_i),(y_j,y_j)} = A_{x_i,y_j} A_{y_i,x_j} = 0$. Hence,

$$n \leq \mathrm{rk}(B) = \mathrm{rk}(A)^2.$$

It is known that, for fields $\mathbb{F}$ with nonzero characteristic, this upper bound is asymptotically attained [6], and for all fields, it is attained up to a multiplicative constant [5]. These results, however, require sophisticated constructions. In this paper, we ask how useful that upper bound is for *typical* functions f.

Put differently, a *fooling-set pattern of size* n is a matrix R with entries in $\{0,1\} \subseteq \mathbb{F}$ with $R_{k,k} = 1$ for all k and $R_{k,\ell} R_{\ell,k} = 0$ whenever $k \neq \ell$. We say that a fooling-set pattern of size n has *density* $p \in \,]0,1]$, if it has exactly $\lceil p\binom{n}{2} \rceil$ off-diagonal 1-entries. So, the density is roughly the quotient $(|R| - n)/\binom{n}{2}$, where $|\cdot|$ denotes the Hamming weight, i.e., the number of nonzero entries. The densest possible fooling-set pattern has $\binom{n}{2}$ off-diagonal ones (density $p = 1$).

For any field $\mathbb{F}$ and $y \in \mathbb{F}$, let $\sigma(y) := 0$, if $y = 0$, and $\sigma(y) := 1$, otherwise. For a matrix (or vector, in case $n = 1$) $M \in \mathbb{F}^{m \times n}$, define the *zero-nonzero pattern of* M, $\sigma(M)$, as the matrix in $\{0,1\}^{m \times n}$ which results from applying σ to every entry of M.

This paper deals with the following question: *For a fooling-set pattern chosen at random, is the minimum rank closer to its lower bound $\sqrt{n}$ or to its trivial upper bound n?* The question turns out to be surprisingly difficult. We give partial results, but we must leave some cases open. The distributions we study are the following:

$Q(n)$ denotes a fooling-set pattern drawn uniformly at random from all fooling-set patterns of size n;

$R(n,p)$ denotes a fooling-set patterns drawn uniformly at random from all fooling-set patterns of size n with density p.

We allow that the density depends on the size of the matrix: $p = p(n)$. From now on, $Q = Q(n)$ and $R = R(n,p)$ will denote these random fooling-set patterns.

Our first result is the following. As customary, we use the terminology "asymptotically almost surely, a.a.s.," to stand for "with probability tending to 1 as n tends to infinity".

Theorem 1. *(a) For every field $\mathbb{F}$, if $p = O(1/n)$, then, a.a.s., the minimum rank of a matrix with zero-nonzero pattern $R(n,p)$ is $\Omega(n)$.*

(b) Let $\mathbb{F}$ be a finite field and $F := |\mathbb{F}|$. (We allow F to grow with n.) If $100 \max(1, \ln \ln F)/n \leq p \leq 1$, then the minimum rank of a matrix over $\mathbb{F}$ with zero-nonzero pattern $R(n,p)$ is

$$\Omega \left(\frac{\log(1/p)}{\log(1/p) + \log(F)}\, n \right) = \Omega(n/\log(F)).$$

(c) For every field $\mathbb{F}$, *if* $p \in]0, 1]$ *is a constant, then the minimum rank of a matrix with zero-nonzero pattern* $R(n, p)$ *is* $\Omega(n)$. *(The same is true for zero-nonzero pattern* $Q(n)$.)*

Since the constant in the big-Ω in Theorem 1(c) tends to 0 with $p \to 0$, the proof technique used for constant p does not work for $p = o(1)$; moreover, the bound in (b) does not give an $\Omega(n)$ lower bound for infinite fields, or for large finite fields, e.g., $\mathrm{GF}(2^n)$. We conjecture that the bound is still true (see Lemma 2 for a lower bound):

Conjecture 1. For every field $\mathbb{F}$ and for all $p = p(n)$, the minimum rank of a fooling-set matrix with random zero-nonzero pattern $R(n, p)$ is $\Omega(n)$.

The bound in Theorem 1(b) is similar to that in [8], but it is better by roughly a factor of $\log n$ if p is (constant or) slowly decreasing, e.g., $p = 1/\log n$. (Their minrank definition gives a lower bound to fooling-set pattern minimum rank.)

The next three sections hold the proofs for Theorem 1.

2   Proof of Theorem 1(a)

It is quite easy to see (using, e.g., Turán's theorem) that in the region $p = O(1/n)$, $R(n, p)$ contains a triangular submatrix with nonzero diagonal entries of order $\Omega(n)$, thus lower bounding the rank over any field. Here, we prove the following stronger result, which also gives a lower bound (for arbitrary fields) for more slowly decreasing p.

Lemma 2. *For* $p(n) = d(n)/n = o(1)$, *if* $d(n) > C$ *for some constant* C, *then zero-nonzero pattern* $R(n, p)$ *contains a triangular submatrix with nonzero diagonal entries of size*

$$\Omega\left(\frac{\ln d}{d} \cdot n\right).$$

We prove the lemma by using the following theorem about the independence number of random graphs in the Erdős-Rényi model. Let $G_{n,q}$ denote the random graph with vertex set $[n]$ where each edge is chosen independently with probability q.

Theorem 2 (Theorem 7.4 in [11]). *Let* $\epsilon > 0$ *be a constant,* $q = q(n)$, *and define*

$$k_{\pm\epsilon} := \left\lfloor \frac{2}{q}(\ln(nq) - \ln\ln(nq) + 1 - \ln 2 \pm \epsilon) \right\rfloor.$$

There exists a constant C_ϵ *such that for* $C_\epsilon/n \le q = q(n) \le \ln^{-2} n$, *a.a.s., the largest independent set in* $G_{n,q}$ *has size between* $k_{-\epsilon}$ *and* $k_{+\epsilon}$.

Proof (of Lemma 2). Construct a graph G with vertex set $[n]$ from the fooling-set pattern matrix $R(n, p)$ in the following way: There is an edge between vertices k and ℓ with $k > \ell$, if and only if $M_{k,\ell} \neq 0$. This gives a random graph $G = G_{n,m,1/2}$ which is constructed by first drawing uniformly at random a graph from all graphs with vertex set $[n]$ and exactly m edges, and then deleting each edge, independently, with probability $1/2$. Using standard results in random graph theory (e.g., Lemma 1.3 and Theorem 1.4 in [7]), this random graph behaves similarly to the Erdős-Rényi graph with $q := p/2$. In particular, since $G_{n,p/2}$ has an independent set of size $\Omega(n)$, so does $G_{n,m,1/2}$.

It is easy to see that the independent sets in G are just the lower-triangular principal submatrices of $R_{n,p}$.

As already mentioned, Theorem 1(a) is completed by noting that for $p < C/n$, an easy application of Turán's theorem (or ad-hoc methods) gives us an independent set of size $\Omega(n)$.

3   Proof of Theorem 1(b)

Let $\mathbb{F}$ be a finite field with $F := |\mathbb{F}|$. As mentioned in Theorem 1, we allow $F = F(n)$ to depend on n. In this section, we need to bound some quantities away from others, and we do that generously.

Let us say that a *tee shape* is a set $T = I \times [n] \cup [n] \times I$, for some $I \subset [n]$. A *tee matrix* is a tee shape T together with a mapping $N : T \to \mathbb{F}$ which satisfies

$$N_{k,k} = 1 \text{ for all } k \in I, \quad \text{and} \quad N_{k,\ell} N_{\ell,k} = 0 \text{ for all } (k, \ell) \in I \times [n], \ k \neq \ell. \quad (1)$$

The *order* of the tee shape/matrix is $|I|$, and the *rank* of the tee matrix is the rank of the matrix $N_{I \times I}$.

For a matrix M and a tee matrix N with tee shape T, we say that M *contains the tee matrix* N, if $M_T = N$.

Lemma 3. *Let M be a matrix with rank $s := \mathrm{rk}\, M$, which contains a tee matrix N of rank s. Then M is the only matrix of rank s which contains N.*

In other words, the entries outside of the tee shape are uniquely determined by the entries inside the tee shape.

Proof. Let $T = I \times [n] \cup [n] \times I$ be the tee shape of a tee matrix N contained in M.

Since $N_{I \times I} = M_{I \times I}$ and $\mathrm{rk}\, N_{I \times I} = s = \mathrm{rk}\, M$, there is a row set $I_1 \subseteq I$ of size $s = \mathrm{rk}\, M$ and a column set $I_2 \subseteq I$ of size s such that $\mathrm{rk}\, M_{I_1 \times I_2} = s$. This implies that M is uniquely determined, among the matrices of rank s, by $M_{T'}$ with $T' := I_1 \times [n] \cup [n] \times I_2 \subseteq T$. (Indeed, since the rows of $M_{I_1 \times [n]}$ are linearly independent and span the row space of M, every row in M is a unique linear combination of the rows in $M_{I_1 \times [n]}$; since the rows in $M_{I_1 \times I_2}$ are linearly independent, this linear combination is uniquely determined by the rows of $M_{[n] \times I_2}$.)

Hence, M is the only matrix M' with $\mathrm{rk}\, M' = s$ and $M'_{T'} = M_{T'}$. Trivially, then, M is the only matrix M' with $\mathrm{rk}\, M' = s$ and $M'_T = M_T = N$.

Lemma 4. *For $r \leq n/5$ and $m \leq 2r(n-r)/3$, there are at most*

$$O(1) \cdot \binom{n}{2r} \cdot \binom{2r(n-r)}{m} \cdot (2F)^m$$

matrices of rank at most r over $\mathbb{F}$ which contain a tee matrix of order $2r$ with at most m nonzeros.

Proof. By the Lemma 3, the number of these matrices is upper bounded by the number of tee matrices (of all ranks) of order $2r$ with at most k nonzeros.

The tee shape is uniquely determined by the set $I \subseteq [n]$. Hence, the number of tee shapes of order $2r$ is

$$\binom{n}{2r}. \tag{$*$}$$

The number of ways to choose the support of a tee matrix. Suppose that the tee matrix has h nonzeros. Due to (1), h nonzeros must be chosen from $\binom{2r}{2} + 2r(n-2r) \leq 2r(n-r)$ opposite pairs. Since $h < 2r(n-r)/2$, we upper bound this by

$$\binom{2r(n-r)}{h}.$$

For each of the h opposite pairs, we have to pick one side, which gives a factor of 2^h. Finally, picking, a number in $\mathbb{F}$ for each of the entries designated as nonzero gives a factor of $(F-1)^h$.

For summing over $h = 0, \ldots, m$, first of all, remember that $\sum_{i=0}^{(1-\varepsilon)j/2} \binom{j}{i} = O_\varepsilon(1) \cdot \binom{j}{(1-\varepsilon)j/2}$ (e.g., Theorem 1.1 in [2], with $p = 1/2$, $u := 1 + \varepsilon$). Since $m \leq 2r(n-r)/3$, we conclude

$$\sum_{h=0}^{m} \binom{2r(n-r)}{h} = O(1) \cdot \binom{2r(n-r)}{m}$$

(with an absolute constant in the big-Oh). Hence, we find that the number of tee matrices (with fixed tee shape) is at most

$$\sum_{h=0}^{m} \binom{2r(n-r)}{h} 2^h (F-1)^h \leq (2F)^m \sum_{h=0}^{m} \binom{2r(n-r)}{h} = O(1) \cdot (2F)^m \cdot \binom{2r(n-r)}{m}.$$

Multiplying by $(*)$, the statement of the lemma follows.

Lemma 5. *Let $r \leq n/5$. Every matrix M of rank at most r contains a tee matrix of order $2r$ and rank $\mathrm{rk}\, M$.*

Proof. There is a row set I_1 of size $s := \mathrm{rk}\, M$ and a column set I_2 of size s such that $\mathrm{rk}\, M_{I_1 \times I_2} = s$. Take I be an arbitrary set of size $2r$ containing $I_1 \cup I_2$, and $T := I \times [n] \cup [n] \times I$. Clearly, M contains the tee matrix $N := M_T$, which is of order $2r$ and rank $s = \mathrm{rk}\, M$.

Lemma 6. Let $100 \max(1, \ln \ln F)/n \leq p \leq 1$, and $n/(1000(\max(1, \ln F)) \leq r \leq n/100$. A.a.s., every tee shape of order $2r$ contained in the random matrix $R(n, p)$ has fewer than $15pr(n - r)$ nonzeros.

Proof. We take the standard Chernoff-like bound for the hypergeometric distribution of the intersection of uniformly random $p\binom{n}{2}$-element subset (the diagonally opposite pairs of $R(n, p)$ which contain a 1-entry) of a $\binom{n}{2}$-element ground set (the total number of diagonally opposite pairs) with a fixed $2r(n - r)$-element subset (the opposite pairs in T) of the ground set:[2] With $\lambda := p2r(n - r)$ (the expected size of the intersection), if $x \geq 7\lambda$, the probability that the intersection has at least x elements is at most e^{-x}.

Hence, the probability that the support of a fixed tee shape of order $2r$ is greater than than $15pr(n - r) \geq 14pr(n - r) + r$ is at most

$$e^{-14pr(n-r)} \leq e^{-r \cdot 14 \cdot 99 \cdot \max(1, \ln \ln F)} \leq e^{-r \cdot 1000 \cdot \max(1, \ln \ln F))}$$

Since the number of tee shapes is

$$\binom{n}{2r} \leq e^{r(1+\ln(n/r))} \leq e^{r(11+\ln \max(1, \ln F))} \leq e^{r(11+\max(1, \ln \ln F))}$$

we conclude that the probability that a dense tee shape exists in $R(n, p)$ is at most $e^{-\Omega(r)}$.

We are now ready for the main proof.

Proof (of Theorem 1(b)). Call a fooling-set matrix M *regular*, if $M_{k,k} = 1$ for all k. The minimum rank over a fooling-set pattern is always attained by a regular matrix (divide every row by the corresponding diagonal element).

Consider the event that there is a regular matrix M over $\mathbb{F}$ with $\sigma(M) = R(n, p)$, and $\mathrm{rk}\, M \leq r := n/(2000 \ln F)$. By Lemma 5, M contains a tee matrix N of order $2r$ and rank $\mathrm{rk}\, M$. If the size of the support of N is larger than $15pr(n - r)$, then we are in the situation of Lemma 6.

Otherwise, M is one of the

$$O(1) \cdot \binom{n}{2r} \cdot \binom{2r(n-r)}{15pr(n-r)} \cdot (2F)^{15pr(n-r)}$$

matrices of Lemma 4.

Hence, the probability of said event is $o(1)$ (from Lemma 6) plus at most an $O(1)$ factor of the following (with $m := pn^2/2$ and $\varrho := r/n$) a constant

---

[2] Specifically, we use Theorem 2.10 applied to (2.11) in [11].

$$\frac{\binom{n}{2r}\cdot\binom{2r(n-r)}{15pr(n-r)}\cdot(2F)^{15pr(n-r)}}{\binom{\binom{n}{2}}{p\binom{n}{2}}2^{p\binom{n}{2}}2^{-O(pn)}}\approx\frac{\binom{n}{2r}\cdot\binom{2r(n-r)}{15pr(n-r)}\cdot(2F)^{15pr(n-r)}}{\binom{n^2/2}{pn^2/2}2^{pn^2/2-O(pn)}}$$

$$=\frac{\binom{n}{2\varrho n}\cdot\binom{4\varrho(1-\varrho)n^2/2}{30p\varrho(1-\varrho)n^2/2}\cdot(2F)^{30p\varrho(1-\varrho)n^2/2}}{\binom{n^2/2}{pn^2/2}2^{pn^2/2-O(pn)}}$$

$$=\frac{\binom{n}{2\varrho n}\cdot\binom{4\varrho(1-\varrho)\,n^2/2}{30\varrho(1-\varrho)\,pn^2/2}\cdot(2F)^{30\varrho(1-\varrho)\,pn^2/2}}{\binom{n^2/2}{pn^2/2}2^{pn^2/2-O(pn)}}=:Q.$$

Abbreviating $\alpha:=30\varrho(1-\varrho)<30\varrho$, denoting $H(t):=-t\ln t-(1-t)\ln(1-t)$, and using

$$\binom{a}{ta}=\Theta\big((ta)^{-1/2}\big)e^{H(t)a},\ \text{for } t\le 1/2\qquad(2)$$

(for a large, "$\le$" holds instead of "$=\Theta$"), we find (the $O(pn)$ exponent comes from replacing $\binom{n}{2}$ by $n^2/2$ in the denominator)

$$\frac{\binom{n}{2\varrho n}2^{30\varrho(1-\varrho)\,pn^2/2}}{2^{pn^2/2-O(pn)}}\le e^{H(2\varrho)n-(\ln 2)(1-\alpha)pn^2/3}$$

$$\le e^{H(2\varrho)n-(\ln 2)(1-\alpha)pn^2/3}$$

$$=e^{n\left(H(2\varrho)-(\ln 2)(1-\alpha)pn/3\right)}$$

$$\le e^{n\left(H(2\varrho)-(\ln 2)33(1-30\varrho)\right)}$$

$$=o(1),$$

as $pn/2\ge 30$ and $1-\alpha>1-30\varrho$, and the expression in the parentheses is negative for all $\varrho\in[0,3/100]$.

For the rest of the fraction Q above, using (2) again, we simplify

$$\frac{\binom{4\varrho(1-\varrho)\,n^2/2}{30\varrho(1-\varrho)\,pn^2/2}F^{30\varrho(1-\varrho)\,pn^2/2}}{\binom{n^2/2}{pn^2/2}}\le\frac{\binom{\alpha\,n^2/2}{\alpha\,pn^2/2}F^{30\varrho(1-\varrho)\,pn^2/2}}{\binom{n^2/2}{pn^2/2}}$$

$$=O(1)\cdot e^{n^2/2\cdot\left((\alpha-1)H(p)+p\alpha\ln F\right)}.$$

Setting the expression in the parentheses to 0 and solving for α, we find

$$\alpha \geq \frac{\ln(1/p)}{\ln(1/p) + \ln F}$$

suffices for $Q = o(1)$; as $\alpha \leq \varrho$, the same inequality with α replaced by ϱ is sufficient.

This completes the proof of the theorem.

4   Proof of Theorem 1(c)

In this section, following the idea of [9], we apply a theorem of Ronyai, Babai, and Ganapathy [15] on the maximum number of zero-patterns of polynomials, which we now describe.

Let $f = (f_j)_{j=1,\dots,h}$ be an h-tuple of polynomials in n variables $x = (x_1, x_2, \cdots, x_n)$ over an arbitrary field $\mathbb{F}$. In line with the definitions above, for $u \in \mathbb{F}^n$, the zero-nonzero pattern of f at u is the vector $\sigma(f(u)) \in \{0,1\}^h$.

Theorem 3 ([15]). *If $h \geq n$ and each f_j has degree at most d then, for all m, the set*

$$\left| \left\{ y \in \{0,1\}^h \;\middle|\; |y| \leq m \text{ and } y = \sigma(f(u)) \text{ for some } u \in \mathbb{F}^n \right\} \right| \leq \binom{n + md}{n}.$$

In other words, the number of zero-nonzero patterns with Hamming weight at most m is at most $\binom{n+md}{n}$.

As has been observed in [9], this theorem is implicit in the proof of Theorem 1.1 of [15]. Since it is not explicitly proven in [15], for the sake of completeness, we repeat the proof here with slight modification of the proof of Theorem 1.1 from Theorem 3 which proves Theorem 3. The only difference between the following proof and that in [15] is where the proof below upper-bounds the degrees of the polynomials g_y.

It has been used in the context of minimum rank problems before (e.g., [9,14]), but our use requires slightly more work.

Proof (of Theorem 3). Consider the set

$$S := \left\{ y \in \{0,1\}^h \;\middle|\; |y| \leq m \text{ and } y = \sigma(f(u)) \text{ for some } u \in \mathbb{F}^n \right\}.$$

For each such y, let $u_y \in \mathbb{F}^n$ be such that $\sigma(f(u_y)) = y$, and let

$$g_y := \prod_{j,y_j=1} f_j.$$

Now define a square matrix A whose row- and column set is S, and whose (y, z) entry is $g_y(u_z)$. We have

$$g_y(u_z) \neq 0 \iff z \geq y,$$

with entry-wise comparison, and "1 > 0". Hence, if the rows and columns are arranged according to this partial ordering of S, the matrix is upper triangular, with nonzero diagonal, so it has full rank, $|S|$. This implies that the g_y, $y \in S$, are linearly independent.

Since each g_y has degree at most $|y| \cdot d \le md$, and the space of polynomials in n variables with degree at most md has dimension $\binom{n+md}{md}$, it follows that S has at most that many elements.

Given positive integers $r < n$, let us say that a *G-pattern* is an $r \times n$ matrix P whose entries are the symbols 0, 1, and $*$, with the following properties.

(1) Every column contains at most one 1, and every column containing a 1 contains no $*$s.
(2) In every row, the leftmost entry different from 0 is a 1, and every row contains at most one 1.
(3) Rows containing a 1 (i.e., not all-zero rows) have smaller row indices than rows containing no 1 (i.e., all-zero rows). In other words, the all-zero rows are at the bottom of P.

We say that an $r \times n$ matrix Y has *G-pattern* P, if $Y_{j,\ell} = 0$ if $P_{j,\ell} = 0$, and $Y_{j,\ell} = 1$ if $P_{j,\ell} = 1$. There is no restriction on the $Y_{j,\ell}$ for which $P_{j,\ell} = *$.

"G" stands for "Gaussian elimination using row operations". We will need the following three easy lemmas.

Lemma 7. *Any $r \times n$ matrix Y' can be transformed, by Gaussian elimination using only row operations, into a matrix Y which has some G-pattern.*

Proof. If Y' has no nonzero entries, we are done. Otherwise start with the leftmost column containing a nonzero entry, say (j, ℓ). Scale row j that entry a 1, permute the row to the top, and add suitable multiples of it to the other rows to make every entry below the 1 vanish.

If all columns $1, \ldots, \ell$ have been treated such that column ℓ has a unique 1 in row, say $j(\ell)$, consider the remaining matrix $\{j(\ell) + 1, \ldots r\} \times \{\ell + 1, \ldots, n\}$. If every entry is a 0, we are done. Otherwise, find the leftmost nonzero entry in the block; suppose it is in column ℓ' and row j'. Scale row j' to make that entry a 1, permute row j' to $j(\ell) + 1$, and add suitable multiples of it to all other rows $\{1, \ldots, r\} \setminus \{j(\ell) + 1\}$ to make every entry below the 1 vanish.

Lemma 8. *For every $r \times n$ G-pattern matrix P, the number of $*$-entries in P is at most $r(n - r/2)$.*

Proof. The G-pattern matrix P is uniquely determined by $c_1 < \cdots < c_s$, the (sorted) list of columns of P which contain a 1. With $c_0 := 0$, for $i = 1, \ldots, s$, if $c_{i-1} < c_i - 1$, then replacing c_i by $c_i - 1$ gives us a G-pattern matrix with one more $*$ entry. Hence, we may assume that $c_i = i$ for $i = 1, \ldots, s$. If $s < r$, then adding $s + 1$ to the set of 1-columns cannot decrease the number of $*$-entries (in fact, it increases the number, unless $s + 1 = n$). Hence, we may assume that $s = r$. The number of $*$-entries in the resulting (unique) G-pattern matrix is

$$n - 1 + \cdots + n - r = rn - r(r+1)/2 \leq r(n - r/2),$$

as promised.

Lemma 9. *Let $\varrho \in]0, .49]$. The number of $n \times \varrho n$ G-pattern matrices is at most*

$$O(1) \cdot \binom{n}{\varrho n}$$

(with an absolute constant in the big-O).

Proof. A G-pattern matrix is uniquely determined by the set of columns containing a 1, which can be between 0 and ϱn. Hence, the number of $n \times \varrho n$ G-pattern matrices is

$$\sum_{j=0}^{\varrho n} \binom{n}{j}. \tag{$*$}$$

From here on, we do the usual tricks. As in the previous section, we use the helpful fact (Theorem 1.1 in [2]) that

$$(*) \leq \frac{1}{1 - \frac{\varrho}{1-\varrho}} \binom{n}{\varrho n}.$$

A swift calculation shows that $1/(1 - \varrho/(1-\varrho)) \leq 30$, which completes the proof.

We are now ready to complete the Proof of Theorem 1(c).

Proof (of Theorem 1(c)). Let M be a fooling-set matrix of size n and rank at most r. It can be factored as $M = XY$, for an $n \times r$ matrix X and an $r \times n$ matrix Y due to rank factorization. By Lemma 7, through applying row operations to Y and corresponding column operations to X, we can assume that Y has a G-pattern.

Now we use Theorem 3, for every G-pattern matrix separately. For a fixed G-pattern matrix P, the variables of the polynomials are

- $X_{k,j}$, where (k, j) ranges over all pairs $[n] \times [r]$; and
- $Y_{j,\ell}$, where (j, ℓ) ranges over all pairs $[r] \times [n]$ with $P_{j,\ell} = *$.

The polynomials are: for every $(k, \ell) \in [n]^2$, with $k \neq \ell$,

$$f_{k,\ell} = \sum_{\substack{j \\ P_{j,\ell}=1}} X_{k,j} + \sum_{\substack{j \\ P_{j,\ell}=*}} X_{k,j} Y_{j,\ell}.$$

Clearly, there are $n(n-1)$ polynomials; the number of variables is $2rn - r^2/2$, by Lemma 8 (and, if necessary, using "dummy" variables which have coefficient 0 always). The polynomials have degree at most 2.

By Theorem 3, we find that the number of zero-nonzero patterns with Hamming weight at most m of fooling-set matrices with rank at most r which result from this particular G-pattern matrix P is at most

$$\binom{2rn - r^2/2 + 2m}{2rn - r^2/2}.$$

Now, take a $\varrho < 1/2$, and let $r := \varrho n$. Summing over all G-pattern matrices P, and using Lemma 9, we find that the number of zero-nonzero patterns with Hamming weight at most m of fooling-set matrices with rank at most ϱn is at most an absolute constant times

$$\binom{n}{\varrho n}\binom{(2\varrho - \varrho^2/2)n^2 + 2m}{(2\varrho - \varrho^2/2)n^2}.$$

Now, take a constant $p \in]0,1]$, and let $m := \lceil p\binom{n}{2}\rceil$. The number of fooling-set patterns of size n with density p is

$$\binom{\binom{n}{2}}{m}2^m, \tag{3}$$

and hence, the probability that the minimum rank of a fooling-set matrix with zero-nonzero pattern $R(n,p)$ has rank at most r is at most

$$\frac{\binom{n}{\varrho n}\binom{(2\varrho - \varrho^2/2)n^2 + 2m}{(2\varrho - \varrho^2/2)n^2}}{\binom{\binom{n}{2}}{m}2^m} \leq \frac{\binom{n}{\varrho n}\binom{(2\varrho - \varrho^2/2)n^2 + 2pn^2/2}{(2\varrho - \varrho^2/2)n^2}}{\binom{n^2/2}{pn^2/2}2^{pn^2/2 + O(pn)}}$$

$$= \frac{\binom{n}{\varrho n}\binom{\alpha n^2 + pn^2}{\alpha n^2}}{\binom{n^2/2}{pn^2/2}2^{pn^2/2 + O(pn)}}$$

where we have set $\alpha := 2\varrho - \varrho^2/2$. As in the previous section, we use (2) to estimate this expression, and we obtain

$$\ln\left(\frac{\binom{n}{\varrho n}\binom{\alpha n^2 + pn^2}{\alpha n^2}}{\binom{n^2/2}{pn^2/2}2^{pn^2/2 + O(pn)}}\right) = nH(\varrho)$$

$$+ n^2\left(\alpha H(\alpha/(\alpha + p)) - \tfrac{1}{2}H(p) - (\ln 2)p/2\right) + O(pn).$$

The dominant term is the one where n appears quadratic. The expression $\tfrac{1}{2}H(p) + (\ln 2)p/2$ takes values in $]0,1[$. For every fixed p, the function $g\colon \alpha \mapsto \alpha H(\alpha/(\alpha + p))$ is strictly increasing on $[0, 1/2]$ and satisfies $g(0) = 0$. Hence, for every given constant p, there exists an α for which the coefficient after the n^2 is negative. (As indicated in the introduction, such an α must tend to 0 with $p \to 0$.)

The proof for the distribution $Q(n)$ follows in the same way, in fact easier: For the uniform distribution, the expression in (3) is replaced by the larger number of all fooling-set patterns, $3^{\binom{n}{2}}$, and it is most convenient to use another theorem directly from Theorem 1.1 in [15] (without going through the version proved in the Theorem 3).

Acknowledgments. The second author would like to thank Kaveh Khoshkhah for discussions on the subject.

References

1. Beasley, L.B., Klauck, H., Lee, T., Theis, D.O.: Communication complexity, linear optimization, and lower bounds for the nonnegative rank of matrices (Dagstuhl Seminar 13082). Dagstuhl Rep. **3**(2), 127–143 (2013)
2. Bollobás, B.: Random Graphs. Cambridge Studies in Advanced Mathematics, 2nd edn., vol. 73. Cambridge University Press, Cambridge (2001)
3. Dietzfelbinger, M., Hromkovič, J., Schnitger, G.: A comparison of two lower-bound methods for communication complexity. Theoret. Comput. Sci. **168**(1), 39–51 (1996). 19th International Symposium on Mathematical Foundations of Computer Science (Košice, 1994)
4. Fiorini, S., Kaibel, V., Pashkovich, K., Theis, D.O.: Combinatorial bounds on nonnegative rank and extended formulations. Discrete Math. **313**(1), 67–83 (2013)
5. Friesen, M., Hamed, A., Lee, T., Theis, D.O.: Fooling-sets and rank. Eur. J. Comb. **48**, 143–153 (2015)
6. Friesen, M., Theis, D.O.: Fooling-sets and rank in nonzero characteristic. In: Nešetřil, J., Pellegrini, M. (eds.) The Seventh European Conference on Combinatorics, Graph Theory and Applications. CRM series, vol.16, pp. 383–390. CRM (2013)
7. Frieze, A., Karoński, M.: Introduction to Random Graphs. Cambridge University Press, Cambridge (2015)
8. Golovnev, A., Regev, O., Weinstein, O.: The minrank of random graphs. Preprint arXiv:1607.04842 (2016)
9. Tracy Hall, T., Hogben, L., Martin, R., Shader, B.: Expected values of parameters associated with the minimum rank of a graph. Linear Algebra Appl. **433**(1), 101–117 (2010)
10. Haviv, I., Langberg, M.: On linear index coding for random graphs. In: 2012 IEEE International Symposium on Information Theory Proceedings (ISIT), pp. 2231–2235. IEEE (2012)
11. Janson, S., Łuczak, T., Rucinski, A.: Random Graphs. Wiley-Interscience Series in Discrete Mathematics and Optimization. Wiley-Interscience, New York (2000)
12. Klauck, H., de Wolf, R.: Fooling one-sided quantum protocols. In: 30th International Symposium on Theoretical Aspects of Computer Science (2013)
13. Kushilevitz, E., Nisan, N.: Communication Complexity. Cambridge University Press, Cambridge (1997)
14. Mallik, S., Shader, B.L.: On graphs of minimum skew rank 4. Linear Multilinear Algebra **64**(2), 279–289 (2016)
15. Rónyai, L., Babai, L., Ganapathy, M.: On the number of zero-patterns of a sequence of polynomials. J. Am. Math. Soc. **14**(3), 717–735 (2001)

On Probabilistic Algorithm for Solving Almost All Instances of the Set Partition Problem

Alexandr V. Seliverstov[(✉)] [iD]

Institute for Information Transmission Problems of the Russian Academy of Sciences
(Kharkevich Institute), Bolshoy Karetny per. 19, build. 1, Moscow 127051, Russia
slvstv@iitp.ru

Abstract. Earlier, I.V. Latkin and the author have shown the set partition problem can be reduced to the problem of finding singular points of a cubic hypersurface. The article focuses on the new link between two different research areas as well as on methods to look for singular points or to confirm the smoothness of the hypersurface. Our approach is based on the description of tangent lines to the hypersurface. The existence of at least one singular point imposes a restriction on the algebraic equation that determines the set of tangent lines passing through the selected point of the space. This equation is based on the formula for the discriminant of a univariate polynomial. We have proposed a probabilistic algorithm for some set of inputs of the set partition problem. The probabilistic algorithm is not proved to have polynomial complexity.

Keywords: Set partition · Cubic hypersurfaces · Smoothness · Tangent line · Polynomial · Discriminant · Computational complexity

1 Introduction

The set partition problem is NP-complete [1]. Let us recall its definition. Given a multiset of positive integers $\{\alpha_0, \ldots, \alpha_n\}$. Can it be partitioned into two subsets with equal sums of elements? Points with coordinates ± 1 are called $(-1, 1)$-points. Obviously, this problem is to recognize whether a $(-1, 1)$-point belongs to the hyperplane given by $\alpha_0 + \alpha_1 x_1 + \cdots + \alpha_n x_n = 0$. So, it is hard to find a $(-1, 1)$-point belonging to the hyperplane in high dimensions. The problem is to solve the system that consists of one linear equation and the set of quadratic equations $x_1^2 = 1, \ldots, x_n^2 = 1$. If there is no solution, then a direct proof of the unsolvability of the system by means of Hilbert's Nullstellensatz requires to produce polynomials of very high degree [2]. The informal explanation is that many $(-1, 1)$-points can lie on a hyperplane. In case $n = 2k$, the number of $(-1, 1)$-points belonging to the hyperplane given by $x_1 + \cdots + x_n = 0$ is equal to $n!/(k!)^2$. The full description of a large number of solutions requires polynomials

The research has been carried out at the expense of the Russian Science Foundation, project no. 14–50–00150.

P. Weil (Ed.): CSR 2017, LNCS 10304, pp. 285–293, 2017.
DOI: 10.1007/978-3-319-58747-9_25

of high degree. There are known randomized algorithms for solving some systems of algebraic equations [3]. But their applicability in this case is doubtful.

There are other methods for solving integer linear programming problems [1,4]. One can find $(-1,1)$-points belonging to the hyperplane given by a linear function with integer coefficients near zero, using dynamic programming [5,6]. There is also the related optimization problem. So, there are well known both fully polynomial time approximation scheme and pseudo-polynomial time algorithm for solving the problem. The obstacle for solving the optimization problem is a large number of values of the linear functional at different $(-1,1)$-points.

In this paper, we focus on an algorithm for solving all but an exponentially small fraction of inputs; these inputs are incorrectly accepted without any warning. In accordance with the Schwartz–Zippel lemma [7], if the stupid algorithm rejects all inputs, then it works correctly on a strongly generic set of inputs [8,9]. But our algorithm can make errors of another type only.

Our method is based on the reduction of the set partition problem to the recognition problem for hypersurface singularities [10,11]. Two viewpoints may clarify each other. Other geometric formulations of related problems have already appeared in the literature [12,13]. For example, maximization of cubic form over the Euclidean ball is NP-hard too. Of course, we consider a very special type of singularities. In general the problem is very hard [14]. Singular points on the variety corresponds to roots of a system of algebraic equations. The best methods for solving the system require at least exponential time in general case [3,15]. A solution to n algebraic equations in n variables can be obtained by a series of hypergeometric type [16]. Methods based on the computation of Gröbner bases are widely used in small dimensions [17–19], but the computational complexity quickly increases in high dimensions [20]. Some examples have been computed by means of the cloud service MathPartner [21].

2   Preliminaries

The binary representation of a positive integer n has the length $\lceil \log_2(n + 1) \rceil$, where $\lceil t \rceil$ is the smallest integer not less than t. We denote by $\mathbb{C}$ and $\mathbb{Q}$ the fields of complex and rational numbers, respectively.

The discriminant Δ_d of a univariate polynomial of degree d is a homogeneous function of its coefficients. The discriminant vanishes if and only if the polynomial has a multiple root. For example, the discriminant Δ_3 of the cubic polynomial $at^3 + bt^2 + pt + q$ is equal to $b^2p^2 - 4ap^3 - 4b^3q - 27a^2q^2 + 18abpq$. Moreover, $\Delta_d(g_0, g_1, \ldots, g_{d-1}, g_d) = \Delta_d(g_d, g_{d-1}, \ldots, g_1, g_0)$. If the leading coefficient vanishes, then the value of the function Δ_d is equal to the discriminant of another polynomial without the constant term. If the degree is equal to $d - 1$, then Δ_d vanishes if and only if the polynomial has a multiple root. If the degree is less than $d - 1$, then $\Delta_d = 0$.

A square-free polynomial is a polynomial that does not have as a factor any square of a polynomial of positive degree. An affine hypersurface is the vanishing locus of a square-free polynomial over the field of complex numbers.

Let us consider an affine hypersurface given by a square-free polynomial f. A straight line passing through the selected point U in n-dimensional affine space is defined as the set of points with coordinates $((x_1-u_1)t+u_1,\ldots,(x_n-u_n)t+u_n)$, where $(u_1,\ldots,u_n)$ are coordinates at U, and t is a parameter. Let us denote by $r(t)$ a univariate polynomial that is the restriction of the polynomial f to the line, and by $D[f,U]$ the discriminant of $r(t)$. If $\deg r(t) < d$, then we use the formula for Δ_d by means of substitution the zero as the leading coefficient. At the general point U the degree of $D[f,U](x_1,\ldots,x_n)$ is equal to d^2-d. If the line is a tangent line to the hypersurface, then the discriminant of the polynomial $r(t)$ vanishes. If U is not a singular point of the hypersurface, then $D[f,U](x_1,\ldots,x_n)$ defines a cone. If U is a smooth point of the hypersurface, the cone is reducible and contains a tangent hyperplane at the point U. If U is singular, then $D[f,U]$ vanishes identically.

If the selected point U is a smooth point of the hypersurface, then let us denote by $B[f,U]$ the discriminant of $r(t)/t$. Since $r(0)=0$, $r(t)/t$ is a polynomial of degree at most $d-1$, where $d=\deg f$. If $\deg r(t) < d-1$, then we use the formula for degree $d-1$ by means of substitution the zero as the leading coefficient. Of course, the polynomial $B[f,U]$ is a divisor of the polynomial $D[f,U]$.

To study generic-case complexity of an algorithm, let us recall the definition of the generic set [8,9]. For every positive n, let B_n denote the set of all inputs of length at most n. Let us define the asymptotic density $\rho(S)$ for S as

$$\rho(S) = \lim_{n\to\infty} \rho_n(S),$$

where

$$\rho_n(S) = \frac{|S \cap B_n|}{|B_n|}.$$

If $\rho(S)=1$, then the subset S is called generic. If in addition $\rho_n(S)$ converges to 1 exponentially fast, then S is called strongly generic.

For example, hard inputs are rare for the simplex algorithm for linear programming [22,23].

3   Results

In this section let us denote

$$f = \alpha_0 + \alpha_1 x_1^3 + \cdots + \alpha_n x_n^3$$
$$h = \alpha_0 + \alpha_1 x_1 + \cdots + \alpha_n x_n,$$

where all coefficients $\alpha_0,\ldots,\alpha_n$ are nonzero. Of course, the hypersurface $f=0$ is smooth. The following theorem is a reformulation of the result from [11].

Theorem 1. *Given a multiset of positive integers $\{\alpha_0,\ldots,\alpha_n\}$, where $n \geq 2$. There exists a one-to-one correspondence between singular points of the affine variety given by two equations $f = h = 0$ and $(-1,1)$-points belonging to the hyperplane given by the equation $h = 0$.*

Proof. If both polynomials f and h vanish simultaneously at a $(-1, 1)$-point, then the hyperplane $h = 0$ is tangent to the hypersurface $f = 0$ at this point. Thus, the hyperplane section is singular.

At a singular point of the section, the hyperplane $h = 0$ coincides with the tangent hyperplane to the hypersurface $f = 0$. Since all the coefficients α_k are nonzero, both gradients ∇f and ∇h can be collinear only at the points whose coordinates satisfy the system of the equations $x_k^2 = x_j^2$ for all indices k and j. All the points are $(-1, 1)$-points. $\square$

The polynomial $D[f, U]$ is equal to the discriminant of a univariate polynomial $at^3 + bt^2 + pt + q$. That is, $D[f, U] = b^2 p^2 - 4ap^3 - 4b^3 q - 27a^2 q^2 + 18abpq$, where the coefficients are sums of univariate polynomials $a = a_1(x_1) + \cdots + a_n(x_n)$, $b = b_1(x_1) + \cdots + b_n(x_n)$, $p = p_0 + p_1 x_1 + \cdots + p_n x_n$, and the constant term q. Each monomial from $D[f, U](x_1, \ldots, x_n)$ is dependent on at most four variables.

The polynomial $B[f, U]$ is equal to the discriminant of a univariate polynomial $at^2 + bt + c$. That is, $B[f, U] = b^2 - 4ac$, where the coefficients are sums of univariate polynomials $a = a_1(x_1) + \cdots + a_n(x_n)$, $b = b_1(x_1) + \cdots + b_n(x_n)$, and $c = c_0 + c_1 x_1 + \cdots + c_n x_n$. Each monomial from $B[f, U](x_1, \ldots, x_n)$ is dependent on at most two variables.

Let us consider the factor ring $\mathbb{C}[x_1, \ldots, x_n]/\langle x_1^2 - 1, \ldots, x_n^2 - 1 \rangle$. It is referred to as the set of multilinear polynomials. In this way, we have a surjective map φ from the set of all polynomials onto the set of multilinear polynomials.

Let us denote by $M[f, U](x_1, \ldots, x_{n-1})$ a multilinear polynomial that is an image of the restriction to the hyperplane $h = 0$ of the multilinear polynomial $\varphi(B[f, U])$. The restriction to the hyperplane $h = 0$ means that we substitute $x_n = -(\alpha_0 + \alpha_1 x_1 + \cdots + \alpha_{n-1} x_{n-1})/\alpha_n$. Unfortunately, it is hard to compute a Gröbner basis of the ideal $\langle h, x_1^2 - 1, \ldots, x_n^2 - 1 \rangle$. Instead, we use computations over the set of multilinear polynomials.

Let us denote by L or $L_{\alpha_0, \ldots, \alpha_n}$ a linear space spanned by all multilinear polynomials $M[f, U](x_1, \ldots, x_{n-1})$, where U belongs to the section $f = h = 0$.

A polynomial vanishes at a $(-1, 1)$-point if and only if its multilinear image vanishes at this point. Thus, if the hyperplane section given by $f = h = 0$ contains a $(-1, 1)$-point, then all multilinear polynomials from L vanish at the point. Contrariwise, if L coincides with the linear space of all multilinear polynomials of degree at most two, then the section does not contain any $(-1, 1)$-point. Of course, all such $(-1, 1)$-points are singular.

Lemma 1. *If $n = 2$ and $\alpha_0 = 1$, then there exist infinitely many values of two coefficients α_1 and α_2 such that the linear space L coincides with the linear space of all multilinear polynomials of degree at most two. In particular, the same is true for all algebraically independent numbers α_1 and α_2.*

Proof. Let us consider a plane curve defined by $f = 3x_1^3 + 2x_2^3 + 1$. The intersection of the line $3x_1 + 2x_2 + 1 = 0$ and the curve $f = 0$ consist of two points $U(-1, 1)$ and $V(\frac{1}{5}, -\frac{4}{5})$. The union of all tangent lines passing through the point U is defined by the polynomial $B[f, U] = -3x_2^4 - 36x_2^3 x_1 - 24x_2^3 -$

$54x_2^2x_1^2 + 36x_2^2 - 36x_2x_1^3 - 24x_2 - 27x_1^4 - 72x_1^3 - 108x_1^2 - 72x_1 - 12$. Its multi-linear image is $\varphi(B[f,U]) = -72x_2x_1 - 48x_2 - 144x_1 - 168$. The substitution $x_2 = -\frac{3x_1+1}{2}$ yields a univariate polynomial $108x_1^2 - 36x_1 - 144$. Its multilinear image $M[f,U] = -36x_1 - 36$. At the second point V the multilinear polynomial

$$M[f,V] = \frac{26172}{3125}x_1 + \frac{428292}{15625}.$$

Two polynomials $M[f,U]$ and $M[f,V]$ together span the whole linear space of univariate linear polynomials. The same is true for almost all cubic curves because the first-order theory of the field of complex numbers admits quantifier elimination. □

Remark 1. Let us consider an affine plane curve defined by $f = x_1^3 + x_2^3 + 1$. The intersection of the curve and the line defined by $h = x_1 + x_2 + 1$ consists of two points $U(0,-1)$ and $V(-1,0)$. The third point does not belong to the affine plane. So, $B[f,U] = -12x_1x_2 - 24x_2 - 12x_1 - 24$; the multilinear polynomial $M[f,U] = 24x_1 + 12$. On the other hand, at the point V the polynomial $B[f,V] = -12x_1x_2 - 12x_2 - 24x_1 - 24$; the multilinear polynomial $M[f,V]$ vanishes identically. Thus, L is a proper subspace in the two-dimensional space of univariate linear polynomials.

Lemma 2. *For all $n \geq 2$, if there exist nonzero numbers $\beta_0, \ldots, \beta_n$ such that the linear space $L_{\beta_0, \ldots, \beta_n}$ coincides with the linear space of all multilinear polynomials of degree at most two, then for almost all nonzero integers $\alpha_0, \ldots, \alpha_n$, the linear space $L_{\alpha_0, \ldots, \alpha_n}$ coincides with the linear space of all multilinear polynomials of degree at most two. Moreover, if for all indices k the numbers $1 \leq \alpha_k \leq S$, then the upper bound on the fraction of the exception set of $(n+1)$-tuples $\{\alpha_0, \ldots, \alpha_n\}$ is equal to $2^{poly(n)}/S$.*

Proof. All coefficients from $M[f,U]$ are continuous functions on the open set $\alpha_0 \neq 0, \ldots, \alpha_n \neq 0$. The matrix determinant is continuous too. Let us consider a set of points $\{U^{(k)}\}$ on the hypersurface $f = 0$ for a set $\{\alpha_0, \ldots, \alpha_n\}$. If all polynomials $\{M[f,U^{(k)}]\}$ are linearly independent, then under a sufficiently small change of α_k there exists a set of points $\{V^{(k)}\}$, such that for all indices $V^{(k)}$ belongs to a small polydisk near $U^{(k)}$, $V^{(k)}$ belongs to the new hypersurface $\check{f} = 0$, and all polynomials $\{M[\check{f},V^{(k)}]\}$ are linearly independent. This property is satisfied on a nonempty open set of $(n+1)$-tuples $\{\alpha_0, \ldots, \alpha_n\}$ because the first-order theory of the field of complex numbers admits quantifier elimination. Thus, $\dim L$ is a lower semi-continuous function.

In accordance with our premise, the fraction of the exception set is less than one. In accordance with Lemma 1, in case $n = 2$, the premise holds.

There exists a nontrivial polynomial $g(\alpha_0, \ldots, \alpha_n)$ of degree at most $2^{poly(n)}$ such that if L does not coincide with the linear space of all multilinear polynomials of degree at most two, then g vanishes. (The converse implication is not necessary true.) Vanishing of the polynomial g is equivalent to inconsistency of a system of $O(n^2)$ algebraic equations, where the degree of each algebraic equations is $poly(n)$. In accordance with [15], the polynomial g can be chosen so that its degree $\deg(g) \leq 2^{poly(n)}$. Thus, in accordance with the Schwartz–Zippel lemma [7], the fraction is less than $2^{poly(n)}/S$.   □

Let us denote by π the projection of the hyperplane section $f = h = 0$ that forgets two coordinates x_{n-1} and x_n. Let us define

$$\lambda(n) = \frac{n(n+1)}{2} + 1$$

that is the upper bound on $\dim L$ for all $n \geq 3$.

Lemma 3. *Given a multiset of positive integers* $\alpha_0, \ldots, \alpha_n$, *and a real* $\varepsilon > 0$. *Let us consider the multilinear polynomials* $m_k = M[f, U^{(k)}]$ *for random points* $U^{(k)}$ *of the hyperplane section given by* $f = h = 0$, *where the index* k *runs the segment* $1 \leq k \leq \lambda(n)$. *If all coordinates of their images* $\pi(U^{(k)})$ *are independent and uniformly distributed on the set of integers from one to* $\lceil 2^{180n^4}/\varepsilon \rceil$, *then the probability of spanning the whole linear space* L *is at least* $1 - \varepsilon$.

Proof. All polynomials $m_1, \ldots, m_{\lambda(n)}$ belong to L. If the polynomials are linearly dependent, then the determinant of the matrix, whose entries are coefficients, vanishes. The order of the matrix is equal to $\lambda(n)$. Each matrix entry is a polynomial of degree at most six. The determinant of the matrix is a polynomial of degree at most $6\lambda(n)$. Let us denote the polynomial by g. The resultant $\mathrm{res}_{x_{n-1}}(g, f(x_1, \ldots, x_{n-1}, -(\alpha_0+\alpha_1 x_1+\cdots+\alpha_{n-1}x_{n-1})/\alpha_n)$ vanishes with probability at most ε. Else it vanishes identically. The resultant degree is less than $(3 + \deg g) \deg g \leq 9n^4 + 18n^3 + 54n^2 + 45n + 54 < 180n^4$. The upper bound on the probability of vanishing the resultant is calculated by the Schwartz–Zippel lemma [7].   □

Remark 2. The enormous integer $\lceil 2^{180n^4}/\varepsilon \rceil$ has a binary representation of polynomial length. But we assume it is only very rough upper bound. Another approach to prove Lemma 3 is briefly discussed in the next section.

Theorem 2. *There exists a function* $S(n)$ *of the type* $2^{poly(n)}$ *such that for any real* $\varepsilon > 0$ *there exists a probabilistic algorithm for solving the set partition problem in certain sense.*

- *The algorithm receives as the input a set of positive integers* $\alpha_0, \ldots, \alpha_n$ *from one to* $S(n)$;
- *The algorithm executes* $O(n^6)$ *arithmetic operations over algebraic numbers as well as square root or cube root extraction operations;*
- *If a solution exists, then the probability of accepting is at least* $1 - \varepsilon$;
- *Else if there exist nonzero numbers* $\beta_0, \ldots, \beta_n$ *such that the linear space* $L_{\beta_0, \ldots, \beta_n}$ *coincides with the linear space of all multilinear polynomials of degree at most two, then the probability of rejecting is at least* $1 - \varepsilon$ *except an exponentially small fraction of inputs, i.e., on a strongly generic set of inputs.*

Proof. Let us consider the cubic hypersurface given by $f = 0$. In accordance with Theorem 1, a singular point of its hyperplane section given by $f = h = 0$ corresponds to a solution to the set partition problem [10,11].

In case $n \leq 1$, the algorithm simply checks all $(-1, 1)$-points.

In case $n \geq 2$, the algorithm picks up $\lambda(n)$ random points on the section. In this way, it picks up a random point P from the coordinate subspace, whose $n - 2$ coordinates are independently and uniformly distributed on the set of integers from one to a large number as in Lemma 3. A preimage $U \in \pi^{-1}(P)$ belongs to the section. Both points P and U have the same $n - 2$ coordinates. Other two coordinates are calculated as a solution of the system of two equations $f = h = 0$. They can be irrational.

If a $(-1, 1)$-point is picked up, then the input is accepted. Else the algorithm calculates a spanning set of the linear space L in accordance with Lemma 3. If L does not coincide with the linear space of all multilinear polynomials of degree at most two, then the input is accepted because a solution gives a linear dependence of polynomials. Else the input is rejected.

The total number of random bits used by the algorithm is bounded by a polynomial in n and $1/\varepsilon$; it does not depend on the values $\alpha_0, \ldots, \alpha_n$. The total number of the arithmetic operations over algebraic numbers is bounded by a polynomial in n.

In accordance with Lemma 2, if there exist nonzero numbers $\alpha_0, \ldots, \alpha_n$ such that the linear space L coincides with the linear space of all multilinear polynomials of degree at most two, then the error probability is small for a generic set of inputs. □

Remark 3. Instead of computation $\dim L$ it is sufficient to check whether a nonzero constant belongs to the linear space L. Moreover, if the linear space L contains a linear polynomial, one can reduce the dimension of the initial task.

4   Discussion

In fact, the algorithm from Theorem 2 computes the determinant of a matrix with irrational entries. Its value is an algebraic number that is result of $poly(n)$ arithmetic operations over roots of cubic polynomials. Unfortunately, there are such algebraic numbers whose both length and degree can be large [24]. On the other hand, if the determinant does not belong to a very small polydisk near zero, then one can use Diophantine approximation to prove that it is nonzero. Thus, we have a sufficient condition over $\mathbb{Q}$ for the absence of any solution for the set partition problem.

In Lemma 3, we pick up a point from the preimage $\pi^{-1}(P)$ containing three points. But we need only one point. Instead, the point on the cubic hypersurface can be computed in more deterministic way using a rational parameterization of the variety. All cubic surfaces as well as hypersurfaces in higher dimensions are unirational over $\mathbb{C}$, although any smooth cubic curve is not unirational. Moreover, such a cubic hypersurface defined over $\mathbb{Q}$ is unirational over $\mathbb{Q}$ if and only if it has a $\mathbb{Q}$-point [25]. Obviously, the same result is true for the hyperplane section $f = h = 0$ that is hypersurface inside the hyperplane. Thus, if the section contain a $\mathbb{Q}$-point, then we have not only a lot of rational points but also a rational map from the set of points with integer coordinates to the variety defined by both polynomials f and h. In this case, one can modify Lemma 3 as well as Theorem 2 to eliminate irrational numbers.

The number of arithmetic operations in the algorithm depends on the computational complexity of a method for solving systems of linear equations. We adopt Gaussian elimination. Some upper bounds can be improved by means of asymptotically more efficient methods [26].

The algorithm works correctly on a strongly generic set of inputs. Maybe the exception set is empty, but this hypothesis is not obvious. Although two smooth hypersurfaces are diffeomorphic each other, their algebraic properties can differ. For example, there exists an exotic smooth complex affine variety which is diffeomorphic to an affine space, but is not algebraically isomorphic to it [27]. In case $n = 2$, see also Remark 1. But in accordance with Lemma 1, the method can be used for smoothness recognition of almost all cubic curves.

In Theorem 2, all $\alpha_0, \ldots, \alpha_n$ are integers with binary representations of length $poly(n)$. One can consider the continuous version, where all $\alpha_0, \ldots, \alpha_n$ are nonzero complex numbers (or algebraic numbers having finite descriptions). In this case, the exception set has measure zero.

The same method can be applied to find additional algebraic equation that vanishes at all singular points of an arbitrary algebraic variety of degree d. In the case, a polynomial of the type $D[f, U]$ can be computed using finitely many tangent lines passing through the selected point U. The approach based on the description of tangent lines to the surface can be useful for solving some problems of machine vision and image recognition.

Acknowledgements. The author would like to thank Mark Spivakovsky, Sergei P. Tarasov, Mikhail N. Vyalyi, and the anonymous reviewers for useful comments.

References

1. Schrijver, A.: Theory of Linear and Integer Programming. Wiley, New York (1986)
2. Margulies, S., Onn, S., Pasechnik, D.V.: On the complexity of Hilbert refutations for partition. J. Symbolic Comput. **66**, 70–83 (2015). doi:10.1016/j.jsc.2013.06.005
3. Herrero, M.I., Jeronimo, G., Sabia, J.: Affine solution sets of sparse polynomial systems. J. Symbolic Comput. **51**, 34–54 (2013). doi:10.1016/j.jsc.2012.03.006
4. Bodur, M., Dash, S., Günlük, O.: Cutting planes from extended LP formulations. Math. Program. **161**(1), 159–192 (2017). doi:10.1007/s10107-016-1005-7
5. Tamir, A.: New pseudopolynomial complexity bounds for the bounded and other integer Knapsack related problems. Oper. Res. Lett. **37**(5), 303–306 (2009). doi:10.1016/j.orl.2009.05.003
6. Claßen, G., Koster, A.M.C.A., Schmeink, A.: The multi-band robust knapsack problem — a dynamic programming approach. Discrete Optimization. **18**, 123–149 (2015). doi:10.1016/j.disopt.2015.09.007
7. Schwartz, J.T.: Fast probabilistic algorithms for verification of polynomial identities. J. ACM **27**(4), 701–717 (1980). doi:10.1145/322217.322225

8. Kapovich, I., Myasnikov, A., Schupp, P., Shpilrain, V.: Generic-case complexity, decision problems in group theory, and random walks. J. Algebra **264**, 665–694 (2003). doi:10.1016/S0021-8693(03)00167-4
9. Rybalov, A.N.: A generic relation on recursively enumerable sets. Algebra Logic **55**(5), 387–393 (2016). doi:10.1007/s10469-016-9410-9
10. Latkin, I.V., Seliverstov, A.V.: Computational complexity of fragments of the theory of complex numbers. Bulletin of University of Karaganda. Ser. Mathematics, vol. 1, pp. 47–55 (2015). (in Russian)
11. Seliverstov, A.V.: On cubic hypersurfaces with involutions. In: Vassiliev, N.N. (ed.) International Conference Polynomial Computer Algebra 2016, St. Petersburg, 18–22 April 2016, pp. 74–77. VVM Publishing, Saint Petersburg (2016)
12. Nesterov, Y.: Random walk in a simplex and quadratic optimization over convex polytopes. CORE Discussion Paper 2003/71 (2003)
13. Hillar, C.J., Lim, L.H.: Most tensor problems are NP-hard. J. ACM **60**(6), 45 (2013). doi:10.1145/2512329
14. Gel'fand, I.M., Zelevinskii, A.V., Kapranov, M.M.: Discriminants of polynomials in several variables and triangulations of Newton polyhedra. Leningrad Math. J. **2**(3), 499–505 (1991)
15. Chistov, A.L.: An improvement of the complexity bound for solving systems of polynomial equations. J. Math. Sci. **181**(6), 921–924 (2012). doi:10.1007/s10958-012-0724-4
16. Kulikov, V.R., Stepanenko, V.A.: On solutions and Waring's formulae for the system of n algebraic equations with n unknowns. St. Petersburg Math. J. **26**(5), 839–848 (2015). doi:10.1090/spmj/1361
17. Bokut, L.A., Chen, Y.: Gröbner-Shirshov bases and their calculation. Bull. Math. Sci. **4**(3), 325–395 (2014). doi:10.1007/s13373-014-0054-6
18. Bardet, M., Faugère, J.-C., Salvy, B.: On the complexity of the F_5 Gröbner basis algorithm. J. Symbolic Comput. **70**, 49–70 (2015). doi:10.1016/j.jsc.2014.09.025
19. Eder, C., Faugère, J.-C.: A survey on signature-based algorithms for computing Gröbner bases. J. Symbolic Comput. **80**(3), 719–784 (2017). doi:10.1016/j.jsc.2016.07.031
20. Mayr, E.W., Ritscher, S.: Dimension-dependent bounds for Gröbner bases of polynomial ideals. J. Symbolic Comput. **49**, 78–94 (2013). doi:10.1016/j.jsc.2011.12.018
21. Malaschonok, G., Scherbinin, A.: Triangular decomposition of matrices in a domain. In: Gerdt, V.P., Koepf, W., Seiler, W.M., Vorozhtsov, E.V. (eds.) CASC 2015. LNCS, vol. 9301, pp. 292–306. Springer, Cham (2015). doi:10.1007/978-3-319-24021-3_22
22. Vershik, A.M., Sporyshev, P.V.: An estimate of the average number of steps in the simplex method, and problems in asymptotic integral geometry. Sov. Math. Dokl. **28**, 195–199 (1983)
23. Smale, S.: On the average number of steps of the simplex method of linear programming. Math. Program. **27**(3), 241–262 (1983). doi:10.1007/BF02591902
24. Dubickas, A., Smyth, C.J.: Length of the sum and product of algebraic numbers. Math. Notes. **77**, 787–793 (2005). doi:10.1007/s11006-005-0079-y
25. Kollár, J.: Unirationality of cubic hypersurfaces. J. Inst. Math. Jussieu. **1**(3), 467–476 (2002). doi:10.1017/S1474748002000117
26. Cenk, M., Hasan, M.A.: On the arithmetic complexity of Strassen-like matrix multiplications. J. Symbolic Comput. **80**(2), 484–501 (2017). doi:10.1016/j.jsc.2016.07.004
27. Hedén, I.: Russell's hypersurface from a geometric point of view. Osaka J. Math. **53**(3), 637–644 (2016)

Dag-Like Communication and Its Applications

Dmitry Sokolov[✉]

St. Petersburg Department of V.A. Steklov Institute of Mathematics
of the Russian Academy of Sciences,
27 Fontanka, St. Petersburg 191023, Russia
sokolov.dmt@gmail.com
http://logic.pdmi.ras.ru/~sokolov

Abstract. In 1990 Karchmer and Widgerson considered the following communication problem Bit: Alice and Bob know a function $f : \{0,1\}^n \to \{0,1\}$, Alice receives a point $x \in f^{-1}(1)$, Bob receives $y \in f^{-1}(0)$, and their goal is to find a position i such that $x_i \neq y_i$. Karchmer and Wigderson proved that the minimal size of a boolean formula for the function f equals the size of the smallest communication protocol for the Bit relation. In this paper we consider a model of dag-like communication complexity (instead of classical one where protocols correspond to trees). We prove an analogue of Karchmer-Wigderson Theorem for this model and boolean circuits. We also consider a relation between this model and communication PLS games proposed by Razborov in 1995 and simplify the proof of Razborov's analogue of Karchmer-Wigderson Theorem for PLS games.

We also consider a dag-like analogue of real-valued communication protocols and adapt a lower bound technique for monotone real circuits to prove a lower bound for these protocols.

In 1997 Krajíček suggested an interpolation technique that allows to prove lower bounds on the lengths of resolution proofs and Cutting Plane proofs with small coefficients (CP*). Also in 2016 Krajíček adapted this technique to "random resolution". The base of this technique is an application of Razborov's theorem. We use real-valued dag-like communication protocols to generalize the ideas of this technique, which helps us to prove a lower bound on the Cutting Plane proof system (CP) and adapt it to "random CP".

Our notion of dag-like communication games allows us to use a Raz-McKenzie transformation [5,17], which yields a lower bound on the real monotone circuit size for the CSP-SAT problem.

1 Introduction

In 1990 Karchmer and Wigderson [9] introduced the following communication problem Bit: Alice receives a point u from a set $U \subseteq \{0,1\}^n$, Bob receives a point v from a set $V \subseteq \{0,1\}^n$, $U \cap V = \emptyset$, and their goal is to find a position i such that $u_i \neq v_i$. There is also a monotone version of this communication problem, called MonBit, in this case the goal of Alice and Bob is to find a position i such that $u_i = 1$ and $v_i = 0$. In [9] Karchmer and Wigderson proved the following

© Springer International Publishing AG 2017
P. Weil (Ed.): CSR 2017, LNCS 10304, pp. 294–307, 2017.
DOI: 10.1007/978-3-319-58747-9_26

Theorem: for every function f, there is a (monotone) boolean formula of size S iff there is a communication protocol of size S for the problem Bit (MonBit), where $U = f^{-1}(1)$ and $V = f^{-1}(0)$. Since then, a lot of results about the formula complexity of functions has been obtained by using this theorem, for example, a lower bound $2^{\Omega(\frac{n}{\log n})}$ on the monotone formula complexity for an explicit function [5], and a lower bound $n^{3-o(1)}$ on the formula complexity in de Morgan basis for an explicit function [4]. Karchmer-Wigderson Theorem gives a characterization of boolean formulas in terms of communication complexity, however, it does not work in the context of boolean circuits.

In 1995 Razborov [18] introduced a model of communication *Polynomial Local Search* games (PLS). He gave a generalization of Karchmer-Wigderson Theorem replacing classical communication protocols by PLS games, and boolean formulas by boolean circuits. In this paper we consider a simplification of communication PLS games that is called boolean communication games (an analogue of this definition was also studied in [16]). We show that for any communication problem there is a boolean communication game of size S iff there is a PLS game of size $\Theta(S)$ for the same communication problem. We also show a simple proof of a generalization of Karchmer-Wigderson result in the case of using boolean communication games and boolean circuits.

Razborov's result about the connection between PLS games and boolean circuits was used in 1997 by Krajíček [10], who introduced a so-called "interpolation technique" for proving lower bounds on the size of propositional proof systems. In order to describe the essence of this technique let us consider a monotone function $f : \{0,1\}^n \to \{0,1\}$ from the class **NP** such that there is a lower bound $T(n)$ on the monotone circuit complexity of f. For example, one can use a function from [1]: let formula $\texttt{Zero}(x,r)$ encode with additional variables r, the fact that $x \in f^{-1}(0)$, and let formula $\texttt{One}(x,q)$ encode with additional variables q, the fact that $x \in f^{-1}(1)$. Krajíček has shown that if a proof system operates with clauses such that the communication complexity of evaluating these clauses (Alice knows the values of a part of variables, and Bob knows the values of the other part of variables) is bounded by parameter t, and in this proof system there is a proof of size S of the unsatisfiable formula $\texttt{Zero}(x,r) \wedge \texttt{One}(x,q)$, then one can create a PLS game of size $S \cdot 2^t$ for the Karchmer-Wigderson problem for function f. If the formulas Zero and One satisfy certain natural properties then this PLS game also solves a monotone version of the Karchmer-Wigderson problem for the function f. Thus we have a lower bound $S \geq \frac{T(n)}{2^t}$. There are proof systems for which lower bounds can be obtained by using this technique, for example, resolution, CP^*, subsystems of LK, OBDD($\exists$, weakening) [12]. However, if we cannot bound the parameter t then this technique does not give us any bounds, in particular we cannot use this technique for the CP proof system (without restrictions on the size of coefficients).

The second important communication problem is a canonical search problem $\texttt{Search}_\phi$ for an unsatisfiable formula $\phi(x,y)$ in CNF [2]: Alice receives values for the variables x, Bob receives values for the variables y, and their goal is to find a clause of ϕ such that it is unsatisfied by this substitution. In the paper [2], the authors present a technique of constructing communication protocols of size poly(S) (in various classical communication models) for the Search$_\phi$ problem, where S is the size of a tree-like proof of ϕ in the proof system Th(k) for fixed k that operates with polynomial inequalities of degree at most k over integer numbers. These proof systems cover a huge class of known proof systems (for example, CP is a special case of Th(1)). In [2,5,8] the authors prove lower bounds on the communication complexity of the Search$_\phi$ problem and, as a corollary, a lower bound on the size of tree-like proofs in Th(k). This technique allows to prove lower bounds only for tree-like versions of proof systems; general lower bounds are still unknown even for Th(2). Also in [5] the authors demonstrate a version of Raz-McKenzie transformation [17] that reduces the problem Search$_\phi$ to the problem MonBit for a certain function SAT$_G$ (see Definition 12). As a corollary the authors obtain a lower bound on the monotone formula complexity of the function SAT$_G$.

Remark 1. Although in Krajíček's paper [10] the problem Search$_\phi$ is not used, in fact all PLS games in that paper with little modification solve this problem. As a corollary, these games also solve the Karchmer-Wigderson problem.

In this paper we also consider real communication games that generalize boolean communication games (which are a dag-like analogue of real-valued classical communication protocols [11]). We prove an analogue of Krajíček's Theorem: we show how to construct a real communication game of size S for the problem Search$_\phi$ from a proof of ϕ in the CP proof system (and, as a corollary, from a proof in any proof system used in Krajíček's paper). Instead of constructing a circuit from a game we directly give a lower bound for real communication protocols. This result generalizes Cook and Haken's result [6] for monotone real circuits. As a corollary of this result we apply a Raz-McKenzie transformation and obtain a lower bound on the monotone real circuit size of the function SAT$_G$.

In [3] the authors introduce a *random resolution* proof system. A δ-random resolution proof distribution for a formula ϕ is a random distribution (π_s, Δ_s) such that Δ_s is a CNF formula, π_s is a resolution proof of $\phi \wedge \Delta_s$, and every fixed truth assignment of all variables satisfies Δ_s with probability at least $1 - \delta$. We can consider a natural generalization of this definition to other proof systems and look at lower bounds for it. The only known technique for proving lower bounds for the CP proof system is the reduction, due to Pudlák [15], to lower bounds on the size of real monotone circuits; Hrubeš [7] generalizes this technique for the semantic version of CP. The exponential lower bounds on these circuits are given in [6,15]. The reduction of lower bounds on the CP proof size to lower bounds on the size of real monotone circuits uses substantially the structure of the initial formula, and so it is unclear how to generalize them for a *random* CP proof system. In this paper we show that lower bounds that are obtained by using real communication games can be generalized for *random* CP by using a technique that has been recently introduced by Krajíček in [13]. Unfortunately, this technique gives us a lower bound only for small values of the parameter δ.

Organization of the paper. In Sect. 2 we give definitions of boolean and real communication games and prove basic properties of these games. In Sect. 3 we define PLS games and prove a relation between PLS games and boolean communication games, also we give a simplification of Razborov's Theorem. In Sect. 4 we consider a construction of communication games from semantic CP proofs. In Sect. 5 we give a lower bound on the size of real communication games. In Sect. 6 we prove a lower bound on *random* CP proof system. In Sect. 7 we give a lower bound on the real circuit complexity of the function SAT_G.

Remark 2. Definition 1 was introduced independently by Pavel Pudlák and Pavel Hrubeš. Also Pavel Pudlák in a private communication announced a proof of the opposite direction of the statement of Lemma 2.

2 Preliminaries

2.1 Games

The following definition has been also independently introduced by Pavel Pudlák and Pavel Hrubeš.

Definition 1. *Let $U, V \in \{0,1\}^n$ be two sets. Let us consider a triple (H, A, B), where H is a directed acyclic graph, $A : H \times U \to \mathbb{R}$ and $B : H \times V \to \mathbb{R}$. We say that a vertex $v \subset H$ is valid for a pair (x, y) iff $A(v, x) > B(v, y)$. We call this triple a real communication game for the pair (U, V) and some relation $N : U \times V \times T \to \{0, 1\}$, where T is a finite set of "possible answers", if the following holds:*

- *H is an acyclic graph and the out-degree of all its vertices is at most 2;*
- *the leaves of H are marked by element of T;*
- *there is a root $s \in H$ with in-degree 0 and this vertex is valid for all pairs from $U \times V$;*
- *if $v \in H$ is valid for (x, y) and v is not a leaf then at least one child of v is valid for (x, y);*
- *if $v \in H$ is valid for (x, y), v is a leaf and v is marked by $t \in T$ then $N(x, y, t) = 1$.*

The size of the game is the size of the graph H.

We call it a boolean communication game if $A : H \times U \to \{0, 1\}$ and $B : H \times V \to \{0, 1\}$. (An analogue of boolean communication games was studied in [16]).

Remark 3. It is useful to think that if we have a boolean communication game (H, A, B) for sets U, V then we mark each vertex $h \in H$ by rectangle $R_h \in U \times V$ of *valid inputs*, where $(x, y) \in R_h$ iff $A(h, x) = 1$ and $B(y, h) = 0$. So, if s is the root then it is marked by the rectangle $U \times V$. If h has two children h' and h'', then $R_h \subseteq R_{h'} \cup R_{h''}$.

Lemma 1. *Let $U, V \in \{0,1\}^n$. If Alice receives $x \in U$, Bob receives $y \in V$ and we have a classical communication protocol of size S for some relation N, then we have a boolean communication game (H, A, B) of size S for sets U, V and relation N. Moreover, H is a tree.*

Proof. Let us consider a tree K that corresponds to a classical communication protocol. Vertices of this tree correspond to the values of transmitted bits. We consider a vertex $k \in K$ and mark it by rectangle $R_k \in U \times V$, where $(x, y) \in R_k$ iff we run protocol on inputs x, y and come to vertex k at some moment. This tree with rectangles defines a boolean communication game (see Remark 3), the root of this tree is the root of the game. All required properties follow from the definition of rectangles R_k. □

Definition 2. *Let $\phi(x, y)$ be an unsatisfiable CNF formula, U be an arbitrary subset of assignments to variables x, and V be an arbitrary subset of assignments to variables y. A canonical search problem (relation) $\mathtt{Search}_\phi : U \times V \times C \to \{0,1\}$, where C is the set of clauses of formula ϕ, contains all triples (u, v, c) such that $c(u, v) = 0$.*

Definition 3. *Let $U, V \subseteq \{0,1\}^n$, $U \cap V = \emptyset$. Relation $\mathtt{Bit}_{U,V} : U \times V \times [n] \to \{0,1\}$ contains all triples (u, v, i) such that $u_i \neq v_i$. If there is a function f such that $U = f^{-1}(1)$ and $V = f^{-1}(0)$ we write $\mathtt{Bit}_f$.*

Let $U, V \subseteq \{0,1\}^n$, $U \cap V = \emptyset$ and $\forall\ x \in U,\ y \in V\ \exists i\ x_i = 1 \wedge y_i = 0$. Relation $\mathtt{MonBit}_{U,V} : U \times V \times [n] \to \{0,1\}$ contains all triples (u, v, i) such that $u_i = 1 \wedge v_i = 0$. If there is a monotone function f such that $U = f^{-1}(1)$ and $V = f^{-1}(0)$ we write $\mathtt{MonBit}_f$.

Lemma 2. *Let $f : \{0,1\}^n \to \{0,1\}$ be a monotone function. If there is a monotone (boolean) real circuit for f of size S then there is a (boolean) real communication game of size S for sets $(f^{-1}(1), f^{-1}(0))$ and relation $\mathtt{MonBit}_f$.*

Proof. A graph H of our real communication game is a graph of the minimal monotone real circuit for function f with inverted edges. $A(e, u)$ returns the value of the gate that corresponds to the vertex e on the input u. We define $B(e, v)$ in the same way. If a leaf $h \in H$ corresponds to an input variable x_i then mark this leaf by i.

Let us check all the required properties:

- H is an acyclic and all leaves are marked;
- the root $s \in H$ corresponds to the output gate of the circuit;
- note that $A(s, f^{-1}(1)) = 1$ and $B(s, f^{-1}(0)) = 0$, hence the root is valid for all pairs from $f^{-1}(1) \times f^{-1}(0)$;
- if $h \in H$ is an inner vertex and $A(h, u) > B(h, v)$, then it has a child h' such that $A(h', u) > B(h', v)$ since a gate that corresponds to h computes a monotone function;
- if $h \in H$ is a leaf with label i then $A(h, u) = u_i$ and $B(h, v) = v_i$. Hence if $A(h, u) > B(h, v)$ then $u_i = 1$ and $v_i = 0$. □

2.2   Semantic Cutting Planes

We consider a semantic version of the Cutting Plane (CP) proof system.

Definition 4 (Hrubeš [7]). *A proof in semantic CP for CNF formula ϕ is a sequence of linear inequalities with real coefficients $C_1, C_2, \ldots, C_k$, such that C_k is the trivially unsatisfiable inequality $0 \geq 1$ and C_i can be obtained by one of the following rules:*

- *C_i is a linear inequality that encodes a clause of formula ϕ;*
- *C_i semantically follows on $\{0,1\}$ values from C_j, C_k where $j, k < i$.*

The size of proof is the number of inequalities k. We say that we have a proof in CP^* if coefficients in the proof are integer and bounded by a polynomial in the number of variables of ϕ.

2.3   Broken Mosquito Screen

Definition 5 (Cook, Haken [6]). *An instance of the Broken Mosquito Screen (BMS) problem encodes a graph with $m^2 - 2$ vertices, where $m \geq 3$ is a convenient parameter for indexing. The graphs are represented in a standard way, as a string of bits that indicates for each pair of vertices whether there is an edge between them, with value 1 for the edge being present and value 0 for the edge being absent.*

The graph is good, or accepted, if there is a partition of its vertices into $m - 1$ sets of size m and one set of size $m - 2$ such that each of these subsets forms a clique. A graph is bad, or rejected if there is a partition of its vertices into $m - 1$ sets of size m and one set of size $m - 2$ such that each of these subsets forms an anticlique.

Lemma 3 (Cook, Haken [6]). *No instance of BMS can be good and bad simultaneously. Furthermore, each element in good set is not less (as a vector) than any element in bad set.*

Definition 6 (Cook, Haken [6]). *Let G_0 be a set of good instances of the BMS problem that are minimal: only the edges that are explicitly needed to meet the acceptance condition are present. Let B_0 be a set of bad instances of the BMS problem that are maximal: all edges are present except those that are explicitly required to be absent to meet the rejection condition.*

Now we describe unsatisfiable formulas that are based on the BMS problem. $BMS(x, q, r) = Part(x, q) \wedge Part(\neg x, r)$, where $x \in \{0,1\}^{(m^2-2)(m^2-3)/2}$ are variables that correspond to a graph, $\neg x$ means that we substitute the negation of the respective literals, $q = \{q_{ijk} \mid i, j \in [m], k \in [m^2 - 2]\}$, $r = \{r_{ijk} \mid i, j \in [m], k \in [m^2 - 2]\}$. $Part(x, y)$ equals true iff x is a good instance of BMS problem, $y_{ijk} = 1$ iff we put a vertex k on the j-th place in the i-th component, and the formula $Part(x, y)$ consists of the following clauses:

- $\forall i, j \in [m], k_1, k_2 \in [m^2 - 2], k_1 \neq k_2 : \quad (\neg y_{ijk_1} \vee \neg y_{ijk_2});$

- $\forall i < m, j \leq m :$ $\displaystyle\bigvee_{k\in[m^2-2]} y_{ijk};$
- $\forall j \leq m - 2 :$ $\displaystyle\bigvee_{k\in[m^2-2]} y_{mjk};$
- $\forall j \in \{m - 1, m\}, k \in [m^2 - 2] :$ $(\neg y_{mjk});$
- $\forall i, j_1 < j_2, k_1 \neq k_2 :$ $(\neg y_{ij_1k_1} \vee \neg y_{ij_2k_2} \vee x_{k_1k_2}).$

We also need a variant of this formula in 3-CNF, denote it by $Part'$. It can be obtained by replacing long clauses by a standard procedure: if we have a clause C of the form $(a \vee b \vee c \vee D)$ then we replace it by two new clauses $(a \vee b \vee \ell)$ and $(\neg \ell \vee D)$, where ℓ is a new variable.

$$\mathrm{BMS}'(x, q, r, z) = Part'(x, q, z) \wedge Part'(\neg x, r, z)$$

3   Bit Relation and Circuits

In this section we prove a generalization of Kachmer-Wigderson Theorem. This Theorem relates the size of classic communication protocols for the relation Bit to the size of boolean formulas. We prove a similar result for boolean communication games and boolean circuits. We also consider a model of PLS communication games [18] with a fixed graph and prove its equivalence to boolean communication games, hence we give a simple proof of Razborov's Theorem about the relation between communication PLS games and boolean circuits.

3.1   PLS Games and Boolean Circuits

We start with a model of PLS games. We use a bit simpler notion of PLS games from Krajíček's paper [10], where the graph of game is fixed.

Definition 7 (Razborov [18], Krajíček [10]). *Let $U, V \in \{0, 1\}^n$ be two sets and let $N : U \times V \times T \to \{0, 1\}$ be a relation, where T is a finite set of "possible answers". A communication PLS game for sets U, V and relation N is a labelled directed graph G satisfying the following four conditions:*

- *G is acyclic and has a root (the in-degree 0 node) denoted $\emptyset$;*
- *each leaf is labelled by some $t \in T$;*
- *there is a function $S(g, x, y)$ (the strategy) that given a node $g \in G$ and a pair $x \in U, y \in V$, outputs the end of an edge leaving the node g;*
- *for every $x \in U, y \in V$, there is a set $F(x, y) \in G$ such that:*
  - *$\emptyset \in F(x, y);$*
  - *if $g \in F(x, y)$ is not a leaf then $S(g, x, y) \in F(x, y);$*
  - *if $g \in F(x, y)$ is a leaf and it is marked by $t \in T$ then $N(x, y, t) = 1$.*

The communication complexity of G is the minimal number t such that for every $g \in G$ the players (one knowing x and g, the other one, y and g) decide whether $g \in F(x, y)$ and compute $S(g, x, y)$ with at most t bits exchanged in the worst case. The size of the game is defined as $|G|$.

Remark 4. We remove the cost function from the original definition in [18] since if a graph is fixed then the cost function can be replaced by the topology number of vertex.

Theorem 1. *Let $U, V \subseteq \{0,1\}^n$, and $N : U \times V \times \mathbb{N} \to \{0,1\}$ be a relation.*

1. *If there is a communication PLS game of size L and communication complexity t for sets U, V and a relation N then there is a boolean communication game of size at most $L \cdot 2^{3t}$ for the same sets and relation.*
2. *If there is a boolean communication game of size L for sets U, V and a relation N then there is a communication PLS game of size L and communication complexity two for the same sets and relation.*

Proof. See full version [20].   □

3.2   Games and Circuits

The proof of the following theorem generalizes a result from [9] and uses a similar proof strategy. The sketch of circuits construction from protocols was given in [16] (Lemma 1), for protocols construction from circuits we will use combination of Razborov's Theorem [18] and Theorem 1.

Theorem 2. *Let $f : \{0,1\}^n \to \{0,1\}$ be a function. There is a boolean communication game for $\mathtt{Bit}_f$ of size S iff there is a circuit for f of size $O(S)$. Moreover there is a boolean communication game for $\mathtt{MonBit}_f$ of size S iff there is a monotone circuit for f of size S.*

Proof. See full version [20].   □

Corollary 1 (Razborov [18]). *Let $f : \{0,1\}^n \to \{0,1\}$ be a function. If there is a PLS communication game for $(\mathtt{MonBit}_f)$ $\mathtt{Bit}_f$ of size S and communication complexity t then there is a (monotone) circuit for f of size $S \cdot 2^{O(t)}$. If there is a (monotone) circuit for f of size S then there is a PLS communication game for $(\mathtt{MonBit}_f)$ $\mathtt{Bit}_f$ of size S and communication complexity two.*

Proof. Follows from Theorems 1 and 2.   □

4   From Proofs to Games

In this section we relate real communication games to proofs in the semantic CP proof system.

At first we consider a connection between $\mathtt{MonBit}$ relation and $\mathtt{Search}_\phi$ problem.

Lemma 4. *Let $U, V \subseteq \{0,1\}^n$, $U \cap V = \emptyset$ and $\forall\, x \in U, y \in V\ \exists i\ x_i = 1 \wedge y_i = 0$. Let $Q(z,q)$ be a boolean CNF formula such that $x \in U$ iff the formula $\exists q\ Q(x,q)$ is true. Let $R(z,r)$ be a boolean CNF formula that satisfies the following properties:*

- *there is at most one variable z in each clause;*
- *all variables z occur with negative signs;*
- *$y \in V$ iff the formula $\exists r\ R(y, r)$ is true.*

For each $x \in U$ one can fix arbitrary q^x such that $Q(x, q^x) = 1$, and for each $y \in V$ fix arbitrary r^y such that $R(y, r^y) = 1$. Let $L = \{(x, q^x) \mid x \in U\}$, and $L' = \{(y, r^y) \mid y \in V\}$. If there is a real (boolean) communication game for sets L, L' and $\mathtt{Search}_{Q(z,q) \wedge R(z,r)}$ of size S then there is a real (boolean) communication game for sets U, V and $\mathtt{MonBit}_{U,V}$ of size S.

Proof. See full version [20].   □

Lemma 5. *Let $\phi(x, y)$ be an unsatisfiable CNF formula, U be an arbitrary subset of substitutions to variables x and V is an arbitrary subset of substitutions to variables y. If there is a semantic CP proof of this formula of size S then there is a real communication game of size S for the sets (U, V) and the canonical search problem $\mathtt{Search}_{\phi}$.*

Proof. Let H be the graph of the semantic CP proof of the formula ϕ with inverted edges. There is a correspondence between vertices and inequalities of the proof. Consider a vertex $h \in H$, this vertex corresponds to inequalities $f(x) + \ell(y) \geq c$, define the functions A, B in the following way $A(h, u) = -f(u)$ and $B(h, v) = \ell(v) - c$. Note that a vertex is valid for pair (u, v) iff $A(h, u) > B(h, v)$, hence $f(u) + \ell(v) < c$, i.e. in this case the inequality is falsified by the substitution (x, y).

The root of our game corresponds to the trivially false inequality $0 \geq 1$, hence the root is valid for any pair $(u, v) \in U \times V$. If a substitution satisfies all inequalities in the children of some vertex $h \in H$ then this substitution satisfies the inequality in h. Thus, if h is valid for some pair then at least one child of h is valid for this pair.

If a leaf h is valid for the pair (u, v) then the inequality in h is falsified by the substitution (u, v).

Lemma 6. *Let $U, V \subseteq \{0, 1\}^n$, $U \cap V = \emptyset$ and $\forall\ x \in U, y \in V\ \exists i\ x_i = 1 \wedge y_i = 0$. Let $Q(z, q)$ be a boolean CNF formula such that $x \in U$ iff the formula $\exists q\ Q(x, q)$ is true. Let $R(z, r)$ be a boolean CNF formula that satisfies the following properties:*

- *there is at most one variable z in each clause;*
- *all variables z occur with negative signs;*
- *$y \in V$ iff the formula $\exists r\ R(y, r)$ is true.*

If there is a proof of formula $Q(z, q) \wedge R(z, r)$ in semantic CP of size S then there is a real communication game for (U, V) and relation $\mathtt{MonBit}_{U,V}$ of size S.

Proof. Follows from Lemmas 4 and 5.

5   Lower Bound

We remind that G_0 is the set of minimal good instances of BMS and B_0 is the set of maximal bad instances of BMS.

Lemma 7 (Cook, Haken [6], Sect. 4.4). $|G_0| = |B_0| = \frac{(m^2-2)!}{(m!)^{m-1}(m-2)!(m-1)!}$.

For the rest of the section we fix some subsets $U_0 \subseteq G_0, V_0 \subseteq B_0$ of size at least $\frac{|G_0|}{2}$, w.l.o.g. $|U_0| = |V_0|$.

Theorem 3. *The size of any real communication game for pair U_0, V_0 and relation* MonBit_{U_0, V_0} *is at least* $\frac{1.8^{\sqrt{m/8}}}{4}$.

Before we prove this theorem we need to present a notion of fences [6]. For the rest of this section we fix some real communication game (H, A, B) for pair (U_0, V_0) and relation MonBit_{U_0, V_0}. Our goal is to construct a partial map μ : $(U_0 \cup V_0) \to H$ such that the domain of μ is big enough and the size of preimage of any element of H is small. We create this map step by step. At the step $i \in 0, 1, \ldots$ (we say that i is the current time) we consider the sets $U_i \subseteq U_0$, $V_i \subseteq V_0$ and pick some element $g \in U_i \cup V_i$ and put it to some vertex from H, after that we increase the time and proceed with sets $U_{i+1} = U_i \setminus \{g\}$ and $V_{i+1} = V_i \setminus \{g\}$. Note that either $U_{i+1} = U_i$ or $V_{i+1} = V_i$.

Definition 8. *Let h be a vertex in a real communication game (H, A, B) and let $g \in U_i$. A fence around graph g in the vertex h at time i is a conjunction $C = z_1 \wedge \cdots \wedge z_q$ where $z_1, \ldots, z_q$ are bits of the input of BMS problem. Furthermore, $C(g) = 1$, and if h is a valid vertex for pair (g, g') for some $g' \in V_i$ then $C(g') = 0$. The length of fence is the number of variables q. A minimal fence around g in h at time i is a fence of minimal length around g in h at time i.*

Dually, a fence around $g \in V_i$ in h at time i is a disjunction $D = z_1 \vee \cdots \vee z_q$, where $z_1, \ldots, z_q$ bits of the input of BMS problem. Furthermore, $D(g) = 0$ and if h is a valid vertex for pair (g', g) for some $g' \in U_i$ then $D(g') = 1$.

Proposition 1. *The length of minimal fence around $g \in U_0 \cup B_0$ in h is not increasing in time.*

Proof. Follows from the definition of fence.

Definition 9 (Cook, Haken [6]). *Let $k = \frac{m}{2}$. We call a fence long if it is longer than $\frac{k}{2}$, otherwise we call it short.*

5.1   Construction of a Mapping μ

Definition 10 (Cook, Haken [6]). *Let us fix some topological sorting of the graph H so that the children of some vertex $h \in H$ have bigger numbers than h. At time i let $h_i \in H$ be the vertex with the maximum topological number such that there is a graph $d_i \in G_i \cup B_i$ such that d_i requires a long fence at h_i at time i. Define $\mu(d_i) = h_i$ and delete d_i from $G_i \cup B_i$ to get $G_{i+1} \cup B_{i+1}$ (if there is more than one such d_i then we choose some from G_i first). This process stops when the remaining graphs have short fences at all gates.*

The following lemmas are proved by analogy with the paper [6].

Lemma 8 (Cook, Haken [6], Lemma 2). *The size of the domain of μ is at least $|U_0|$.*

Proof. See full version [20]. □

The next lemma is an analogue of Lemma 4 from [6].

Lemma 9. *The number of graphs from $U_0 \cup V_0$ that can be mapped by μ to any single $h \in H$ is at most*

$$2\frac{(km)^{r/2}(m^2 - m)^{r/2}(m^2 - 2 - r)!}{(m!)^{m-1}(m-2)!(m-1)!},$$

where r is the greatest even number that is less or equal to $\sqrt{\frac{m}{2}}$.

Proof. See full version [20]. □

Proof (Proof of Theorem 3). The size of real communication game is at least the size of the domain of μ divided by the maximum size of preimage of the elements in the image of μ, hence from Lemmas 8 and 9 we conclude that the size is at least

$$2\frac{|U_0|(m!)^{m-1}(m-2)!(m-1)!}{(km)^{r/2}(m^2-m)^{r/2}(m^2-2-r)!} \geq \frac{|G_0|(m!)^{m-1}(m-2)!(m-1)!}{(km)^{r/2}(m^2-m)^{r/2}(m^2-2-r)!} \geq \frac{1.8^{\sqrt{m/8}}}{4}.$$

The last inequality follows from [6], Sect. 4.6.

Corollary 2. *Let $Q(z,q)$ be a boolean CNF formula that $x \in G_0$ iff the formula $\exists q\, Q(x,q)$ is true. Let $R(z,r)$ be a boolean CNF formula that satisfy the following properties:*

- *there is at most one variable z in each clause;*
- *all variables z occur with negative signs;*
- *$y \in B_0$ iff the formula $\exists r\, R(y,r)$ is true.*

Let L be a set of substitution to variable z, q and L' be a set of substitution to variable r. The size of any real communication game for the pair L, L' and the relation $\mathrm{Search}_{Q(z,y) \wedge R(z,r)}$ is at least $\frac{1.8^{\sqrt{m/8}}}{4}$.

Proof. Follows from Theorem 3 and Lemma 4.

6 Random Cutting Planes

Definition 11. *A δ-random CP proof distribution of formula ϕ is a random distribution (π_s, Δ_s) such that Δ_s is a CNF formula, π_s is a CP proof of $\phi \wedge \Delta_s$, and every fixed truth assignments of all variables satisfies the formula Δ_s with probability at least $1 - \delta$.*
    The size of distribution is the maximum size of π_s.

Theorem 4. *Let* (π_s, Δ_s) *be a* δ-random CP *proof distribution of the formula* BMS *for a convenient parameter* m. *Let* d *be the maximum number of clauses in formulas* Δ_s. *If* $d\sqrt{\delta} \leq \frac{1}{2}$ *then the size of this distribution is at least* $(1 - d\sqrt{\delta})\frac{1.8\sqrt{m/8}}{4}$.

For $(g, h) \in G_0 \times B_0$ define $w(g, h) = (g, q^g, r^h)$ such that $Part(g, q^g) = 1$ and $Part(\neg h, r^h) = 1$. Let us assume that w is an injective map (since G_0 and B_0 are extremal instances we can choose w in such a way).

Let (π_s, Δ_s) be an arbitrary δ-random CP proof. Denote the size of π_s by k. For a sample s define a set $Bad_s \subseteq G_0 \times B_0$ to be the set of all pairs (g, h) such that $w(g, h)$ falsifies Δ_s.

Lemma 10 (Krajíček [13], Lemma 2.1). *There exists a sample s such that* $|Bad_s| \leq \delta|G_0 \times B_0|$.

Let us fix s from this Lemma. Let d be the number of clauses in Δ_s.

Lemma 11 (Krajíček [13], Lemma 2.2). *There exist subsets* $U \subseteq G_0$ *and* $V \subseteq B_0$ *such that*

- $U \times V \cap Bad_s = \emptyset$;
- $|U| \geq (1 - d\sqrt{\delta})|G_0|$;
- $|V| \geq (1 - d\sqrt{\delta})|B_0|$.

Lemma 12. *Consider a pair* (U, V) *from Lemma 6. There is a real communication game for* (U, V) *and relation* MonBit$_{U,V}$ *of size that equals the size of* π_s.

Proof. See full version [20].   □

Proof (Proof of Theorem 4). Consider s from Lemma 10. By Lemma 12 and Lemma we have a real communication game for sets (U, V) and relation MonBit$_{U,V}$ of size that equals the size of π_s where $U \subseteq G_0$, $V \subseteq B_0$ and $|U| \geq \frac{|G_0|}{2}$, $|V| \geq \frac{|B_0|}{2}$. Hence the statement of the Theorem follows from Theorem 3.

7   Monotone CSP-SAT

In this section we consider a monotone function called CSP-SAT. This function was defined in [5,14]; in [19] the authors gave a fully exponential lower bound on the size of monotone boolean formulas for this function. We prove that this function requires an exponential monotone real circuit size.

Definition 12 (Göös, Pitassi [5]). *The function* CSP-SAT *is defined relative to some finite alphabet* Σ *and a fixed constraint topology given by a bipartite graph* G *with left vertices* V *(variable nodes) and right vertices* U *(constraint nodes). We think of each* $v \in V$ *as a variable taking on values from* Σ, *an edge*

$(v, u) \in E(G)$ *indicates that variable v is involved in constraint node u. Let d be the maximum degree of a node in U. We define* $\mathrm{SAT} = \mathrm{SAT}_{G,\Sigma} : \{0,1\}^N \to \{0,1\}$ *on $N \leq |U| \cdot |\Sigma|^d$ bits as follows. An input $\alpha \in \{0,1\}^N$ describes a* CSP *instance by specifying, for each constraint node $u \in U$, its truth table: a list of at most $|\Sigma|^d$ bits that record which assignments to the variables involved in u satisfy u. Then* $\mathrm{SAT}(\alpha) := 1$ *iff the* CSP *instance described by α is satisfiable. This encoding of* CSP *satisfiability is indeed monotone: if we flip any 0 in a truth table of a constraint into a 1, we are only making the constraint easier to satisfy.*

The proof of the following theorem use a simplification of analogy of reduction from [5,17].

Theorem 5. *Let Φ be an unsatisfiable d-CNF formula on n variables and m clauses with the variables splitted into sets X, Y. Let G be a constraint topology of Φ. If there is a real (boolean) communication game of size S for sets* $\mathrm{SAT}^{-1}_{G,\{0,1\}}(1), \mathrm{SAT}^{-1}_{G,\{0,1\}}(0)$ *and* $\mathrm{MonBit}_{\mathrm{SAT}_{G,\{0,1\}}}$ *relation then there is a real (boolean) communication game of size S for sets $\{0,1\}^{|X|}, \{0,1\}^{|Y|}$ (sets of all possible substitution to variables X and Y) and* Search_ϕ *relation.*

Proof. See full version [20]. □

Corollary 3. *Let G be a constraint topology of* BMS$'$*. The size of any monotone real circuit that computes* $\mathrm{SAT}_{G,\{0,1\}} : \{0,1\}^N \to \{0,1\}$ *is at least $2^{\Omega(N^{1/8})}$.*

Proof. Follows from Theorem 5 and Corollary 2.

Acknowledgements. This research is supported by Russian Science Foundation (project 16-11-10123).

The author is grateful to Pavel Pudlák and Dmitry Itsykson for fruitful discussions. The author also thanks Edward Hirsch, Dmitry Itsykson and anonymous reviewers for error correction.

References

1. Alon, N., Boppana, R.B.: The monotone circuit complexity of boolean functions. Combinatorica **7**(1), 1–22 (1987). http://dx.doi.org/10.1007/BF02579196
2. Beame, P., Pitassi, T., Segerlind, N.: Lower bounds for Lovász-Schrijver systems and beyond follow from multiparty communication complexity. SIAM J. Comput. **37**(3), 845–869 (2007). http://dx.doi.org/10.1137/060654645
3. Buss, S.R., Kolodziejczyk, L.A., Thapen, N.: Fragments of approximate counting. J. Symb. Log. **79**(2), 496–525 (2014). http://dx.doi.org/10.1017/jsl.2013.37
4. Dinur, I., Meir, O.: Toward the KRW composition conjecture: Cubic formula lower bounds via communication complexity. In: 31st Conference on Computational Complexity, CCC 2016, May 29 to June 1, 2016, Tokyo, Japan, pp. 3:1–3:51 (2016). http://dx.doi.org/10.4230/LIPIcs.CCC.2016.3
5. Göös, M., Pitassi, T.: Communication lower bounds via critical block sensitivity. In: Symposium on Theory of Computing, STOC 2014, New York, NY, USA, 31 May–03 June, 2014, pp. 847–856 (2014). http://doi.acm.org/10.1145/2591796.2591838

6. Haken, A., Cook, S.A.: An exponential lower bound for the size of monotone real circuits. J. Comput. Syst. Sci. **58**(2), 326–335 (1999)
7. Hrubeš, P.: A note on semantic cutting planes. Electron. Colloquium Comput. Complex. (ECCC) **20**, 128 (2013). http://eccc.hpi-web.de/report/2013/128
8. Huynh, T., Nordström, J.: On the virtue of succinct proofs: amplifying communication complexity hardness to time-space trade-offs in proof complexity. In: Proceedings of the 44th Symposium on Theory of Computing Conference, STOC 2012, New York, NY, USA, 19–22 May 2012, pp. 233–248 (2012). http://doi.acm.org/10.1145/2213977.2214000
9. Karchmer, M., Wigderson, A.: Monotone circuits for connectivity require super-logarithmic depth. SIAM J. Discrete Math. **3**(2), 255–265 (1990). http://dx.doi.org/10.1137/0403021
10. Krajíček, J.: Interpolation theorems, lower bounds for proof systems, and independence results for bounded arithmetic. J. Symb. Log. **62**(2), 457–486 (1997). http://dx.doi.org/10.2307/2275541
11. Krajíček, J.: Interpolation by a game. Math. Log. Q. **44**, 450–458 (1998). http://dx.doi.org/10.1002/malq.19980440403
12. Krajíček, J.: An exponential lower bound for a constraint propagation proof system based on ordered binary decision diagrams. J. Symb. Log. **73**(1), 227–237 (2008). http://dx.doi.org/10.2178/jsl/1208358751
13. Krajíček, J.: A feasible interpolation for random resolution. CoRR abs/1604.06560 (2016). http://arxiv.org/abs/1604.06560
14. Oliveira, I.: Unconditional lower bounds in complexity theory. Ph.D. thesis, Columbia university (2015)
15. Pudlák, P.: Lower bounds for resolution and cutting plane proofs and monotone computations. J. Symb. Log. **62**(3), 981–998 (1997). http://dx.doi.org/10.2307/2275583
16. Pudlák, P.: On extracting computations from propositional proofs (a survey). In: IARCS Annual Conference on Foundations of Software Technology and Theoretical Computer Science, FSTTCS 2010, 15–18 December 2010, Chennai, India, pp. 30–41 (2010). http://dx.doi.org/10.4230/LIPIcs.FSTTCS.2010.30
17. Raz, R., McKenzie, P.: Separation of the monotone NC hierarchy. Combinatorica **19**(3), 403–435 (1999). http://dx.doi.org/10.1007/s004930050062
18. Razborov, A.A.: Unprovability of lower bounds on circuit size in certain fragments of bounded arithmetic. Izvestiya RAN. Ser. Mat. **59**, 201–224 (1995)
19. Robert, R., Pitassi, T.: Strongly exponential lower bounds for monotone computation. ECCC Report: TR16-188 (2016)
20. Sokolov, D.: Dag-like communication and its applications. Electronic Colloquium on Computational Complexity (ECCC) (2016). http://eccc.hpi-web.de/report/2016/202

The Descriptive Complexity of Subgraph Isomorphism Without Numerics

Oleg Verbitsky[1]([⊠]) and Maksim Zhukovskii[2]

[1] Institut für Informatik, Humboldt-Universität zu Berlin,
Unter den Linden 6, 10099 Berlin, Germany
verbitsk@informatik.hu-berlin.de
[2] Department of Discrete Mathematics,
Moscow Institute of Physics and Technology (State University),
Dolgoprudny, Moscow Region, Russia

Abstract. Let F be a connected graph with ℓ vertices. The existence of a subgraph isomorphic to F can be defined in first-order logic with quantifier depth no better than ℓ, simply because no first-order formula of smaller quantifier depth can distinguish between the complete graphs K_ℓ and $K_{\ell-1}$. We show that, for some F, the existence of an F subgraph in *sufficiently large* connected graphs is definable with quantifier depth $\ell - 3$. On the other hand, this is never possible with quantifier depth better than $\ell/2$. If we, however, consider definitions over connected graphs with *sufficiently large treewidth*, the quantifier depth can for some F be arbitrarily small comparing to ℓ but never smaller than the treewidth of F.

We also prove that any first-order definition of the existence of an *induced* subgraph isomorphic to F requires quantifier depth strictly more than the density of F, even over highly connected graphs. From this bound we derive a succinctness result for existential monadic second-order logic: A usage of just one monadic quantifier sometimes reduces the first-order quantifier depth at a super-recursive rate.

1 Introduction

For a fixed graph F on ℓ vertices, let $\mathcal{S}(F)$ denote the class of all graphs containing a subgraph isomorphic to F. The decision problem for $\mathcal{S}(F)$ is known as SUBGRAPH ISOMORPHISM problem. It is solvable in time $O(n^\ell)$ on n-vertex input graphs by exhaustive search. Nešetřil and Poljak [15] showed that $\mathcal{S}(F)$ can be recognized in time $O(n^{(\omega/3)\ell+2})$, where $\omega < 2.373$ is the exponent of fast square matrix multiplication. Moreover, the color-coding method by Alon, Yuster and Zwick [2] yields the time bound

$$2^{O(\ell)} \cdot n^{tw(F)+1} \log n,$$

The first author was supported by DFG grant VE 652/1-2. He is on leave from the IAPMM, Lviv. The second author was supported by grants No. 15-01-03530 and 16-31-60052 of Russian Foundation for Basic Research.

© Springer International Publishing AG 2017
P. Weil (Ed.): CSR 2017, LNCS 10304, pp. 308–322, 2017.
DOI: 10.1007/978-3-319-58747-9_27

where $tw(F)$ denotes the treewidth of F. On the other hand, the decision problem for $\mathcal{S}(K_\ell)$, that is, the problem of deciding if an input graph contains a clique of ℓ vertices, cannot be solved in time $n^{o(\ell)}$ unless the Exponential Time Hypothesis fails.

We here are interested in the *descriptive complexity* of SUBGRAPH ISOMOR-PHISM. A sentence Φ *defines* a class of graphs $\mathcal{C}$ if

$$G \models \Phi \iff G \in \mathcal{C}, \tag{1}$$

where $G \models \Phi$ means that Φ is true on G. For a logic $\mathcal{L}$, we let $D_\mathcal{L}(\mathcal{C})$ (resp. $W_\mathcal{L}(\mathcal{C})$) denote the minimum quantifier depth (resp. variable width) of $\Phi \in \mathcal{L}$ defining $\mathcal{C}$. Note that $W_\mathcal{L}(\mathcal{C}) \leq D_\mathcal{L}(\mathcal{C})$. We simplify notation by writing

$$W_\mathcal{L}(F) = W_\mathcal{L}(\mathcal{S}(F)) \text{ and } D_\mathcal{L}(F) = D_\mathcal{L}(\mathcal{S}(F)). \tag{2}$$

We are primarily interested in the first-order logic of graphs with relation symbols for adjacency and equality of vertices, that will be denoted by FO. We suppose that the vertex set of any n-vertex graph is $\{1, \ldots, n\}$. Seeking the adequate logical formalism for various models of computation, descriptive complexity theory considers also more expressive logics involving numerical relations over the integers. Given a set $\mathcal{N}$ of such relations, FO[$\mathcal{N}$] is used to denote the extension of FO whose language contains symbols for each relation in $\mathcal{N}$. Of special interest are FO[<], FO[+, ×], and FO[Arb], where Arb indicates that arbitrary relations are allowed. It is known [10, 14] that FO[Arb] and FO[+, ×] capture (non-uniform) AC^0 and DLOGTIME-uniform AC^0 respectively.

We will simplify the notation (2) further by writing $D(F) = D_{FO}(F)$ and $W(F) = W_{FO}(F)$. Dropping FO in the subscript, we also use notation like $D_<(F)$ or $W_{Arb}(F)$. In this way we obtain two hierarchies of width and depth parameters. In particular,

$$W_{Arb}(F) \leq W_<(F) \leq W(F) \text{ and } D_{Arb}(F) \leq D_<(F) \leq D(F).$$

The relation of FO[Arb] to circuit complexity implies that $\mathcal{S}(F)$ is recognizable by bounded-depth unbounded-fan-in circuits of size $n^{W_{Arb}(F)+o(1)}$; see [10, 18]. The interplay between the two areas has been studied in [12, 13, 18, 19]. Noteworthy, the parameters $W_{Arb}(F)$ and $D_{Arb}(F)$ admit combinatorial upper bounds

$$W_{Arb}(F) \leq tw(F) + 3 \text{ and } D_{Arb}(F) \leq td(F) + 2 \tag{3}$$

in terms of the treewidth and treedepth of F; see [20].[1]

The focus of our paper is on FO without any background arithmetical relations. Our interest in this weakest setting is motivated by the prominent problem

---

[1] In his presentation [20], Benjamin Rossman states upper bounds $W_{FO}(F) \leq tw(F) + 1$ and $D_{FO}(F) \leq td(F)$ for the *colorful* version of SUBGRAPH ISOMORPHISM studied in [13]. It is not hard to observe that the auxiliary color predicates can be defined in FO[Arb] at the cost of two extra quantified variables by the color-coding method developed in [2]; see also [3, Theorem 4.2].

on the power of encoding-independent computations; see, e.g., [9]. It is a long-standing open question in finite model theory as to whether there exists a logic capturing polynomial time on finite relational structures. The existence of a logic capturing polynomial time would mean that any polynomial-time computation could be made, in a sense, independent of the input encoding. If this is true, are the encoding-independent computations necessarily slower than the standard ones? This question admits the following natural variation. Suppose that a decision problem a priori admits an encoding-independent polynomial-time algorithm, say, being definable in FO, like SUBGRAPH ISOMORPHISM for a fixed pattern graph F. Is it always true that the running time of this algorithm can be improved in the standard encoding-dependent Turing model of computation?

A straightforward conversion of an FO sentence defining $\mathcal{S}(F)$ into an algorithm recognizing $\mathcal{S}(F)$ results in the time bound $O(n^{D(F)})$ for SUBGRAPH ISOMORPHISM, which can actually be improved to $O(n^{W(F)})$; see [14, Proposition 6.6]. The same applies to FO[<]. The last logic is especially interesting in the context of *order-invariant definitions*. It is well known [14,21] that there are properties of (unordered) finite structures that can be defined in FO[<] but not in FO. Even if a property, like $\mathcal{S}(F)$, is definable in FO, one can expect that in FO[<] it can be defined much more succinctly. As a simple example, take F to be the star graph $K_{1,s}$ and observe that $D_<(K_{1,s}) \leq \log_2 s + 3$ and $W_<(K_{1,s}) \leq 3$ while $W(K_{1,s}) = s + 1$.

The main goal we pose in this paper is examining abilities and limitations of the "pure" FO in succinctly defining SUBGRAPH ISOMORPHISM. Actually, if a pattern graph F has ℓ vertices, then the trivial upper bound $D(F) \leq \ell$ cannot be improved. We have $W(F) = \ell$ simply because no first-order formula with less than ℓ variables can distinguish between the complete graphs K_ℓ and $K_{\ell-1}$. Is this, however, the only reason preventing more succinct definitions of $\mathcal{S}(F)$? How succinctly can $\mathcal{S}(F)$ be defined on large enough graphs? The question can be formalized as follows. We say that a sentence Φ defines $\mathcal{S}(F)$ on *sufficiently large connected graphs* if there is k such that (1) with $\mathcal{C} = \mathcal{S}(F)$ is true for all connected G with at least k vertices. Let $W_v(F)$ (resp. $D_v(F)$) denote the minimum variable width (resp. quantifier depth) of such Φ.

Throughout the paper, we assume that the fixed **pattern graph F is connected**. Therefore, F is contained in a host graph G if and only if it is contained in a connected component of G. By this reason, the decision problem for $\mathcal{S}(F)$ efficiently reduces to its restriction to connected input graphs. Since it suffices to solve the problem only on all sufficiently large inputs, $\mathcal{S}(F)$ is still recognizable in time $O(n^{W_v(F)})$, while $W_v(F) \leq W(F)$.

A further relaxation is motivated by Courcelle's theorem [6] saying that every graph property definable by a sentence in monadic second-order logic can be efficiently decided on graphs of bounded treewidth. More precisely, for SUBGRAPH ISOMORPHISM Courcelle's theorem implies that $\mathcal{S}(F)$ is decidable in time $f(\ell, tw(G)) \cdot n$, which means linear time for any class of input graphs having bounded treewidth.

Now, we say that a sentence Φ defines $\mathcal{S}(F)$ on *connected graphs with sufficiently large treewidth* if there is k such that (1) with $\mathcal{C} = \mathcal{S}(F)$ is true for all connected G with treewidth at least k. Denote the minimum variable width (resp. quantifier depth) of such Φ by $W_{tw}(F)$ (resp. $D_{tw}(F)$). Fix k that ensures the minimum value $W_{tw}(F)$ and recall that, by Courcelle's theorem, the subgraph isomorphism problem is solvable on graphs with treewidth less than k in linear time. Note that, for a fixed k, whether or not $tw(G) < k$ is also decidable in linear time [4]. It follows that $\mathcal{S}(F)$ is recognizable even in time $O(n^{W_{tw}(F)})$, while $W_{tw}(F) \leq W_v(F)$.

The above discussion shows that the parameters $W_v(F)$, $D_v(F)$, $W_{tw}(F)$, and $D_{tw}(F)$ have clear algorithmic meaning. Analyzing this setting, we obtain the following results.

- We demonstrate that non-trivial definitions over sufficiently large graphs are possible by showing that $D_v(F) \leq v(F) - 3$ for some F, where $v(F)$ denotes the number of vertices in F. On the other hand, we show limitations of this approach by proving that $W_v(F) \geq (v(F) - 1)/2$ for all F.
- The last barrier (as well as any lower bound in terms of $v(F)$) can be overcome by definitions over graphs with sufficiently large treewidth. Specifically, for every ℓ and $a \leq \ell$ there is an ℓ-vertex F such that $D_{tw}(F) \leq a$ and, moreover, $tw(F) = a - 1$. On the other hand, $W_{tw}(F) \geq tw(F)$ for all F. Note that, along with (3), this implies that $W_{Arb}(F) \leq W_{tw}(F) + 3$.

We also address the descriptive complexity of the INDUCED SUBGRAPH ISOMORPHISM problem. Let $\mathcal{I}(F)$ denote the class of all graphs containing an *induced* subgraph isomorphic to F. The state-of-the-art of the algorithmics for INDUCED SUBGRAPH ISOMORPHISM is different from SUBGRAPH ISOMORPHISM. Floderus et al. [8] collected evidence in favor of the conjecture that $\mathcal{I}(F)$ for F with ℓ vertices cannot be recognized faster than $\mathcal{I}(K_{c\ell})$, where $c < 1$ is a constant.

Similarly to $D(F)$, we define $D[F] = D(\mathcal{I}(F))$, where the square brackets indicate that the case of induced subgraphs is considered. The trivial argument showing that $D(F) = v(F)$ does not work anymore unless F is a complete graph (whereas K_ℓ contains every ℓ-vertex F as a subgraph, it contains no *induced* copy of F unless $F = K_\ell$). Proving or disproving that $D[F] = v(F)$ seems to be a subtle problem. Our results on INDUCED SUBGRAPH ISOMORPHISM are as follows.

- We prove a general lower bound $D[F] > e(F)/v(F)$, where $e(F)$ denotes the number of edges in F. In fact, the bound holds true even for $D_{tw}[F]$.
- From this bound we derive a succinctness result for existential monadic second-order logic: A usage of just one monadic quantifier sometimes reduces the FO quantifier depth at a super-recursive rate. More precisely, let $D_{\exists MSO}[F]$ denote the minimum quantifier depth of a second-order sentence with a single existential monadic quantifier that defines $\mathcal{I}(F)$. Then $D_{\exists MSO}[F]$ can sometimes be so small comparing to $D[F] = D_{FO}[F]$ that there is no total recursive function f such that $f(D_{\exists MSO}[F]) \geq D[F]$ for all F.

2   Preliminaries

First-Order Complexity of Graph Properties. We consider first-order sentences about graphs in the language containing the adjacency and the equality relations. Let $\mathcal{C}$ be a first-order definable class of graphs and π be a graph parameter. Let $D_\pi^k(\mathcal{C})$ denote the minimum quantifier depth of a first-order sentence Φ such that, for every connected graph G with $\pi(G) \geq k$, Φ is true on G exactly when G belongs to $\mathcal{C}$. Note that $D_\pi^k(\mathcal{C}) \geq D_\pi^{k+1}(\mathcal{C})$, and define $D_\pi(\mathcal{C}) = \min_k D_\pi^k(\mathcal{C})$. In other words, $D_\pi(\mathcal{C})$ is the minimum quantifier depth of a first-order sentence defining $\mathcal{C}$ over connected graphs with sufficiently large values of π.

The *variable width* of a first-order sentence Φ is the number of first-order variables used to build Φ; different occurrences of the same variable do not count. By $W_\pi(\mathcal{C})$ we denote the minimum variable width of Φ defining $\mathcal{C}$ over connected graphs with sufficiently large π. Note that $W_\pi(\mathcal{C}) \leq D_\pi(\mathcal{C})$.

Recall that a graph is *k-connected* if it has more than k vertices, is connected, and remains connected after removal of any $k-1$ vertices. The *connectivity* $\kappa(G)$ of G is equal to the maximum k such that G is k-connected. We will consider the depth $D_\pi(\mathcal{C})$ and the width $W_\pi(\mathcal{C})$ for three parameters π, namely the number of vertices $v(G)$, the treewidth $tw(G)$, and the connectivity $\kappa(G)$. It is not hard to see that

$$D_v(\mathcal{C}) \geq D_{tw}(\mathcal{C}) \geq D_\kappa(\mathcal{C}) \text{ and } W_v(\mathcal{C}) \geq W_{tw}(\mathcal{C}) \geq W_\kappa(\mathcal{C}).$$

As it was discussed in Sect. 1, the values of $D_v(\mathcal{C})$ and $D_{tw}(\mathcal{C})$, as well as $W_v(\mathcal{C})$ and $W_{tw}(\mathcal{C})$, are related to the time complexity of the decision problem for $\mathcal{C}$. Consideration of $D_\kappa(\mathcal{C})$ and $W_\kappa(\mathcal{C})$ is motivated by the fact that some lower bounds we are able to show for $D_v(\mathcal{C})$ and $D_{tw}(\mathcal{C})$ actually hold for $D_\kappa(\mathcal{C})$ or even for $W_\kappa(\mathcal{C})$, and it is natural to present them in this stronger form.

Recall that $\mathcal{S}(F)$ denotes the class of graphs containing a subgraph isomorphic to F. Simplifying the notation, we write $D_v(F) = D_v(\mathcal{S}(F))$, $W_v(F) = W_v(\mathcal{S}(F))$, etc.

Given two non-isomorphic graphs G and H, let $D(G, H)$ (resp. $W(G, H)$) denote the minimum quantifier depth (resp. variable width) of a sentence that is true on one of the graphs and false on the other.

Lemma 1. *1. $D_\pi(\mathcal{C}) \geq d$ if there are connected graphs $G \in \mathcal{C}$ and $H \notin \mathcal{C}$ with arbitrarily large values of $\pi(G)$ and $\pi(H)$ such that $D(G, H) \geq d$.*
2. $W_\pi(\mathcal{C}) \geq d$ if there are connected graphs $G \in \mathcal{C}$ and $H \notin \mathcal{C}$ with arbitrarily large values of $\pi(G)$ and $\pi(H)$ such that $W(G, H) \geq d$.
3. $D_\pi(\mathcal{C}) \leq d$ if $D(G, H) \leq d$ for all connected graphs $G \in \mathcal{C}$ and $H \notin \mathcal{C}$ with sufficiently large values of $\pi(G)$ and $\pi(H)$.

Lemma 1 reduces estimating $D_\pi(\mathcal{C})$ to estimating $D(G, H)$ over connected $G \in \mathcal{C}$ and $H \notin \mathcal{C}$ with large values of π. Also, proving lower bounds for $W_\pi(\mathcal{C})$ reduces to proving lower bounds for $W(G, H)$. For estimating $D(G, H)$ and $W(G, H)$ we will use the well-known characterization of these parameters in terms of the k-pebble Ehrenfeucht-Fraïssé game [10]:

1. $D(G, H)$ is equal to the minimum k such that Spoiler has a winning strategy in the k-round k-pebble game on G and H.
2. $W(G, H)$ is equal to the minimum k such that, for some d, Spoiler has a winning strategy in the d-round k-pebble game on G and H.

Graph-Theoretic Preliminaries. Recall that $v(G)$ denotes the number of vertices in a graph G. The treewidth of G is denoted by $tw(G)$. The *neighborhood* $N(v)$ of a vertex v consists of all vertices adjacent to v. The number $\deg v = |N(v)|$ is called the *degree* of v. The vertex of degree 1 is called *pendant*.

We use the standard notation K_n for complete graphs, P_n for paths, and C_n for cycles on n vertices. Furthermore, $K_{a,b}$ denotes the complete bipartite graph whose vertex classes have a and b vertices. In particular, $K_{1,n-1}$ is the star graph on n vertices. The subscript in the name of a graph will almost always denote the number of vertices. If a graph is indexed by two parameters, their sum is typically equal to the total number of vertices in the graph.

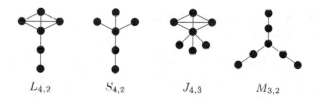

$L_{4,2}$ $\qquad$ $S_{4,2}$ $\qquad$ $J_{4,3}$ $\qquad$ $M_{3,2}$

Fig. 1. Special graph families: Lollipops, sparklers, jellyfishes, and megastars.

The following definitions are illustrated in Fig. 1. Let $a \geq 3$ and $b \geq 1$. The *lollipop graph* $L_{a,b}$ is obtained from K_a and P_b by adding an edge between an end vertex of P_b and a vertex of K_a. We also make a natural convention that $L_{a,0} = K_a$. Furthermore, the *sparkler graph* $S_{a,b}$ is obtained from $K_{1,a-1}$ and P_b by adding an edge between an end vertex of P_b and the central vertex of $K_{1,a-1}$. The *jellyfish graph* $J_{a,b}$ is the result of attaching b pendant vertices to a vertex of K_a. Finally, the *megastar* graph $M_{s,t}$ is obtained from the star $K_{1,s}$ by subdividing each edge into P_{t+1}; thus $v(M_{s,t}) = st + 1$.

3   Definitions over Sufficiently Large Graphs

Our first goal is to demonstrate that non-trivial definitions over large connected graphs are really possible. The lollipop graphs $L_{a,1}$ give simple examples of pattern graphs F with $D_v(F) \leq v(F) - 1$. Though not so easily, the same can be shown for the path graphs P_ℓ. We are able to show better upper bounds using sparkler graphs.

Theorem 2. *There is a graph F with $D_v(F) \leq v(F) - 3$. Specifically, $D_v(S_{4,4}) = 5$.*

For the proof we need two technical lemmas.

Lemma 3. *Suppose that a connected graph H contains the 4-star $K_{1,4}$ as a subgraph but does not contain any subgraph $S_{4,4}$. Then H contains a vertex of degree more than $(v(H)/2)^{1/7}$.*

Proof. H cannot contain P_{15} because, together with $K_{1,4}$, it would give an $S_{4,4}$ subgraph. Consider an arbitrary spanning tree T in H and denote its maximum vertex degree by d and its radius by r. Note that $v(T) \leq 1 + d + d(d-1) + \ldots + d(d-1)^{r-1}$. Since T contains no P_{15}, we have $r \leq 7$. It follows that $v(H) = v(T) < 2d^7$.   □

Let $\sim$ denote the adjacency relation.

Lemma 4. *Let $y_0 \in V(H)$ and assume that*

- *H is a sufficiently large connected graph,*
- *H does not contain $S_{4,4}$,*
- *$\deg y_0 \geq 4$,*
- *$y_0 y_1 y_2 y_3 y_4$ is a path in H.*

Then (see Fig. 2)

1. *$\deg y_0 = 4$,*
2. *$y_0 \sim y_2$, $y_0 \nsim y_3$, $y_0 \nsim y_4$,*
3. *if $N(y_0) = \{y_1, y_2, y', y''\}$, then $y_1 \nsim y'$ and $y_1 \nsim y''$.*

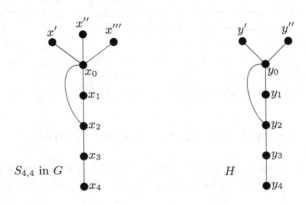

Fig. 2. Proof of Theorem 2.

Proof. By Lemma 3 we know that H must contain a vertex z of large degree, namely $\deg z \geq 7$. We have $y_0 \nsim y_4$ for else H would contain a cycle C_5 and, together with z, this would give us a subgraph $S_{4,4}$ (because, by connectedness of H, we would have a path P_5 emanating from z). Therefore, y_0 has a neighbor $y' \notin \{y_1, y_2, y_3, y_4\}$. Furthermore, $y_0 \nsim y_3$ for else, considering a path from z to one of the vertices $y', y_0, y_1, y_2, y_3, y_4$, we get a P_5 emanating from z and, hence, an $S_{4,4}$. Therefore, y_0 has another neighbor $y'' \notin \{y', y_1, y_2, y_3, y_4\}$. Furthermore, $y_0 \sim y_2$ for else y_0 would have three neighbors y', y'', y''' different from y_1, y_2, y_3, y_4, which would give $S_{4,4}$. By the same reason, y_0 has no other neighbors, that is, $N(y_0) = \{y_1, y_2, y', y''\}$ and $\deg y_0 = 4$. Note that $z \in \{y_0, y_1, y_2, y_3, y_4\}$ for else we easily get an $S_{4,4}$ by considering a path from z to one of these vertices. It is also easy to see that $z \neq y_0, y_4, y_3, y_1$ (for example, if $\deg y_1 \geq 7$, then it would give an $S_{4,4}$ with tail $y_1 y_0 y_2 y_3 y_4$). Therefore, $z = y_2$. If $y_1 \sim y'$ or $y_1 \sim y''$, we would have an $S_{4,4}$ with tails $y_2 y_1 y' y_0 y''$ or $y_2 y_1 y'' y_0 y'$ respectively.   □

Proof (of Theorem 2). We are now ready to prove the upper bound $D_v(S_{4,4}) \leq 5$. Consider sufficiently large connected graphs G and H and suppose that G contains an $S_{4,4}$, whose vertices are labeled as in Fig. 2, and H contains no copy of $S_{4,4}$. We describe a winning strategy for Spoiler in the game on G and H.

1st round. Spoiler pebbles x_0. Denote the response of Duplicator in H by y_0. Assume that $\deg y_0 \geq 4$ for else Spoiler wins in the next 4 moves. Assume that $x_0 \sim x_2$ for else Spoiler wins by pebbling x_1, x_2, x_3, x_4 (if Duplicator responds with a path $y_0 y_1 y_2 y_3 y_4$, she loses by Condition 2 in Lemma 4).

2nd round. Spoiler pebbles x_1. Denote the response of Duplicator in H by y_1. Assume that there is a path $y_0 y_1 y_2 y_3 y_4$ for else Spoiler wins in the next 3 moves.

Case 1: x_1 is adjacent to any of the vertices x', x'', x''', say, to x'. Spoiler pebbles x_2 and x' and wins. Indeed, Duplicator has to respond with two vertices in H both in $N(y_0) \cap N(y_1)$, which is impossible by Conditions 1 and 3 of Lemma 4.

Case 2: $x_1 \nsim x', x_1 \nsim x'', x_1 \nsim x'''$. Spoiler wins by pebbling x', x'', x'''. Duplicator has to respond with three vertices in $N(y_0) \setminus N(y_1)$, which is impossible by Conditions 1 and 2 of Lemma 4.

This completes the proof of the upper bound. On the other hand, we have $D_v(S_{4,4}) > 4$ by considering the jellyfish graphs $G = J_{5,n}$ and $H = J_{4,n}$.   □

We now show general lower bounds for $D_v(F)$ and $W_v(F)$. For this, we need some definitions. Let $v_0 v_1 \dots v_t$ be an induced path in a graph G. We call it *pendant* if $\deg v_0 \neq 2$, $\deg v_t = 1$ and $\deg v_i = 2$ for all $1 \leq i < t$. Furthermore, let S be an induced star $K_{1,s}$ in G with the central vertex v_0. We call S *pendant* if all its pendant vertices are pendant also in G, and in G there are no more than s pendant vertices adjacent to v_0. The definition ensures that a pendant path (or star) cannot be contained in a larger pendant path (or star). As an example, note that the sparkler graph $S_{s+1,t}$ has a pendant P_{t+1} and a pendant $K_{1,s}$.

Let $p(F)$ denote the maximum t such that F has a pendant path P_{t+1}. Similarly, let $s(F)$ denote the maximum s such that F has a pendant star $K_{1,s}$. If F has no pendant vertex, then we set $p(F) = 0$ and $s(F) = 0$.

Theorem 5. $D_v(F) \geq (v(F) + 1)/2$ and $W_v(F) \geq (v(F) - 1)/2$ for every connected F unless $F = P_2$ or $F = P_3$.

Proof. Denote

$$\ell = v(F), \ t = p(F) \text{ and } s = s(F).$$

We begin with noticing that

$$D_v(F) \geq \ell - t \text{ and } W_v(F) \geq \ell - t - 1. \tag{4}$$

Indeed, this is obvious if F is a path, that is, $F = P_{t+1}$. If F is not a path, we consider lollipop graphs $G = L_{\ell-t,n}$ and $H = L_{\ell-t-1,n}$ for each $n \geq t$ (note that $\ell \geq t+3$ and, if $\ell = t+3$, then $H = L_{2,n} = P_{n+2}$). Obviously, G contains F, and H does not. It remains to note that $D(G, H) \geq \ell - t$ and $W(G, H) \geq \ell - t - 1$.

We also claim that

$$D_v(F) \geq \ell - s \text{ and } W_v(F) \geq \ell - s - 1. \tag{5}$$

This is obvious if F is a star, that is, $F = K_{1,s}$. If F is not a star, we consider jellyfish graphs $G = J_{\ell-s,n}$ and $H = J_{\ell-s-1,n}$ for each $n \geq s$ (note that $\ell \geq s+3$ and, if $\ell = s + 3$, then $H = J_{2,n} = K_{1,n+1}$). Clearly, G contains F, and H does not. It remains to observe that $D(G, H) \geq \ell - s$ and $W(G, H) \geq \ell - s - 1$.

Let $F = K_{1,\ell-1}$ or $F = P_\ell$, where $\ell \geq 4$. Using (4) and (5) respectively, we get $D_v(F) \geq \ell - 1 \geq \frac{\ell+1}{2}$ and, similarly, $W_v(F) \geq \ell - 2 \geq \frac{\ell-1}{2}$. Assume, therefore, that F is neither a star nor a path. In this case we claim that

$$t + s < \ell. \tag{6}$$

This is obviously true if F has no pendant vertex, that is, $t = s = 0$. Suppose that F has a pendant vertex and, therefore, both $t > 0$ and $s > 0$. Consider an arbitrary spanning tree T of F and note that T contains all pendant paths and stars of F. Fix a longest pendant path P and a largest pendant star S in F. If P and S share at most one common vertex, we readily get (6). If they share two vertices, then $S = K_{1,1}$, i.e., $s = 1$, and $t + 1 < \ell$ follows from the assumption that F is not a path.

The theorem readily follows from (4)–(6).   □

4   Definitions over Graphs of Sufficiently Large Treewidth

Theorem 5 poses limitations on the succinctness of definitions over sufficiently large graphs. We now show that there are no such limitations for definitions over connected graphs with sufficiently large treewidth.

The Grid Minor Theorem says that every graph of large treewidth contains a large grid minor; see [7]. The strongest version of this result belongs to Chekuri and Chuzhoy [5] who proved that, for some $\epsilon > 0$, every graph G of treewidth

k contains the $m \times m$ grid as a minor with $m = \Omega(k^\epsilon)$. If $m > 2b$, then G must contain $M_{3,b}$ as a subgraph. This applies also to all subgraphs of $M_{3,b}$. The following result is based on the fact that a graph of large treewidth contains a long path.

Theorem 6. *For all a and ℓ such that $3 \le a \le \ell$ there is a graph F with $v(F) = \ell$ and $tw(F) = a - 1$ such that $D_{tw}(F) \le a$. Specifically, $D_{tw}(L_{a,b}) = W_\kappa(L_{a,b}) = a$ if $a \ge 3$ and $b \ge 0$.*

Note for comparison that $W_v(L_{a,b}) \ge a + b - 2$, as follows from the bound (5) in the proof of Theorem 5.

Proof. We first prove the upper bound $D_{tw}(L_{a,b}) \le a$. If a connected graph H of large treewidth does not contain $L_{a,b}$, it cannot contain even K_a for else K_a could be combined with a long path to give $L_{a,b}$. Therefore, Spoiler wins on $G \in \mathcal{S}(L_{a,b})$ and such H in a moves.

For the lower bound $W_\kappa(L_{a,b}) \ge a$, consider $G = K(a, n)$ and $H = K(a - 1, n)$, where $K(a, n)$ denotes the complete a-partite graph with each part having n vertices. Note that this graph is $(a - 1)n$-connected. If $n > b$, then G contains $L_{a,b}$, while H for any n does not contain even K_a. It remains to note that $W(G, H) \ge a$ if $n \ge a - 1$. □

We now prove a general lower bound for $W_{tw}(F)$ in terms of the treewidth $tw(F)$. Using the terminology of [11, Chap. 5], we define the *core* F_0 of F to be the graph obtained from F by removing, consecutively and as long as possible, vertices of degree at most 1. If F is not a forest, then F_0 is nonempty; it consists of all cycles of F and the paths between them.

We will use the well-known fact that there are cubic graphs of arbitrary large treewidth. This fact dates back to Pinsker [17] who showed that a random cubic graph with high probability has good expansion properties, implying linear treewidth.

Theorem 7. *If F is connected, then*

1. $W_{tw}(F) \ge v(F_0)$, *and*
2. $W_{tw}(F) \ge tw(F) + 1$ *unless F is contained in some 3-megastar $M_{3,b}$.*

Note that the bound in part 2 of Theorem 7 is tight by Theorem 6.

Proof. 1. Denote $v(F) = \ell$ and $v(F_0) = \ell_0$. If F is a tree, then $\ell_0 = 0$, and the claim is trivial. Suppose, therefore, that F is not a tree. In this case, $\ell_0 \ge 3$.

We begin with a cubic graph B of as large treewidth $tw(B)$ as desired. Let $(B)_\ell$ denote the graph obtained from B by subdividing each edge by ℓ new vertices. Since B is a minor of $(B)_\ell$, we have $tw((B)_\ell) \ge tw(B)$; see [7].

Next, we construct a gadget graph A as follows. By a *k-uniform tree* we mean a tree of even diameter where every non-leaf vertex has degree k and all distances between a leaf and the central vertex are equal. The graph A is obtained by merging the ℓ-uniform tree of radius ℓ and $(B)_\ell$; merging is done by identifying one leaf of the tree and one vertex of $(B)_\ell$.

We now construct G by attaching a copy of A to each vertex of K_{ℓ_0}. Specifically, a copy A_u of A is created for each vertex u of K_{ℓ_0}, and u is identified with the central vertex of (the tree part of) A_u. Let H be obtained from G by shrinking its clique part to K_{ℓ_0-1}. Since both G and H contain copies of $(B)_\ell$, these two graphs have treewidth at least as large as $tw(B)$.

The clique part of G is large enough to host the core F_0, and the remaining tree shoots of F fit into the A-parts of G. Therefore, G contains F as a subgraph. On the other hand, the clique part of H is too small for hosting F_0, and no cycle of F fits into any A-part because A has larger girth than F. Therefore, H does not contain F. It remains to notice that $W(G, H) \geq \ell_0$.

2. Suppose first that F is not a tree. By part 1, we then have

$$W_{tw}(F) \geq v(F_0) \geq tw(F_0) + 1 = tw(F) + 1.$$

If F is a tree not contained in any 3-megastar, then there are connected graphs of arbitrarily large treewidth that do not contain F as a subgraph (for example, consider $(B)_\ell$ for a connected cubic graph B as in part 1). Trivially, there are also connected graphs of arbitrarily large treewidth that contain F as a subgraph. Since one pebble is not enough for Spoiler to distinguish the latter from the former, we have $W_{tw}(F) \geq 2 = tw(F) + 1$ in this case. $\square$

5 Induced Subgraphs: Trading Super-Recursively Many First-Order Quantifiers for a Single Monadic One

By $\mathcal{I}(F)$ we denote the class of all graphs containing an *induced* subgraph isomorphic to F. Similarly to $D(F)$, we use the notation $D[F] = D(\mathcal{I}(F))$, where the square brackets indicate that only induced subgraphs are considered. In the same vein, $D_\kappa[F] = D_\kappa(\mathcal{I}(F))$.

Unlike the case of (not necessarily induced) subgraphs, where the equality $D(F) = v(F)$ is trivial, determining and estimating the parameter $D[F]$ seems to be a subtle problem. In this section we prove a lower bound for $D[F]$ in terms of the density of F; this bound actually holds for $D_\kappa[F]$. The proof will use known facts about random graphs in the Erdős-Rényi model $G(n, p)$, collected below. It should be stressed that, whenever the term *subgraph* stands alone, it refers to a *not necessarily induced* subgraph. *With high probability* means that the probability approaches 1 as $n \to \infty$.

The *density* of a graph K is defined to be the ratio $\rho(K) = e(K)/v(K)$. The maximum $\rho(K)$ over all subgraphs K of a graph F will be denoted by $\rho^*(F)$. The following fact from the random graph theory was used also in [13] for proving average-case lower bounds on the AC^0 complexity of SUBGRAPH ISOMORPHISM.

Lemma 8 (Subgraph Threshold, see [11, Chap. 3]).

1. *If $\alpha = 1/\rho^*(F)$, then the probability that $G(n, n^{-\alpha})$ contains F as a subgraph converges to a limit different from 0 and 1 as $n \to \infty$.*
2. *If $\alpha > 1/\rho^*(F)$, then with high probability $G(n, n^{-\alpha})$ does not contain F as a subgraph.*

Let $\alpha > 0$. Given a graph S and its subgraph K, we define $f_\alpha(S, K) = v(S) - v(K) - \alpha(e(S) - e(K))$.

Lemma 9 (Generic Extension, see [1, Chap. 10]). *Let F be a graph with vertices $v_1, \ldots, v_\ell$ and K be a subgraph of F with vertices $v_1, \ldots, v_k$. Assume that $f_\alpha(S, K) > 0$ for every subgraph S of F containing K as a proper subgraph. Then with high probability every sequence of pairwise distinct vertices $x_1, \ldots, x_k$ in $G(n, n^{-\alpha})$ can be extended with pairwise distinct $x_{k+1}, \ldots, x_\ell$ such that $x_i \sim x_j$ if and only if $v_i \sim v_j$ for all $i \leq \ell$ and $k < j \leq \ell$.*

Lemma 10 (Zero-One d-Law [23]). *Let $0 < \alpha < \frac{1}{d-2}$, and Ψ be a first-order statement of quantifier depth d. Then the probability that Ψ is true on $G(n, n^{-\alpha})$ converges either to 0 or to 1 as $n \to \infty$.*

We are now ready to prove our result.

Theorem 11. *If $e(F) > v(F)$, then $D_\kappa[F] \geq \frac{e(F)}{v(F)} + 2$ and $D_\kappa(F) \geq \frac{e(F)}{v(F)} + 2$.*

Proof. We prove the bound for $D_\kappa[F]$. The same proof works as well for $D_\kappa(F)$ (and is even simpler as the equality (7) below is only needed in the induced case).

Set $\alpha = 1/\rho^*(F)$ and denote $\mathbb{G}_n = G(n, n^{-\alpha})$. We begin with proving that

$$P[\mathbb{G}_n \in \mathcal{I}(F)] = P[\mathbb{G}_n \subset \mathcal{S}(F)] - o(1). \tag{7}$$

Let K be a maximal subgraph of F with $\rho(K) = \rho^*(F)$. Note that K is an induced subgraph of F. Note also that, if F is balanced, i.e., $\rho^*(F) = \rho(F)$, then $K = F$. The graph K has less than $\binom{v(K)}{2}$ supergraphs K' obtainable by adding an edge to K, and every K' has density strictly larger than K, that is, $\rho(K') > 1/\alpha$. By part 2 of Lemma 8, each such K' appears as a subgraph in $\mathbb{G}_n$ with probability $o(1)$. It follows that

$$P[\mathbb{G}_n \in \mathcal{I}(K)] = P[\mathbb{G}_n \in \mathcal{S}(K)] - o(1). \tag{8}$$

which readily implies (7) in the case that F is balanced.

Suppose now that F is not balanced. In this case, for every subgraph S of F containing K properly we have $v(S)/e(S) > \alpha$, which implies $f_\alpha(S, K) > 0$. Lemma 9 ensures that, with probability $1 - o(1)$, every induced copy of K in $\mathbb{G}_n$ extends to an induced copy of F. Therefore,

$$P[\mathbb{G}_n \in \mathcal{S}(F)] \geq P[\mathbb{G}_n \in \mathcal{I}(F)] \geq P[\mathbb{G}_n \in \mathcal{I}(K)] - o(1)$$
$$\geq P[\mathbb{G}_n \in \mathcal{S}(K)] - o(1) \geq P[\mathbb{G}_n \in \mathcal{S}(F)] - o(1), \tag{9}$$

where the last but one inequality is due to (8). Equality (7) is proved.

By part 1 of Lemma 8, $\lim_{n\to\infty} P[\mathbb{G}_n \in \mathcal{S}(F)]$ exists and equals neither 0 nor 1. It follows from (7) that $P[\mathbb{G}_n \in \mathcal{I}(F)]$ converges to the same limit, different from 0 and 1.

Now, assume that a first-order sentence Φ of quantifier depth d defines $\mathcal{S}(F)$ over k-connected graphs for all $k \geq k_0$. We have to prove that $d \geq \frac{e(F)}{v(F)} + 2$, whatever k_0.

By the assumption of the theorem, $\rho^*(F) \geq \rho(F) > 1$. Fix k such that $1 + 1/k < \rho(F)$ and $k \geq k_0$. Lemma 9 implies that with high probability every two vertices in $\mathbb{G}_n$ can be connected by k vertex-disjoint paths (of length k each). Therefore, $\mathbb{G}_n$ is k-connected with high probability.

Since Φ correctly decides the existence of an induced copy of F on all k-connected graphs,

$$\mathsf{P}[\mathbb{G}_n \models \Phi] = \mathsf{P}[\mathbb{G}_n \in \mathcal{I}(F)] + o(1).$$

Therefore, $\mathsf{P}[\mathbb{G}_n \models \Phi]$ converges to the same limit as $\mathsf{P}[\mathbb{G}_n \in \mathcal{I}(F)]$, which, as we have seen, is different from 0 and 1. By Lemma 10, this implies that $\alpha \geq \frac{1}{d-2}$. From here we conclude that

$$d \geq \rho^*(F) + 2 \geq \frac{e(F)}{v(F)} + 2,$$

as required. $\qquad\square$

We now turn to *existential monadic second-order logic*, denoted by $\exists$MSO, whose formulas are of the form

$$\exists X_1 \ldots \exists X_m\, \Phi, \tag{10}$$

where a first-order subformula Φ is preceded by (second-order) quantification over unary relations (that is, we are now allowed to use existential quantifiers over subsets of vertices $X_1, X_2, \ldots$). The second-order quantifiers contribute to the *quantifier depth* as well as the first-order ones. Thus, the quantifier depth of the sentence (10) is larger by m than the quantifier depth of the formula Φ. If a graph property $\mathcal{C}$ is definable in $\exists$MSO, the minimum quantifier depth of a defining formula will be denoted by $D_{\exists\mathrm{MSO}}(\mathcal{C})$. Furthermore, we define $D_{\exists\mathrm{MSO}}[F] = D_{\exists\mathrm{MSO}}(\mathcal{I}(F))$.

It is very well known that $\exists$MSO is strictly more expressive than first-order logic. For example, the properties of a graph to be disconnected or to be bipartite are expressible in $\exists$MSO but not in FO. We now show that $\exists$MSO is also much more succinct than FO, which means that some properties of graphs that are expressible in FO can be expressed in $\exists$MSO with significantly smaller quantifier depth. In fact, this can be demonstrated by considering the properties of containing a fixed induced subgraph. It turns out that, if we are allowed to use just one monadic second-order quantifier, the number of first-order quantifiers can sometimes be drastically reduced.

Theorem 12. *There is no total recursive function f such that*

$$f(D_{\exists\mathrm{MSO}}[F]) \geq D[F]$$

for all graphs F. Moreover, this holds true even for the fragment of $\exists$MSO where exactly one second-order quantifier is allowed.

The proof, which can be found in a long version of this paper [22], is based on Theorem 11 and [16, Theorem 4.2].

Acknowledgements. We would like to thank Tobias Müller for his kind hospitality during the Workshop on Logic and Random Graphs in the Lorentz Center (August 31–September 4, 2015), where this work was originated. We also thank the anonymous referee who provided us with numerous useful comments on the manuscript.

References

1. Alon, N., Spencer, J.H.: The Probabilistic Method. Wiley, Chichester (2016)
2. Alon, N., Yuster, R., Zwick, U.: Color-coding. J. ACM **42**(4), 844–856 (1995)
3. Amano, K.: k-Subgraph isomorphism on AC^0 circuits. Comput. Complex. **19**(2), 183–210 (2010)
4. Bodlaender, H.L.: A linear-time algorithm for finding tree-decompositions of small treewidth. SIAM J. Comput. **25**(6), 1305–1317 (1996)
5. Chekuri, C., Chuzhoy, J.: Polynomial bounds for the grid-minor theorem. In: Proceedings of the 46th ACM Symposium on Theory of Computing (STOC 2014), pp. 60–69 (2014)
6. Courcelle, B.: The monadic second-order logic of graphs I. Recognizable sets of finite graphs. Inf. Comput. **85**(1), 12–75 (1990)
7. Diestel, R.: Graph Theory. Springer, New York (2000)
8. Floderus, P., Kowaluk, M., Lingas, A., Lundell, E.: Induced subgraph isomorphism: are some patterns substantially easier than others? Theor. Comput. Sci. **605**, 119–128 (2015)
9. Grädel, E., Grohe, M.: Is polynomial time choiceless? In: Beklemishev, L., Blass, A., Dershowitz, N., Finkbeiner, B., Schulte, W. (eds.) Fields of Logic and Computation II. LNCS, vol. 9300, pp. 193–209. Springer, Cham (2015). doi:10.1007/978-3-319-23534-9_11
10. Immerman, N.: Descriptive Complexity. Springer, New York (1999)
11. Janson, S., Łuczak, T., Ruciński, A.: Random Graphs. Wiley, New York (2000)
12. Koucký, M., Lautemann, C., Poloczek, S., Thérien, D.: Circuit lower bounds via Ehrenfeucht-Fraïssé games. In: Proceedings of the 21st Annual IEEE Conference on Computational Complexity (CCC 2006), pp. 190–201 (2006)
13. Li, Y., Razborov, A.A., Rossman, B.: On the AC^0 complexity of subgraph isomorphism. In: Proceedings of the 55th IEEE Annual Symposium on Foundations of Computer Science (FOCS 2014), pp. 344–353. IEEE Computer Society (2014)
14. Libkin, L.: Elements of Finite Model Theory. Springer, Berlin (2004)
15. Nešetřil, J., Poljak, S.: On the complexity of the subgraph problem. Commentat. Math. Univ. Carol. **26**, 415–419 (1985)
16. Pikhurko, O., Spencer, J., Verbitsky, O.: Succinct definitions in the first order theory of graphs. Ann. Pure Appl. Log. **139**(1–3), 74–109 (2006)
17. Pinsker, M.S.: On the complexity of a concentrator. In: Proceedings of the 7th Annual International Teletraffic Conference, vol. 4, pp. 1–318 (1973)
18. Rossman, B.: On the constant-depth complexity of k-clique. In: Proceedings of the 40th Annual ACM Symposium on Theory of Computing (STOC 2008), pp. 721–730. ACM (2008)
19. Rossman, B.: An improved homomorphism preservation theorem from lower bounds in circuit complexity (2016). Manuscript

20. Rossman, B.: Lower bounds for subgraph isomorphism and consequences in first-order logic. In: Talk in the Workshop on Symmetry, Logic, Computation at the Simons Institute, Berkeley, November 2016. https://simons.berkeley.edu/talks/benjamin-rossman-11-08-2016
21. Schweikardt, N.: A short tutorial on order-invariant first-order logic. In: Bulatov, A.A., Shur, A.M. (eds.) CSR 2013. LNCS, vol. 7913, pp. 112–126. Springer, Heidelberg (2013). doi:10.1007/978-3-642-38536-0_10
22. Verbitsky, O., Zhukovskii, M.: The descriptive complexity of subgraph isomorphism without numerics. E-print (2016). http://arxiv.org/abs/1607.08067
23. Zhukovskii, M.: Zero-one k-law. Discrete Math. **312**, 1670–1688 (2012)

On a Generalization of Horn Constraint Systems

Piotr Wojciechowski[1], R. Chandrasekaran[2], and K. Subramani[1(✉)]

[1] LDCSEE, West Virginia University, Morgantown, USA
pwjociec@mix.wvu.edu, k.subramani@mail.wvu.edu
[2] Computer Science and Engineering, University of Texas at Dallas, Richardson, USA
chandra@utdallas.edu

Abstract. In this paper, we study linear constraint systems in which each constraint is a *fractional Horn constraint*. A constraint is fractional horn, if it can be written in the form: $\sum_{i=1}^{n} a_i \cdot x_i \geq c$, where the a_i and c are integral, and at most one of the $a_i > 0$ and all negative a_i are equal to -1. A conjunction of fractional Horn constraints is called a Fractional Horn Systems (FHS). FHSs generalize a number of specialized constraint systems such as Difference Constraint Systems and Horn Constraint Systems. We show that the problem of checking linear feasibility in these systems is in **P**, whereas the problem of checking lattice point feasibility is **NP-complete**. We then study a sub-class of fractional horn systems called Binary fractional horn systems (BFHS) in which each constraint has at most two non-zero coefficients with at most one being positive. In this case, we show that the problem of lattice point feasibility is in **P**.

Keywords: Fractional horn · Integer feasibility · Linear constraints

1 Introduction

In this paper, we focus on analyzing Fractional Horn Systems (FHS) from the perspectives of linear and integer feasibilities. A constraint is fractional Horn, if it can be written in the form: $\sum_{i=1}^{n} a_i \cdot x_i \geq c$, where the a_i and c are integral, and at most one of the $a_i > 0$ and all negative a_i are equal to -1. A conjunction of such constraints is called a Fractional Horn System. FHSs generalize several commonly occurring linear constraint systems such as Difference constraint systems [2] and Horn constraint systems [3].

FHSs find applications in a number of domains, but primarily in program verification (abstract interpretation). Our goal in this paper is to establish boundaries between sub-classes of FHSs, which are computationally hard and those

---

P. Wojciechowski—This research was supported in part by the National Science Foundation through Award CCF-1305054.

K. Subramani—This work was supported by the Air Force Research Laboratory under US Air Force contract FA8750-16-3-6003. The views expressed are those of the authors and do not reflect the official policy or position of the Department of Defense or the U.S. Government.

© Springer International Publishing AG 2017
P. Weil (Ed.): CSR 2017, LNCS 10304, pp. 323–336, 2017.
DOI: 10.1007/978-3-319-58747-9_28

which are solvable in polynomial time. Towards this end, we study Binary fractional horn systems (BFHS). These are Fractional Horn Systems, in which there are at most two non-zero entries per constraint. Our work is closely related to Lattice programming. Lattice programming is concerned with predicting the direction of change in global optima and equilibria resulting from changing conditions based on problem structure alone without data gathering or computation [11].

Associated with the polyhedra discussed in this paper are two related problems, viz., the linear feasibility problem (**LF**) and the integer feasibility problem (**IF**). In case of Difference Constraint systems (DCS) and Horn Constraint Systems (HCS), these two problems coincide. However, in case of Fractional Horn Systems, these problems are not identical in that there exist instances for which the **LF** problem is feasible, while the **IF** problem is not. Note that if the **IF** problem is feasible the **LF** problem is trivially feasible.

The **LF** problem can be solved in polynomial time by use of an algorithm for linear programming problems. Finding a strongly polynomial combinatorial algorithm for this problem is a known open problem at this time and would constitute a very significant achievement in the progress towards obtaining such an algorithm for general linear programs – a longstanding open problem of importance.

The principal contributions of this paper are as follows:

1. Showing that the **IF** problem in FHSs is **NP-hard** (see Sect. 4).
2. Designing a polynomial time **combinatorial** algorithm for solving the **IF** problem in BFHS (see Subsect. 5.2).
3. Proving several properties of the greatest integer point (GIP) of a bounded BFHS (see Subsect. 5.2).
4. Proving several properties of the least point (LLP) and least integer point (LIP) of a BFHS (see Subsect. 5.3).

The rest of this paper is organized as follows: The problems under consideration are formally described in Sect. 2. The motivation for our work and related approaches in the literature are described in Sect. 3. The computational complexity of the **IF** problem for FHS is detailed in Sect. 4. In Sect. 5, we discuss the **IF** problem for BFHS. Finally, in Sect. 6, we summarize our results and discuss avenues for future research.

2   Statement of Problems

We now define the various specialized constraint systems discussed in this paper.
Let

$$\mathbf{A}\mathbf{x} \geq \mathbf{c} \tag{1}$$

denote a polyhedral system. In System (1), $\mathbf{A}$ is an $m \times n$ integral matrix, $\mathbf{c}$ is an integral m-vector and $\mathbf{x} = [x_1,\ x_2,\ \ldots x_n]^T$ is a variable n-vector. We use $\mathbf{X}$ to denote the set of all the variables in System (1), i.e., $\mathbf{X} = \{x_1, x_2, \ldots, x_n\}$.

Definition 1. *System (1) is said to be a Difference Constraint System (DCS), if each row of* **A** *contains at most one positive entry which is a 1 and at most one negative entry which is a −1.*

Definition 2. *System (1) is said to be a Horn Constraint system or a Horn polyhedron if*

1. *The entries in* **A** *belong to the set* $\{0, 1, -1\}$.
2. *Each row contains at most one positive entry.*

Definition 3. *System (1) is said to be a Fractional Horn Constraint system (FHS) or a Fractional Horn polyhedron if*

1. *Each row contains at most one positive entry which is an integer.*
2. *All negative entries of the matrix are −1.*

Example 1. The following is a Fractional Horn Constraint.

$$5 \cdot x_1 - x_2 - x_3 - x_5 \geq 9$$

Definition 4. *System (1) is said to be a Binary Fractional Horn Constraint System (BFHS), if each row of* **A** *contains at most one positive entry (which might be an arbitrary positive integer) and at most one negative entry which is a −1.*

Example 2. The following is a Binary Fractional Horn Constraint.

$$5 \cdot x_1 - x_5 \geq 9$$

Definition 5. *System (1) is said to be an Extended Difference Constraint System (EDCS), if each row of* **A** *contains at most one positive entry, which is 1, and at most one negative entry, which can be an arbitrary negative integer.*

Example 3. The following is an Extended Difference Constraint.

$$x_1 - 5 \cdot x_5 \geq 9$$

Definition 6. *Let U represent the set of solutions to System (1).* **x** *is the least point (LLP) of System (1) if,* $\mathbf{x} \in U$ *and* $\forall \mathbf{y} \in U$ $\mathbf{x} \leq \mathbf{y}$. **x** *is the least integer point (LIP) of System (1) if,* $\mathbf{x} \in U \cap \mathbb{Z}^n$ *and* $\forall \mathbf{y} \in U \cap \mathbb{Z}^n$ $\mathbf{x} \leq \mathbf{y}$. *We can similarly define the greatest point (GLP) and greatest integer point (GIP) of System (1).*

FHS systems do have least points when they are linearly feasible but these need not be integral. Moreover, it is possible to have an FHS that has a fractional feasible solution but no integer feasible solution. The following example illustrates this property.

Example 4. Consider the system:

$$x_1 - x_2 \geq 1 \quad -x_1 + x_2 \geq -1$$
$$x_3 - x_4 \geq 1 \quad -x_3 + x_4 \geq -1$$
$$2 \cdot x_1 - x_3 - x_4 \geq 0 \quad -x_1 - x_2 + 2 \cdot x_4 \geq 0 \tag{2}$$

It is easy to check that $\mathbf{x} = [1, 0, \frac{3}{2}, \frac{1}{2}]$ is a linear solution to System (2). We now show that there is no integer feasible solution.

The first and second constraints of System (2) imply:

$$x_1 - x_2 = 1.$$

Similarly, the third and fourth constraints imply:

$$x_3 - x_4 = 1.$$

Combining these results with the fifth and sixth constraints, we get:

$$2 \cdot (1 + x_2) - (1 + x_4) - x_4 \geq 0$$
$$-(1 + x_2) - x_2 + 2 \cdot x_4 \geq 0.$$

This in turn implies:

$$2 \cdot x_2 - 2 \cdot x_4 \geq -1$$
$$-2 \cdot x_2 + 2 \cdot x_4 \geq 1.$$

Thus, we get:

$$2 \cdot x_2 - 2 \cdot x_4 = -1$$
$$x_2 - x_4 = -\frac{1}{2}.$$

This implies that there is no integer feasible solution to System (2).

This paper is concerned with the **LF** and **IF** problems in FHS and BFHS.

3   Motivation and Related Work

Since the **IF** problem is **NP-hard** for general polyhedra $(\mathbf{A} \cdot \mathbf{x} \geq \mathbf{c})$ a fair amount of research has been devoted towards the design of polynomial time algorithms for various special cases, restricting the structure of $\mathbf{A}$.

It is well-known that, if the constraint matrix $\mathbf{A}$ is Totally Unimodular (TUM) and the vector $\mathbf{b}$ is integral, then the system $\mathbf{A} \cdot \mathbf{x} \geq \mathbf{c}$ has integral extreme point solutions [8]. Recall that a matrix $\mathbf{A}$ is totally unimodular if the determinant of every square sub-matrix of $\mathbf{A}$ is 0, 1, or −1. Difference constraint systems are a sub-class of TUM systems, in which each constraint has at most one positive entry and one negative entry, with the positive entry being 1 and the negative entry being −1.

A related constraint system is the Unit Two Variables per Inequality (UTVPI) system in which both sum and difference relationships can be expressed. The **IF** feasibility problem for this class was shown to be in **P** [5]. Unlike DCSs though, in a UTVPI system the answers to the **LF** and **IF** problems do not coincide in that such a system could be linear feasible but not integer feasible. UTVPI systems find applications in a host of verification-related problems such as abstract interpretation and array-bounds checking [1,7].

Horn Constraint Systems generalize difference constraints in that multiple negative unity entries are permitted in a row. It is easy to see that Horn systems are not TUM. However, a Horn constraint system always has a least element (if it is feasible) and the least element of a Horn system is always integral. It follows that the **LF** and **IF** problems coincide in case of Horn Constraint systems [3]. Veinott [9,10] has a non-polynomial algorithm for the **LF** problem of Horn type programs where the positive and negative elements can take any value.

Our work is closely related to Lattice programming. Lattice programming is concerned with predicting the direction of change in global optima and equilibria resulting from changing conditions based on problem structure alone without data gathering or computation. Rooted in the theory of lattices, this work is also useful for characterizing the form of optimal and equilibrium policies, improving the efficiency of computation and suggesting desirable properties of heuristics. Applications range widely over dynamic programming, statistical decision-making, cooperative and noncooperative games, economics, network flows, Leontief substitution systems, production and inventory management, project planning, scheduling, marketing, and reliability and maintenance [11].

In this paper, we focus on yet another generalization of DCS; viz., Fractional Horn systems. Here the positive entry does not have to be unity, although the negative entries are −1. In this case, the **LF** and **IF** problems do not coincide; indeed, the **IF** problem is **NP-hard**. We also analyze a special sub-class of FHSs and show that the **IF** problem can be decided in polynomial time in this case.

4   Integer Feasibility of Fractional Horn Constraints

We now show that the problem of finding an integer solution to an FHS is **NP-complete**. We do this via a reduction from monotone TVPI.

Definition 7. *System (1) is said to be a monotone* **two variables per inequality** *(TVPI), if each row of* **A** *contains at most one positive and at most one negative entry (both of which can be arbitrary integers).*

It is known that finding an integer solution to a monotone TVPI system is **NP-complete** [4,6].

Theorem 1. *Finding an integer solution to an FHS is* **NP-complete**.

Proof. First observe that the problem of finding an integer solution to a FHS is in **NP**. This is because this problem is a special case of integer programming and it is well-known that integer programming is in **NP** [8].

We show **NP-hardness** by reducing **IF** for Monotone TVPI to **IF** for FHS. A system of Monotone TVPI constraints consists of constraints of the form $a_k \cdot x_i - b_k \cdot x_j \geq c_k$ where $a_k, b_k \in \mathbb{Z}^+$ and $c_k \in \mathbb{Z}$. Let $M = \lfloor \max \log_2 b_k \rfloor$, for each x_i in the monotone TVPI system we construct $(2 \cdot M + 2)$ variables. Let $x_{i,l,1}$ and $x_{i,l,2}$ for $l = 0, \ldots, M$ be these variables.

We add the following constraints:

$$x_{i,l,1} - x_{i,l,2} \geq 0 \ \ l = 0, \ldots, M$$
$$x_{i,l,2} - x_{i,l,1} \geq 0 \ \ l = 0, \ldots, M$$
$$x_{i,l+1,1} - x_{i,l,1} - x_{i,l,2} \geq 0 \ \ l = 0, \ldots, M - 1$$
$$2 \cdot x_{i,l,1} - x_{i,l+1,1} \geq 0 \ \ l = 0, \ldots, M - 1$$

From the first two groups of constraints we get:

$$x_{i,l,1} = x_{i,l,2} \ \ l = 0, \ldots, M.$$

Combining this result with the third group of constraints yields:

$$x_{i,l+1,1} - 2 \cdot x_{i,l,1} \geq 0 \ \ l = 0, \ldots, M - 1.$$

Combining this result with the fourth group of constraints yields:

$$x_{i,l+1,1} = 2 \cdot x_{i,l,1} \ \ l = 0, \ldots, M - 1.$$

This is equivalent to:

$$x_{i,l,1} = 2^l \cdot x_{i,0,1} \ \ l = 0, \ldots, M$$

Let us consider the constraint $a_k \cdot x_i - b_k \cdot x_j \geq c_k$. Let $S_k \subseteq \mathbb{Z}$ be such that

$$\sum_{l \in S_k} 2^l = b_k.$$

Note that, S_k represents the binary expansion of b_k. We now add the following constraint to the FHS:

$$a_k \cdot x_{i,0,1} - \sum_{l \in S_k} x_{j,l,1} \geq c_k.$$

When we simplify this constraint, we get:

$$a_k \cdot x_{i,0,1} - \sum_{l \in S_k} x_{j,l,1} = a_k \cdot x_{i,0,1} - \sum_{l \in S_k} 2^l \cdot x_{j,0,1} = a_k \cdot x_{i,0,1} - b_k \cdot x_{j,0,1} \geq c_k.$$

Thus, if the FHS is integer feasible we can satisfy the original monotone TVPI system by setting $x_i = x_{i,0,1}$ for $i = 1, \ldots, n$. Similarly, if the monotone TVPI system is integer feasible we can satisfy the FHS by setting $x_{i,l,1} = x_{i,l,2} = 2^l \cdot x_i$ for $i = 1, \ldots, n$ and $l = 0, \ldots, M$. $\qquad \square$

Despite **IF** being **NP-hard** for FHS, if an FHS has an integral solution then it is unbounded or it has a least integer point. This is a direct consequence of Theorem 2.

Definition 8. *Let* **y** *and* **z** *be two n dimensional vectors. We have that* **w** $=$ $\min[\mathbf{y}, \mathbf{z}]$ *if* $w_i = \min[y_i, z_i]$ *for each* $i = 1 \ldots n$. *Similarly,* **w** $= \max[\mathbf{y}, \mathbf{z}]$ *if* $w_i = \max[y_i, z_i]$ *for each* $i = 1 \ldots n$.

Theorem 2. *If* **y** *and* **z** *are feasible solutions to the FHS system*

$$\mathbf{A} \cdot \mathbf{x} \geq \mathbf{c},$$

then so is **w** $= \min[\mathbf{y}, \mathbf{z}]$.

Proof. Consider an arbitrary constraint

$$a \cdot x_i - \sum x_j \geq c.$$

We have that w_i is either equal to y_i or equal to z_i. If $w_i = y_i$, then

$$a \cdot w_i - \sum w_j = a \cdot y_i - \sum w_j \geq a \cdot y_i - \sum y_j \geq c$$

Similarly, if $w_i = z_i$, then

$$a \cdot w_i - \sum w_j = a \cdot z_i - \sum w_j \geq a \cdot z_i - \sum z_j \geq c$$

Thus, **w** is feasible.   □

5   Binary Fractional Horn Constraints

We now study systems of binary fractional horn constraints. First, we will provide an algorithm for solving BFHSs which are unbounded from above (Subsect. 5.1). Then, we will extend this algorithm to BFHSs which have upper bounds (Subsect. 5.2). Finally, we will prove several properties of least points of BFHSs (Subsect. 5.3).

5.1   BFHS Systems Unbounded from Above

Let us first consider BFHS systems in which there is no row of the matrix **A** that has -1 as its only nonzero element. This means that there is no constraint of the form $-x_j \geq c_k$ in the system.

Theorem 3. *A BFHS with no row whose only nonzero element is -1, is either unbounded or infeasible.*

Proof. Given the conditions on the matrix $\mathbf{A}$, it follows that $\mathbf{A} \cdot \mathbf{e} \geq 0$ where $\mathbf{e} = (1, 1, ..., 1) \in \mathbb{R}^n$. It follows that $\mathbf{A}$ and $\mathbf{c}$ can be written after permuting their rows as:

$$\mathbf{A} = \boxed{\begin{array}{c} \mathbf{A}_1 \\ \hline \mathbf{A}_2 \end{array}}, \quad \mathbf{c} = \boxed{\begin{array}{c} \mathbf{c}^1 \\ \hline \mathbf{c}^2 \end{array}}$$

such that the following relations hold:

$$\mathbf{A}_1 \cdot \mathbf{e} > 0$$
$$\mathbf{A}_2 \cdot \mathbf{e} = 0$$

Hence, given any vector $\mathbf{x}$ such that $\mathbf{A}_2 \cdot \mathbf{x} \geq \mathbf{c}^2$, letting $\mathbf{y} = \mathbf{x} + M \cdot \mathbf{e}$ yields:

$$\mathbf{A}_1 \cdot \mathbf{y} = \mathbf{A}_1 \cdot (\mathbf{x} + M \cdot \mathbf{e}) = \mathbf{A}_1 \cdot \mathbf{x} + M \cdot \mathbf{A}_1 \cdot \mathbf{e} \geq \mathbf{c}^1$$
$$\mathbf{A}_2 \cdot \mathbf{y} = \mathbf{A}_2 \cdot (\mathbf{x} + M \cdot \mathbf{e}) = \mathbf{A}_2 \cdot \mathbf{x} + M \cdot \mathbf{A}_2 \cdot \mathbf{e} = \mathbf{A}_2 \cdot \mathbf{x} \geq \mathbf{c}^2$$

for sufficiently large positive values of M. Hence, $\mathbf{y}$ is a feasible solution for large values of M. If the vector $\mathbf{x}$ in the above is integral, then, by choosing M to be an integer, we can guarantee that $\mathbf{y}$ is also integral. Moreover, if the system is feasible it is unbounded. Checking whether there exists an integral vector $\mathbf{x}$ such that $\mathbf{A}_2 \cdot \mathbf{x} \geq \mathbf{c}^2$ can be done in $\mathbf{P}$ since this is a DCS system. A process for determining an exact value for M can be found in Subsect. 5.2. $\square$

5.2   BFHS Systems with Upper Bounds

Now we consider systems that do have constraints of the form $-x_j \geq c_k$. This is an integral upper bound for the variable x_j. This in turn may imply integral upper bounds for other variables. For example, if there is a constraint of the form $a_{k'} \cdot x_j - x_i \geq c_{k'}$, then combining the two constraints yields an integral upper bound for x_i.

Thus, we can divide the variables into two sets – those with a derivable upper bound and those without. Let S be the set of variables which have no derivable upper bound.

We will show that any BFHS, $\mathbf{A} \cdot \mathbf{x} \geq \mathbf{c}$, can be solved in the following manner:

1. Determine the set S of variables with no derivable upper bound.
2. If every variable is in S, solve using the method detailed in Subsect. 5.1.
3. Otherwise, solve the subsystem of constraints involving variables not in S. Then, extend to a full solution.

First, we need to prove several theorems.

Lemma 1. *There are no constraints of the form:*

$$a_k \cdot x_i - x_j \geq c_k \qquad i \notin S; j \in S$$

Proof. Since $x_i \notin S$, we can derive an upper bound on x_i. Let x_i^* be this upper bound. Thus we can derive the constraint $-x_i \geq -x_i^*$. Together with the constraint $a_k \cdot x_i - x_j \geq c_k$, this lets us derive the constraint $-x_j \geq c_k - a_k \cdot x_i^*$. However, this places an upper bound on x_j, which contradicts the fact that $x_j \in S$. □

Theorem 4. *Every variable in S can be increased uniformly without violating any constraints.*

Proof. Let $\mathbf{x}'$ be a feasible solution to the BFHS and $\mathbf{x}''$ be the vector obtained from $\mathbf{x}'$ by adding 1 to every variable in S. We show that $\mathbf{x}''$ is also a solution to the BFHS.

Let x_i, x_j, and x_l be variables such that $x_i, x_j \in S$ and $x_l \notin S$. Thus, $x_i'' = x_i' + 1$, $x_j'' = x_j' + 1$ and $x_l'' = x_l'$. From Lemma 1, we know that there are no constraints of the form $a_k \cdot x_l - x_i \geq c_k$ or $a_k \cdot x_l - x_j \geq c_k$. We also know that there are no constraints of the form $-x_i \geq c_k$ or $-x_j \geq c_k$. Otherwise, x_i or x_j would not be in S.

Thus, we only need to consider the following constraints.

1. $a_k \cdot x_i - x_j \geq c_k$ where $a_k \geq 1$: We have that

$$a_k \cdot x_i'' - x_j'' = a_k \cdot (x_i' + 1) - (x_j' + 1) = a_k \cdot x_i' - x_j' + (a_k - 1) \geq a_k \cdot x_i' - x_j' \geq c_k$$

2. $a_k \cdot x_i - x_l \geq c_k$: We have that

$$a_k \cdot x_i'' - x_l'' = a_k \cdot (x_i' + 1) - x_l' = a_k \cdot x_i' - x_l' + a_k \geq a_k \cdot x_i' - x_l' \geq c_k$$

Thus, $\mathbf{x}''$ is still a valid solution to the BFHS. This means that we can increase each variable in S uniformly without bound. □

From the preceding theorem, we have that if a variable has no derivable upper bound, then it is unbounded.

This allows us to rewrite the system in the following form:

$$\begin{array}{|c|c|} \hline \mathbf{A}_1 & 0 \\ \hline \mathbf{A}_2 & 0 \\ \hline \mathbf{B} & \mathbf{A}_3 \\ \hline \end{array} \cdot \begin{array}{|c|} \hline \mathbf{x}^1 \\ \hline \mathbf{x}^2 \\ \hline \end{array} \geq \begin{array}{|c|} \hline \mathbf{c}^1 \\ \hline \mathbf{c}^2 \\ \hline \mathbf{c}^3 \\ \hline \end{array} \tag{3}$$

such that:

1. The variables in $\mathbf{x}^1$ are in the set S and those in $\mathbf{x}^2$ are not.
2. The sum of the elements in each row of $\mathbf{A}_1$ is positive.
3. The sum of the elements in each row of $\mathbf{A}_2$ is zero.
4. Each row of $\mathbf{B}$ may have a positive entry but no negative entries.

From Lemma 1, we know that there is no constraint such that a variable in $\mathbf{x}^1$ has a negative coefficient and a variable in $\mathbf{x}^2$ has a non-zero coefficient. Thus, $\mathbf{A} \cdot \mathbf{x} \geq \mathbf{c}$ can be divided as described.

We already know how to solve the problem:

$$\boxed{\begin{array}{c} \mathbf{A}_1 \\ \hline \mathbf{A}_2 \end{array}} \cdot \mathbf{x}^1 \geq \boxed{\begin{array}{c} \mathbf{c}^1 \\ \hline \mathbf{c}^2 \end{array}}$$

From Theorem 3, if this system is feasible, then it is unbounded. Thus, what remains is the problem:

$$\mathbf{B} \cdot \mathbf{x}^1 + \mathbf{A}_3 \cdot \mathbf{x}^2 \geq \mathbf{c}^3$$

Since $\mathbf{x}^1$ is unbounded, the constraints corresponding to rows of $\mathbf{B}$ which have a positive entry can eventually be made feasible by uniformly increasing all components of $\mathbf{x}^1$. Thus, we can remove these constraints from the system. This yields a subsystem of the form

$$\mathbf{D} \cdot \mathbf{x}^2 \geq \mathbf{c}^4 \tag{4}$$

where $\mathbf{D}$ is a row submatrix of $\mathbf{A}_3$ and $\mathbf{c}^4$ is the corresponding part of $\mathbf{c}^3$. In this system, every variable is bounded from above. We now show how to solve this particular type of system.

Theorem 5. *If a BFHS system is bounded from above, then the system is feasible if and only if it has an integer valued greatest point.*

Proof. Let $\mathbf{x}^*$ be the least upper bound on $\mathbf{x}$ that can be derived by combining the inequalities in System (4). Since the only negative entries in $\mathbf{A}$ are -1, $\mathbf{x}^*$ is integral. If $\mathbf{x}^*$ is feasible, then

$$\mathbf{D} \cdot \mathbf{x}^* \geq \mathbf{c}^4.$$

Thus, the system has integer solution and $\mathbf{x}^*$ is the greatest point since we can derive the constraints:

$$-\mathbf{x} \geq -\mathbf{x}^*$$

If $\mathbf{x}^*$ is not feasible, then there is a constraint of the form

$$a_k \cdot x_i - x_j \geq c_k$$

such that

$$a_k \cdot x_i^* - x_j^* < c_k$$

which implies that

$$x_j^* > a_k \cdot x_i^* - c_k$$

But we can derive the constraint:

$$-x_i \geq -x_i^*$$

Combining this with constraint

$$a_k \cdot x_i - x_j \geq c_k$$

we get

$$-x_j \geq c_k - a_k \cdot x_i^*$$

Thus, $a_k \cdot x_i^* - c_k$ is a derivable upper bound on x_j. But x_j^* is the least upper bound for x_j. Hence we have

$$x_j^* \leq a_k \cdot x_i^* - c_k$$

which is a contradiction. Thus, $\mathbf{x}^*$ is feasible.

If $\mathbf{x}^*$ does not exist, then for some variable there is no least upper bound which implies that the system is not feasible whether we seek integer or linear solutions. Thus, if the system is feasible, there exists an integer solution which is the greatest point. □

The existence of a greatest point, but not its integrality, is also a direct result of Theorem 6.

Theorem 6. *If* $\mathbf{y}$ *and* $\mathbf{z}$ *are feasible solutions to the BFHS system* $\mathbf{Ax} \geq \mathbf{c}$, *then so is* $\mathbf{w} - \max[\mathbf{y}, \mathbf{z}]$.

Proof. Consider an arbitrary constraint

$$a \cdot x_i - x_j \geq c.$$

We have that w_j is either equal to y_j or equal to z_j. If $w_j = y_j$, then

$$a \cdot w_i - w_j = a \cdot w_i - y_j \geq a \cdot y_i - y_j \geq c$$

Similarly, if $w_j = z_j$, then

$$a \cdot w_i - w_j = a \cdot w_i - z_j \geq a \cdot z_i - z_j \geq c$$

Thus, $\mathbf{w}$ is feasible. Moreover, if $\mathbf{y}, \mathbf{z}$ are integral, then so is $\mathbf{w}$. □

Remark 1. Unlike Theorem 2, Theorem 6 does not hold for general FHS systems. For example consider the constraint

$$5x_1 - x_2 - x_3 \geq 0.$$

Both vectors $[1, 0, 5]$ and $[1, 5, 0]$ are feasible but $[1, 5, 5] = \max\{[1, 0, 5], [1, 5, 0]\}$ is not feasible.

This allows us to obtain an integer solution to a general BFHS system.

1. First, we determine the variables in S as follows.
   (a) Start with all variables $x_i \in S$.

(b) Remove each x_i such that $-x_i \geq c_k$ is a constraint in the BFHS.
(c) Remove each x_j such that $a_k \cdot x_i - x_j \geq c_k$ is a constraint in the BFHS and $x_i \notin S$.
(d) Continue until no further variables can be removed. This can be done by a breadth-first search from each $x_i \notin S$.
2. Construct $\mathbf{A}_1$, $\mathbf{A}_2$, $\mathbf{A}_3$, $\mathbf{B}$, $\mathbf{c}^1$, $\mathbf{c}^2$, $\mathbf{c}^3$, $\mathbf{c}^4$ and $\mathbf{D}$ as specified in Systems (3) and (4).
3. Let $\mathbf{x'}^1$ be an integer solution to the DCS $\mathbf{A}_2 \cdot \mathbf{x}^1 \geq \mathbf{c}^2$.
4. Let $\mathbf{x'}^2$ be the greatest point of $\mathbf{D} \cdot \mathbf{x}^2 \geq \mathbf{c}^4$.
5. Let M be an integer such that:
   (a) For all constraints $a_k \cdot x_i - x_j \geq c_k$ where $x_i, x_j \in S$ and $a_k > 1$ (this constraint is in $\mathbf{A}_1 \cdot \mathbf{x}^1 \geq \mathbf{c}^1$)

$$M \geq \frac{c_k - a_k \cdot x'_i + x'_j}{a_k - 1}.$$

   Note that, this guarantees $a_k \cdot (x'_i + M) - (x'_j + M) \geq c_k$.
   (b) For all constraints $a_k \cdot x_i - x_j \geq c_k$ where $x_i \in S$, $x_j \notin S$, and $a_k \geq 1$ (this constraint is in $\mathbf{B} \cdot \mathbf{x}^1 + \mathbf{A}_3 \cdot \mathbf{x}^2 \geq \mathbf{c}^3$)

$$M \geq \frac{c_k - a_k \cdot x'_i + x'_j}{a_k}.$$

   Note that, this guarantees $a_k \cdot (x'_i + M) - x'_j \geq c_k$.
6. Thus,

$$\mathbf{x} = \boxed{\begin{array}{c} \mathbf{x'}^1 + M \cdot \mathbf{e} \\ \hline \mathbf{x'}^2 \end{array}}$$

is a feasible solution to the BFHS.

Note that, each part of this process can be performed in polynomial time.

1. Finding S: Accomplished using a breadth first search starting from the explicitly bounded variables. This can be done in $O(m)$ time.
2. Determining the subformulas: Once S is known, this can be done by checking whether each constraint belongs to:
   (a) $\mathbf{A}_1 \cdot \mathbf{x}^1 \geq \mathbf{c}^1$,
   (b) $\mathbf{A}_2 \cdot \mathbf{x}^1 \geq \mathbf{c}^2$,
   (c) $\mathbf{B} \cdot \mathbf{x}^1 - \mathbf{A}_3 \cdot \mathbf{x}^2 \geq \mathbf{c}^3$,
   (d) or $\mathbf{D} \cdot \mathbf{x}^2 \geq \mathbf{c}^4$.
   This can be done in $O(m)$ time.
3. Solving the DCS, $\mathbf{A}_2 \cdot \mathbf{x}^1 \geq \mathbf{c}^2$: This can be done in $O(m \cdot n)$ time.
4. Finding the greatest point of $\mathbf{D} \cdot \mathbf{x}^2 \geq \mathbf{c}^4$: Letting $\mathbf{y} = -\mathbf{x}^2$ changes this system to $-\mathbf{D} \cdot \mathbf{y} \geq \mathbf{c}^4$ and this system is an EDCS system which has a least point $\mathbf{y}^*$. It should be clear that $(\mathbf{x}^2)^* = -\mathbf{y}^*$ is the required greatest point of $\mathbf{D} \cdot \mathbf{x}^2 \geq \mathbf{c}^4$. In [3], the lifting algorithm is shown to produce this point in $O(m \cdot n)$ time.
5. Finding M: This is done by performing constant time checks on every constraint. This can be done in $O(m)$ time.

Thus, this entire procedure runs in $O(m \cdot n)$ time.

5.3   Least Point

We now see what happens when a BFHS is bounded from below. We show that while the system does have a least point, the least point is not necessarily integral. However, the set of integral solutions does have a least point solution. First, we show that any feasible solution to a BFHS can be rounded up to obtain a feasible lattice point solution.

Theorem 7. *Let* $\mathbf{A} \cdot \mathbf{x} \geq \mathbf{c}$ *represent a BFHS. If* $\hat{\mathbf{x}}$ *is a feasible solution, then so is* $\lceil \hat{\mathbf{x}} \rceil$.

Proof. Let us look at each type of constraint separately:

1. If the constraint is of the form: $-x_i \geq c$, then $-\hat{x}_i \geq c$ which implies that $-\lceil \hat{x}_i \rceil \geq c$.
2. If the constraint is of the form $a \cdot x_i - x_j \geq c$, we have

$$a \cdot \lceil \hat{x}_i \rceil \geq \lceil a \cdot \hat{x}_i \rceil \geq \lceil c + \hat{x}_j \rceil = c + \lceil \hat{x}_j \rceil$$

Thus, $\lceil \hat{\mathbf{x}} \rceil$ satisfies all constraints that $\hat{\mathbf{x}}$ satisfies. Thus, if the system is feasible, it has an integer feasible solution.

□

Theorem 8. *If a BFHS system is bounded from below, then it is feasible and has a least point.*

Proof. If the system is bounded from below, then for each variable x_i we can derive a greatest lower bound $\hat{x}_i$. Consider the point $\hat{\mathbf{x}}$. If $\hat{\mathbf{x}}$ is feasible, it is the least point. Suppose $\hat{\mathbf{x}}$ is not feasible; there are two cases to consider:

1. There is a constraint of the form $-x_i \geq c$ such that $-\hat{x}_i < c$. However, we can derive the constraint $x_i \geq \hat{x}_i$ which when combined with the violated constraint yields $0 \geq \hat{x}_i + c$ which is equivalent to $-\hat{x}_i \geq c$ which leads to a contradiction.
2. There is a constraint of the form $a \cdot x_i - x_j \geq c$ with $a \geq 1$ such that $a \cdot \hat{x}_i - \hat{x}_j < c$. This is equivalent to $\hat{x}_i < \frac{c + \hat{x}_j}{a}$. However, we can derive the constraint $x_j \geq \hat{x}_j$ which when combined with the violated constraint yields $a \cdot x_i \geq c + \hat{x}_j$. Thus, $\frac{c + \hat{x}_j}{a}$ is a derivable lower bound on x_i and this means $\hat{x}_i \geq \frac{c + \hat{x}_j}{a}$. This also leads to a contradiction.

Hence, $\hat{\mathbf{x}}$ is a feasible solution to the system and also the least point.   □

$\hat{\mathbf{x}}$ may not be integral in general. But, Theorem 7 shows that $\lceil \hat{\mathbf{x}} \rceil$ is integral and feasible and a least integer point if the system is bounded from below.

6   Conclusion

In this paper, we studied a class of constraints called *Fractional Horn Constraints*. These constraints generalize difference constraints and are closely related to horn constraints. While in Horn constraint systems, the linear feasibility and integer feasibility coincide, the situation is quite different in FHSs. In case of FHSs, the **LF** problem is solvable in polynomial time, whereas the **IF** problem is **NP-complete**. Despite **IF** being **NP-hard** for FHS, if an FHS has an integral solution then it has a least integer point. We also analyzed a sub-class of FHSs called *Binary Fractional Horn constraints* and showed that the **IF** problem is in **P**, in this case.

References

1. Bagnara, R., Hill, P.M., Zaffanella, E.: Weakly-relational shapes for numeric abstractions: improved algorithms and proofs of correctness. Formal Methods Syst. Des. **35**(3), 279–323 (2009)
2. Cormen, T.H., Leiserson, C.E., Rivest, R.L., Stein, C.: Introduction to Algorithms. MIT Press, Cambridge (2001)
3. Chandrasekaran, R., Subramani, K.: A combinatorial algorithm for horn programs. Discrete Optim. **10**, 85–101 (2013)
4. Hochbaum, D.S., (Seffi) Naor, J.: Simple and fast algorithms for linear and integer programs with two variables per inequality. SIAM J. Comput. **23**(6), 1179–1192 (1994)
5. Jaffar, J., Maher, M.J., Stuckey, P.J., Yap, R.H.C.: Beyond finite domains. In: Borning, A. (ed.) PPCP 1994. LNCS, vol. 874, pp. 86–94. Springer, Heidelberg (1994). doi:10.1007/3-540-58601-6_92
6. Lagarias, J.C.: The computational complexity of simultaneous Diophantine approximation problems. SIAM J. Comput. **14**(1), 196–209 (1985)
7. Lahiri, S.K., Musuvathi, M.: An efficient decision procedure for UTVPI constraints. In: Gramlich, B. (ed.) FroCoS 2005. LNCS (LNAI), vol. 3717, pp. 168–183. Springer, Heidelberg (2005). doi:10.1007/11559306_9
8. Schrijver, A.: Theory of Linear and Integer Programming. Wiley, New York (1987)
9. Veinott, A.F.: Representation of general and polyhedral subsemilattices and sublattices of product spaces. Linear Algebra Appl. **114**(115), 681–704 (1989)
10. Veinott, A.F., LiCalzi, M.: Subextremal functions and lattice programming, July 1992. Unpublished Manuscript
11. Veinott, A.F., Wagner, H.M.: Optimal capacity scheduling: Parts i and ii. Oper. Res. **10**, 518–547 (1962)

Author Index

Printed in the United States
By Bookmasters